Centrifugal Partition Chromatography

CHROMATOGRAPHIC SCIENCE SERIES

A Series of Monographs

Editor: JACK CAZES
Cherry Hill, New Jersey

1. Dynamics of Chromatography, *J. Calvin Giddings*
2. Gas Chromatographic Analysis of Drugs and Pesticides, *Benjamin J. Gudzinowicz*
3. Principles of Adsorption Chromatography: The Separation of Nonionic Organic Compounds, *Lloyd R. Snyder*
4. Multicomponent Chromatography: Theory of Interference, *Friedrich Helfferich and Gerhard Klein*
5. Quantitative Analysis by Gas Chromatography, *Josef Novák*
6. High-Speed Liquid Chromatography, *Peter M. Rajcsanyi and Elisabeth Rajcsanyi*
7. Fundamentals of Integrated GC-MS (in three parts), *Benjamin J. Gudzinowicz, Michael J. Gudzinowicz, and Horace F. Martin*
8. Liquid Chromatography of Polymers and Related Materials, *Jack Cazes*
9. GLC and HPLC Determination of Therapeutic Agents (in three parts), *Part 1 edited by Kiyoshi Tsuji and Walter Morozowich, Parts 2 and 3 edited by Kiyoshi Tsuji*
10. Biological/Biomedical Applications of Liquid Chromatography, *edited by Gerald L. Hawk*
11. Chromatography in Petroleum Analysis, *edited by Klaus H. Altgelt and T. H. Gouw*
12. Biological/Biomedical Applications of Liquid Chromatography II, *edited by Gerald L. Hawk*
13. Liquid Chromatography of Polymers and Related Materials II, *edited by Jack Cazes and Xavier Delamare*
14. Introduction to Analytical Gas Chromatography: History, Principles, and Practice, *John A. Perry*
15. Applications of Glass Capillary Gas Chromatography, *edited by Walter G. Jennings*
16. Steroid Analysis by HPLC: Recent Applications, *edited by Marie P. Kautsky*
17. Thin-Layer Chromatography: Techniques and Applications, *Bernard Fried and Joseph Sherma*
18. Biological/Biomedical Applications of Liquid Chromatography III, *edited by Gerald L. Hawk*
19. Liquid Chromatography of Polymers and Related Materials III, *edited by Jack Cazes*
20. Biological/Biomedical Applications of Liquid Chromatography, *edited by Gerald L. Hawk*
21. Chromatographic Separation and Extraction with Foamed Plastics and Rubbers, *G. J. Moody and J. D. R. Thomas*

22. Analytical Pyrolysis: A Comprehensive Guide, *William J. Irwin*
23. Liquid Chromatography Detectors, *edited by Thomas M. Vickrey*
24. High-Performance Liquid Chromatography in Forensic Chemistry, *edited by Ira S. Lurie and John D. Wittwer, Jr.*
25. Steric Exclusion Liquid Chromatography of Polymers, *edited by Josef Janča*
26. HPLC Analysis of Biological Compounds: A Laboratory Guide, *William S. Hancock and James T. Sparrow*
27. Affinity Chromatography: Template Chromatography of Nucleic Acids and Proteins, *Herbert Schott*
28. HPLC in Nucleic Acid Research: Methods and Applications, *edited by Phyllis R. Brown*
29. Pyrolysis and GC in Polymer Analysis, *edited by S. A. Liebman and E. J. Levy*
30. Modern Chromatographic Analysis of the Vitamins, *edited by André P. De Leenheer, Willy E. Lambert, and Marcel G. M. De Ruyter*
31. Ion-Pair Chromatography, *edited by Milton T. W. Hearn*
32. Therapeutic Drug Monitoring and Toxicology by Liquid Chromatography, *edited by Steven H. Y. Wong*
33. Affinity Chromatography: Practical and Theoretical Aspects, *Peter Mohr and Klaus Pommerening*
34. Reaction Detection in Liquid Chromatography, *edited by Ira S. Krull*
35. Thin-Layer Chromatography: Techniques and Applications. Second Edition, Revised and Expanded, *Bernard Fried and Joseph Sherma*
36. Quantitative Thin-Layer Chromatography and Its Industrial Applications, *edited by Laszlo R. Treiber*
37. Ion Chromatography, *edited by James G. Tarter*
38. Chromatographic Theory and Basic Principles, *edited by Jan Åke Jönsson*
39. Field-Flow Fractionation: Analysis of Macromolecules and Particles, *Josef Janča*
40. Chromatographic Chiral Separations, *edited by Morris Zief and Laura J. Crane*
41. Quantitative Analysis by Gas Chromatography, Second Edition, Revised and Expanded, *Josef Novák*
42. Flow Perturbation Gas Chromatography, *N. A. Katsanos*
43. Ion-Exchange Chromatography of Proteins, *Shuichi Yamamoto, Kazuhiro Nakanishi, and Ryuichi Matsuno*
44. Countercurrent Chromatography: Theory and Practice, *edited by N. Bhushan Mandava and Yoichiro Ito*
45. Microbore Column Chromatography: A Unified Approach to Chromatography, *edited by Frank J. Yang*
46. Preparative-Scale Chromatography, *edited by Eli Grushka*
47. Packings and Stationary Phases in Chromatographic Techniques, *edited by Klaus K. Unger*
48. Detection-Oriented Derivatization Techniques in Liquid Chromatography, *edited by Henk Lingeman and Willy J. M. Underberg*
49. Chromatographic Analysis of Pharmaceuticals, *edited by John A. Adamovics*
50. Multidimensional Chromatography: Techniques and Applications, *edited by Hernan Cortes*
51. HPLC of Biological Macromolecules: Methods and Applications, *edited by Karen M. Gooding and Fred E. Regnier*

52. Modern Thin-Layer Chromatography, edited by Nelu Grinberg
53. Chromatographic Analysis of Alkaloids, Milan Popl, Jan Fähnrich, and Vlastimil Tatar
54. HPLC in Clinical Chemistry, I. N. Papadoyannis
55. Handbook of Thin-Layer Chromatography, edited by Joseph Sherma and Bernard Fried
56. Gas-Liquid-Solid Chromatography, V. G. Berezkin
57. Complexation Chromatography, edited by D. Cagniant
58. Liquid Chromatography-Mass Spectrometry, W. M. A. Niessen and Jan van der Greef
59. Trace Analysis with Microcolumn Liquid Chromatography, Miloš Krejčí
60. Modern Chromatographic Analysis of Vitamins: Second Edition, edited by André P. De Leenheer, Willy E. Lambert, and Hans J. Nelis
61. Preparative and Production Scale Chromatography, edited by G. Ganetsos and P. E. Barker
62. Diode Array Detection in HPLC, edited by Ludwig Huber and Stephan A. George
63. Handbook of Affinity Chromatography, edited by Toni Kline
64. Capillary Electrophoresis Technology, edited by Norberto A. Guzman
65. Lipid Chromatographic Analysis, edited by Takayuki Shibamoto
66. Thin-Layer Chromatography: Techniques and Applications, Third Edition, Revised and Expanded, Bernard Fried and Joseph Sherma
67. Liquid Chromatography for the Analyst, Raymond P. W. Scott
68. Centrifugal Partition Chromatography, edited by Alain P. Foucault

ADDITIONAL VOLUMES IN PREPARATION

Handbook of Size Exclusion Chromatography, edited by Chi-San Wu

Centrifugal Partition Chromatography

edited by
Alain P. Foucault
Laboratoire de Bioorganique et Biotechnologies
Centre National de la Recherche Scientifique
Paris, France

Marcel Dekker, Inc.　　　New York • Basel • Hong Kong

Library of Congress Cataloging-in-Publication Data

Centrifugal partition chromatography / edited by Alain P. Foucault.
    p.   cm. — (Chromatographic science series ; v. 68)
    Includes bibliographical references and index.
    ISBN 0-8247-9257-2
    1. Countercurrent chromatography.   I. Foucault, Alain P.
    II. Series: Chromatographic science ; v. 68.
    QP519.9.C68C46   1994
    543'.0894—dc20   94-27246
                                                                    CIP

The publisher offers discounts on this book when ordered in bulk quantities. For more information, write to Special Sales/Professional Marketing at the address below.

This book is printed on acid-free paper.

Copyright © 1995 by MARCEL DEKKER, INC.   All Rights Reserved.

Neither this book nor any part may be reproduced or transmitted in any form or by any means, electronic or mechanical, including photocopying, microfilming, and recording, or by any information storage and retrieval system, without permission in writing from the publisher.

MARCEL DEKKER, INC.
270 Madison Avenue, New York, New York   10016

Current printing (last digit):
10 9 8 7 6 5 4 3 2 1

PRINTED IN THE UNITED STATES OF AMERICA

Preface

Modern countercurrent chromatography (CCC) originates from the pioneering studies of Y. Ito et al. [1], who first constructed, in Japan, an apparatus designed to differentiate particles in suspension or solutes in solution in a solvent system subjected to a centrifugal acceleration field. This first machine opened the way in two main directions: one, pursued by Y. Ito in the United States, is based on a wide variety of "CCC apparatuses" (countercurrent chromatographic apparatuses), most of the recent ones using a variable gravity field produced by a two-axis gyration mechanism and rotary seal–free arrangement for the column; the other, pursued by K. Nunogaki in Japan, is based on the "CPC apparatus" (centrifugal partition chromatographic apparatus) and uses a constant gravity field produced by a single-axis rotation mechanism, and two rotary seal joints for inlet and outlet of the mobile phase.

The historical linkage between countercurrent distribution (Jantzen, Watanabe, Van Dijek, Martin and Synge, Craig, among others) [2] and countercurrent chromatography is responsible for the name *countercurrent chromatography* in which a strong gravitational field is used to keep a liquid stationary phase in a steady immobilized state while the mobile phase is pumped through. With technological improvements, the performance of today's instruments is much closer to that of liquid–liquid chromatography using a solid support to hold the liquid stationary phase.

The goal of this volume is to provide a forum for scientists who are already using centrifugal partition chromatographs in their research to share with others their personal knowledge in this specific field of chromatography.

Two books [3,4] appeared recently, devoted mainly to the CCC apparatus (two-axis gyration mechanism). These books contain very useful information on the principles of CCC and its applications, and the interested reader should obtain at least one of them to complement the material found in this volume, which is devoted exclusively to the CPC apparatus (single-axis rotation mechanism).

CPC and HPLC (high-performance liquid chromatography) are similar in several respects. They share the same fundamental mechanism (partitioning of solutes), the same goal (separation, purification), and the same ancillary techniques (pumps, injectors, detectors).

Figure 1 is a schematic of HPLC and CPC. For convenience, we have chosen a reversed phase chromatographic process. With HPLC (a), the stationary phase is an organic moiety bonded to silica and solvated mainly by the organic solvent of the aqueous organic mobile phase. The volume ratio of the stationary phase is only ≈5% of the total volume of the column, and it will not change significantly as the mobile phase composition varies.

With CPC (b), the silica is replaced with a strong gravitational field that

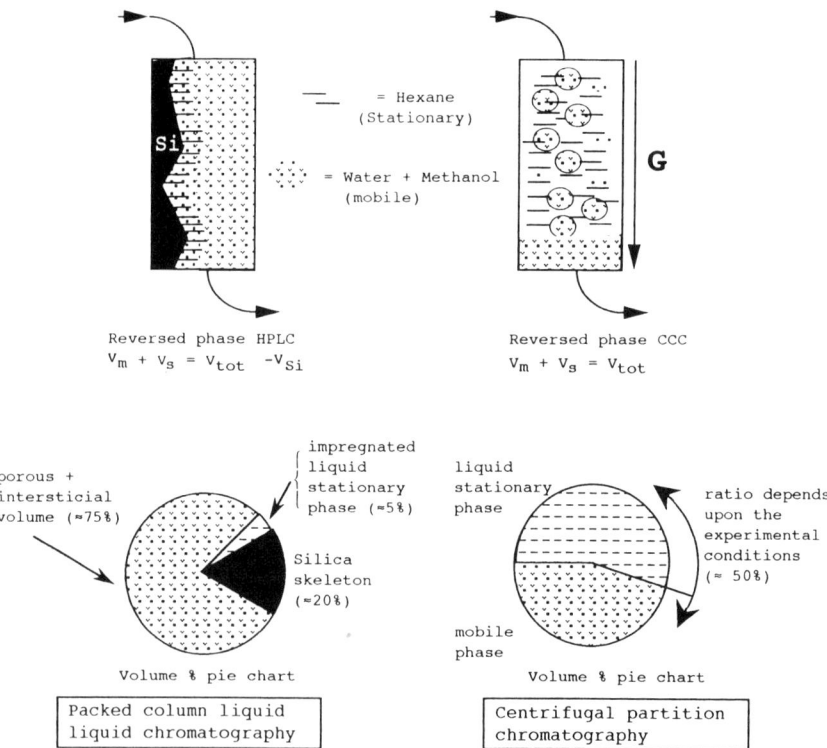

Figure 1 Schematic comparison between conventional reversed phase HPLC (a) and reversed phase CPC (b).

"frees" 20–30% of the total volume. (The porosity factor, ϵ_c, defined as the ratio of the void volume to the total volume of a column, is ≈ 0.7–0.8 for porous silica and ≈ 0.35 for solid beads.) Instead of an organic moiety bonded to a solid support, we now simply use an immiscible organic solvent (e.g., hexane) as the stationary phase; the droplets of the water–organic mixture pass through the stationary phase. The volume ratio of the stationary phase is 50–80% for most cases; the actual value depends on the instrument, the experimental conditions, and the composition of the ternary mixture used to prepare the mobile and stationary phases.

An advantage of the centrifugal system is that the volume ratio of the stationary phase to the total volume is much higher than that of the solid support system. Direct consequences of this better balance between the two phases in CPC are (1) the range of partition coefficients useful for CPC purification is centered around 1, and (2) the capacity of the CPC column is high when compared with an HPLC column with the same total volume.

The chapters of this book are totally independent, and they cover four main topics:

Instrumentation and theory
Octanol–water partition coefficient
CPC in natural product chemistry
CPC in inorganic chemistry

Most work presented in this book has been performed with the CPC Model LLN, which was the only analytical/semiprep CPC instrument available until 1992. The HPCPC (high-performance CPC), recently introduced on the market, and which is definitely more efficient, will be preferred to the LLN by anyone who would like to start using centrifugal partition chromatography.

REFERENCES

1. Y. Ito, M. Weinstein, I. Aoki, R. Harada, E. Kimura, and K. Nunogaki, *Nature, 212*: 985 (1966).
2. N. B. Mandava and J. M. Ruth, in *Countercurrent Chromatography*, Marcel Dekker, Inc., New York, p. 27 and references therein (1988).
3. N. B. Mandava and Y. Ito, *Countercurrent Chromatography, Theory and Practice*, Marcel Dekker, Inc., New York (1988).
4. W. D. Conway, ed., *Countercurrent Chromatography*, VCH Publishers, New York (1990).

Contents

Preface		*iii*
Contributors		*ix*
1.	Operating the Centrifugal Partition Chromatograph *Alain Berthod, Chau-Dung Chang, and Daniel W. Armstrong*	1
2.	Theory of Centrifugal Partition Chromatography *Alain P. Foucault*	25
3.	Pressure Drop in Centrifugal Partition Chromatography *M. J. van Buel, L. A. M. van der Wielen, and K. Ch. A. M. Luyben*	51
4.	Solvent Systems in Centrifugal Partition Chromatography *Alain P. Foucault*	71
5.	Fractionation of Plant Polyphenols *Takuo Okuda, Takashi Yoshida, and Tsutomu Hatano*	99
6.	Centrifugal Partition Chromatography in Assay-Guided Isolation of Natural Products: A Case Study of Immunosuppressive Components of *Tripterygium wilfordii* *Jan A. Glinski and Gary O. Caviness*	133
7.	Liquid–Liquid Partition Coefficients: The Particular Case of Octanol–Water Coefficients *Alain Berthod*	167

8.	Centrifugal Partition Chromatography for the Determination of Octanol–Water Partition Coefficients *Steven J. Gluck, Eric Martin, and Marguerite Healy Benko*	199
9.	Mutual Separation of Lanthanoid Elements by Centrifugal Partition Chromatography *Kenichi Akiba*	219
10.	Separator-Aided Centrifugal Partition Chromatography *Takeo Araki*	241
11.	Centrifugal Partition Chromatographic Separations of Metal Ions *S. Muralidharan and H. Freiser*	317
12.	Preparative Centrifugal Partition Chromatography *Rodolphe Margraff*	331

Appendix I: Various Ways to Fill a CPC   351
Appendix II: CPC Instrumentation   355
Appendix III: Ternary Diagrams   363

Index   *409*

Contributors

Kenichi Akiba, Ph.D. Professor, Institute for Advanced Materials Processing, Tohoku University, Sendai, Japan

Takeo Araki, Ph.D. Professor, Department of Polymer Science and Engineering, Kyoto Institute of Technology, Matsugasaki, Kyoto, Japan

Daniel W. Armstrong, Ph.D. Professor, Department of Chemistry, University of Missouri—Rolla, Rolla, Missouri

Marguerite Healy Benko, Ph.D. Senior Research Chemist, Discovery Research, DowElanco, Indianapolis, Indiana

Alain Berthod, Ph.D. Professeur Agrégé, Directeur de Recherche, Laboratoire des Sciences Analytiques, Centre National de la Recherche Scientifique, Université de Lyon 1, Villeurbanne, France

Gary O. Caviness, M.S. Scientist III, Pharmaceutics, Boehringer Ingelheim Pharmaceuticals, Inc., Ridgefield, Connecticut

Chau-Dung Chang, Ph.D.[*] Research Scientist, Department of Chemistry, University of Missouri—Rolla, Rolla, Missouri

Alain P. Foucault, Ph.D. Senior Research Scientist, Laboratoire de Bioorganique et Biotechnologies, Centre National de la Recherche Scientifique, Paris, France

H. Freiser, Ph.D. Professor, Strategic Metals Recovery Research Facility, Department of Chemistry, University of Arizona, Tucson, Arizona

---

[*]*Current affiliation*: Research Scientist, Chemistry, Genzyme Corporation, Cambridge, Massachusetts

Jan A. Glinski, Ph.D. Principal Scientist, Pharmaceutics, Boehringer Ingelheim Pharmaceuticals, Inc., Ridgefield, Connecticut

Steven J. Gluck, M.S. Research Leader, Analytical Sciences Laboratory, Dow Chemical, Midland, Michigan

Tsutomu Hatano, Ph.D. Associate Professor, Faculty of Pharmaceutical Sciences, Okayama University, Tsushima, Okayama, Japan

K. Ch. A. M. Luyben, IR. Professor and Head, Department of Biochemical Engineering, Delft University of Technology, Delft, The Netherlands

Rodolphe Margraff, Ph.D. Research Advisor, Biotechnologies, Rhône-Poulenc Rorer, S.A., Vitry-sur-Seine, France

Eric Martin, Ph.D. Senior Scientist, Chemical Therapeutics, Chiron Corporation, Emeryville, California

S. Muralidharan, Ph.D. Senior Scientist, Strategic Metals Recovery Research Facility, Department of Chemistry, University of Arizona, Tucson, Arizona

Takuo Okuda, Ph.D. Emeritus Professor, Faculty of Pharmaceutical Sciences, Okayama University, Tsushima, Okayama, Japan

M. J. van Buel, IR. Department of Biochemical Engineering, Delft University of Technology, Delft, The Netherlands

L. A. M. van der Wielen, IR. Assistant Professor, Department of Biochemical Engineering, Delft University of Technology, Delft, The Netherlands

Takashi Yoshida, Ph.D. Professor, Faculty of Pharmaceutical Sciences, Okayama University, Tsushima, Okayama, Japan

Centrifugal Partition Chromatography

1
Operating the Centrifugal Partition Chromatograph

Alain Berthod
Laboratoire des Sciences Analytiques, Centre National de la Recherche Scientifique, Université de Lyon 1, Villeurbanne, France

Chau-Dung Chang* and Daniel W. Armstrong
University of Missouri—Rolla, Rolla, Missouri

I. INTRODUCTION

To some individuals, countercurrent chromatography (CCC) is a complicated technique that involves two liquid phases flowing in opposite directions. Indeed, this is indicated by the very name of the technique. In fact, however, CCC is analogous to other chromatographic methods in that there is a stationary phase and a mobile phase that results in a relative countercurrent motion. CCC is a liquid chromatography (LC) technique with a liquid mobile phase and a stationary phase *that is also a liquid*. It is important to realize that *only the liquid mobile phase moves* in the CCC technique. CCC uses the same basic components, including pumps, valve injectors, and detectors, as classical LC. Only the column is different, because it is much more difficult to keep a liquid motionless than a solid.

In this chapter, we first try to show the nonspecialist CCC is a useful technique with an important role to play in prepurification, sample preparation, and preparative chromatography. Second, we demonstrate several ways we routinely use centrifugal partition chromatography by describing different experiments. The two first parts of the chapter describe the important and sometimes unique factors and parameters used in CCC. The preparation and use of the CCC "column" are described. The advantages of a liquid stationary phase are pointed

---

*Current affiliation: Genzyme Corporation, Cambridge, Massachusetts

Table 1 A Comparison of the Features of the Two Main High-Speed CCC Apparatuses

Feature	Multilayer coil separator extractor, high-speed counter-current chromatograph, HSCC	Centrifugal partition chromatograph, CPC
Type	Hydrodynamic	Hydrostatic
Liquid is retained in	Coiled Teflon tubes	Channels
Liquid retention	Variable	Good
Efficiency	Up to 4 plates per tube turn or 50 plates per ml or less	Up to 1 plate per channel or 20 plates per ml or less
Internal volume adjustment	Changing spool or tubing	Sanki LLN: changing the cartridge number
Pressure	Low, 0.1–10 kg/cm^2	Medium, 2–70 kg/cm^2
Maintenance	Connecting tubing to change every ~50 hours	Rotating seals to lubricate every ~50 hours
Other	Noisy gear assembly, no control on temperature	Quiet centrifuge, very good temperature control on the Sanki LLN

out. In the last part, three examples are selected to illustrate the capabilities of the technique.

Two types of CCC apparatuses are used in high-speed countercurrent chromatography: (1) apparatuses containing coiled tubes, the hydrodynamic CCC machines, and (2) those with rotary seals and channels, the hydrostatic CCC machines [1–3]. The hydrodynamic Ito machines will be called *high-speed countercurrent chromatographs*, or HSCCCs. The hydrostatic machines will be called *centrifugal partition chromatographs*, or CPCs, which is a trade name of Sanki Engineering of Japan. Table 1 lists the general features of the two types of modern high-speed CCC apparatuses. Only the operating mode of the centrifugal partition chromatographs (CPC) is described here. For a complete description of these devices, Refs. 1–3 are recommended.

II. CHROMATOGRAPHIC PROPERTIES

A. Solute Retention

An important advantage of CCC over LC is that solute retention depends on one parameter only, the liquid–liquid partition coefficient **P**:

$$V_R = V_M + \mathbf{P}V_S \tag{1}$$

where V_R, V_M, and V_S are the solute retention volume, the mobile phase volume, and the stationary phase volume inside the CCC "column," respectively. The sum

$$V_M + V_S = V_T \tag{2}$$

is the internal volume V_T of the CCC "column," or apparatus.

The capacity factor k' is the parameter most commonly used in LC to study solute retention. It is defined as

$$k' = \frac{V_R - V_M}{V_M} \tag{3}$$

The partition coefficient **P** is defined as the solute affinity for the stationary phase:

$$\mathbf{P} = \frac{[\text{solute concentration in the stationary phase}]}{[\text{solute concentration in the mobile phase}]} \tag{4}$$

The partition coefficient depends on which solvent or solvent mixture is used as the stationary phase. If the phase role is reversed, the mobile phase becomes the stationary phase and the partition coefficient **P** is replaced by $\mathbf{P}' = 1/\mathbf{P}$. This leads to one of the big advantages of CCC compared with LC: No solute can be irreversibly retained in the column. A high retention volume means a high partition coefficient **P**. Reversing the phases will produce a low **P**′ value ($\mathbf{P}' = 1/\mathbf{P}$) and a low retention volume.

The capacity factor k' can be related to the solute partition coefficient by

$$k' = \frac{\mathbf{P} V_S}{V_M} \tag{5}$$

The difference in retention between two solutes is measured by the selectivity factor α:

$$\alpha = \frac{k'_2}{k'_1} = \frac{\mathbf{P}_2}{\mathbf{P}_1} \tag{6}$$

and α is always higher or equal to 1.

B. Peak Efficiency

The efficiency N is related to the peak width. Efficiency is measured by plate numbers. The classical equation for a Gaussian-shaped peak is

$$N = 4\left(\frac{V_R}{W_{0.6h}}\right)^2 \tag{7}$$

in which $W_{0.6h}$ is the peak width at 60% of the peak height. $W_{0.6h}$ corresponds exactly to 2σ, the variance, if the peak is perfectly Gaussian. The Gaussian peak base width W_b is equal to 4σ, or $2W_{0.6h}$.

Efficiency is related to the solute mass transfer between the two liquid phases. In general, it is observed that the efficiency increases when the viscosity of the liquid phases decreases. Also, the efficiency is directly proportional to the number of channels. Consequently, more channels, that is, more cartridges, give more plates and sharper peaks, but also higher pressure drops. Efficiency is also related to the mobile phase flow rate and the stationary phase volume retention, in a more complex manner. This is reviewed by Foucault in Chapter 2.

C. Solute Resolution

1. Definition

The resolution Rs is a parameter measuring the quality of a separation. A 1.5 resolution value means that two adjacent peaks are separated with a baseline return in between. A resolution factor lower than 1.5 means there is some peak overlap. Resolution factors higher than 1.5 mean there is space in between the two peaks. Rs is expressed as the ratio of the distance between the two peak maxima to the mean value of the peak width W at the baseline:

$$Rs = \frac{V_2 - V_1}{(W_2 + W_1)/2} \tag{8}$$

Using Eqs. 1 and 6 and assuming the efficiency N is constant, Rs can be rewritten as

$$Rs = \tfrac{1}{4}\sqrt{N}\,\frac{P_2 - P_1}{\dfrac{V_M}{V_S} + \dfrac{P_2 + P_1}{2}} \tag{9}$$

The resolution factor can be expressed with capacity factors instead of partition coefficients. Substituting Eq. 5, one obtains

$$Rs = \tfrac{1}{4}\sqrt{N}\,\frac{k'_2 - k'_1}{1 + \dfrac{k'_2 + k'_1}{2}} \tag{10}$$

Using the selectivity factor (Eq. 6) and making the approximation $k'_1 \approx (k'_2 + k'_1)/2$, we obtain the classical resolution equation used in LC [4]:

$$Rs = \tfrac{1}{4}(\alpha - 1)\sqrt{N}\,\frac{k'_1}{1 + k'_1} \tag{11}$$

2. LC and CCC

CCC is not a technique that will supplant LC. CCC is not the best technique for analytical purposes; the efficiency is too low. CCC may, however, be a good alternative to the classical preparative LC technique. It is important to realize that CCC and preparative LC do not operate in the same manner. The resolution equation, Eq. 11, is valid for both techniques. But Eq. 11 conceals the critical role played by the phase volume ratio. Equation 9 shows that the volume ratio V_M/V_S acts on the resolution factor. The minimum partition coefficient difference between two solutes that is necessary to obtain a baseline resolution (Rs = 1.5) depends on the V_S/V_M ratio as illustrated by Fig. 1.

The CCC and LC V_S/V_M ranges are indicated by arrows. In CCC, a minimum of 10% stationary phase retention is required to perform a separation; the V_S/V_M ratio is 0.10, or $\log V_S/V_M = -1$. With a well-retained liquid system, the stationary phase retention can reach 80%, producing a V_S/V_M ratio of 4 ($\log V_S/V_M = 0.6$). In LC, the solute interacts with the surface of the solid stationary phase. In reversed phase LC, the average bonding density is 3 μmol/m², with an average 200 m²/g surface area, the estimated bonding material amount is about 6×10^{-4} mol/g. A heavy commonly bonded moiety is octadecane, C18, with a molar weight of 253; it gives 0.15 g of bonded phase per g of packing. This is a volume of about 0.19 cm³ per gram of LC packing, which can be taken as the solute-accessible stationary phase volume. The mobile phase volume per gram of LC packing is usually in the cm³ range [5]. V_S/V_M ratio is 0.19 ($\log V_S/V_M = -0.7$).

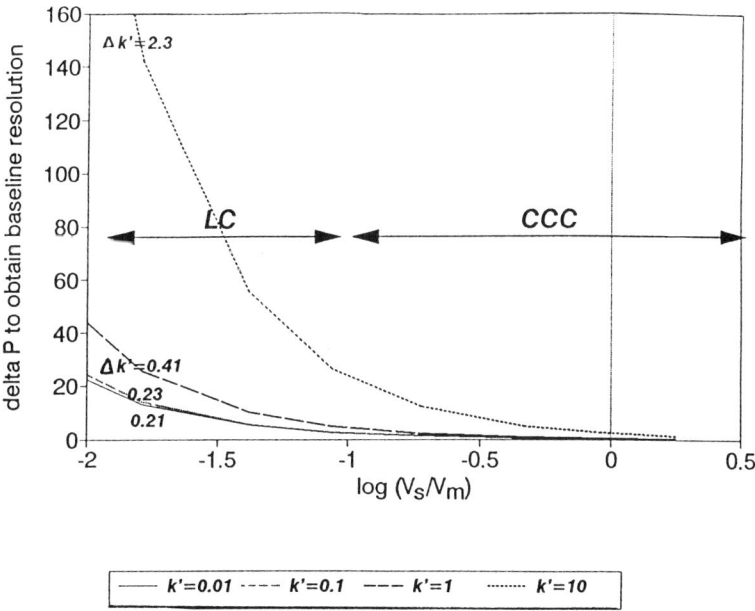

Figure 1 Plot of the partition coefficient difference required to obtain a baseline resolution (Rs = 1.5) versus the phase ratio V_S/V_M for four different sets of solutes. The $\Delta k'$ difference does not depend on the phase ratio; its value is indicated on the figure. The LC and CCC operating ranges are located by arrows.

With a LC packing less densely grafted and/or using a lighter moiety (C10, C4, C1), the V_S/V_M ratio can be as low as 0.01 (log $V_S/V_M = -2$). In normal phase LC, V_S corresponds to the superficial polar layer of unbounded porous silica. An unknown amount of water and/or polar molecules are tightly bound to the surface silanol groups, making the exact estimate of the V_S volume impossible, but very low anyway.

Figure 1 shows that the partition coefficient difference needed to obtain a baseline resolution (Rs = 1.5) can be higher than 100 in LC with two solutes in the $k' = 10$ range. The corresponding **P** difference is around 3 in CCC. In other words, a $\Delta k'$ difference of 2.3 allows one to obtain baseline resolution between two solutes, $k'_1 = 10$ and $k'_2 = 12.3$, in both LC and CCC methods. This 2.3 $\Delta k'$ difference is obtained with a Δ**P** difference of 55 units in LC with a 0.04 V_S/V_M ratio, and Δ**P** is only 1.3 units in CCC because of the 1.8 V_S/V_M ratio (Fig. 1). It should be pointed out that the same 4000 plate efficiency was used for the calculations of Fig. 1. LC compensates for the low V_S/V_M ratio with higher efficiency. Small changes in a molecule, however, induce small partition coefficient differences that may render a separation possible in CCC but not in LC.

D. Pressure Drop

We demonstrated that the observed pressure drop, Pd, in operating a CPC machine was the sum of two terms [6]:

$$Pd = n(\Delta d \omega^2 R h + \nu \gamma F) \qquad (12)$$

in which n is the channel number, Δd is the phase density difference, ω is the rotor spin rate, R is the rotor radius, h is the stationary phase height in a channel that also contains the mobile phase, ν is the mobile phase viscosity, γ is a constant depending on the type or geometry of the apparatus used, and F is the mobile phase flow rate. Assuming the volume of the ducts is small compared with the channel volume, the stationary phase height h can be taken as the physical height of a channel multiplied by the ratio V_S/V_T. The first term of Eq. 12 is the hydrostatic term; the second term is the hydrodynamic, or viscous, term. Figure 2 illustrates the pressure drop changes with the rotor spin rate. If the density difference between the two liquid phases is low ($\Delta d \leq 0.1$ g/cm^2), the maximum operating pressure, 70 kg/cm^2, is not reached. With a density difference of 0.4 g/cm^3 (e.g., hexane–water, $\Delta d \leq 0.34$ g/cm^3), the pressure drop increases dramatically with the rotation speed (Fig. 2).

Equation 12 shows that the pressure drop depends on h, the average height of the stationary phase crossed by the mobile phase inside the channels. This means, at the beginning of the equilibration of the CPC apparatus, the pressure drop is low, whatever liquid system is used, due only to the hydrodynamic viscous term. Then, it increases linearly with time as more channels are equilibrated [6]. It is important

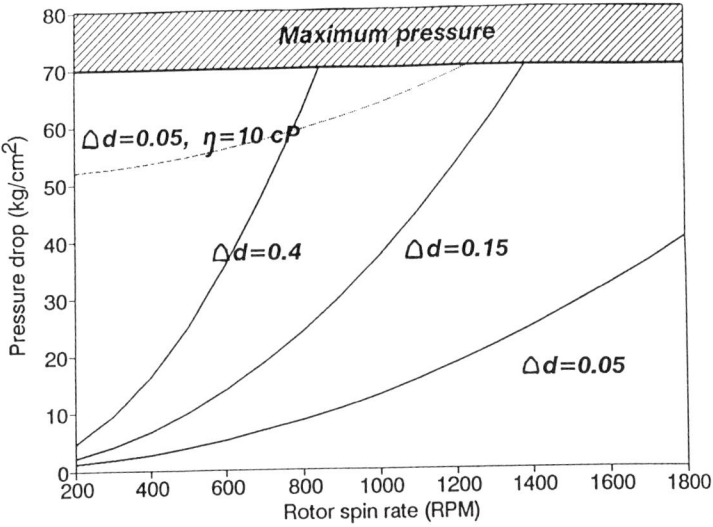

Figure 2 CPC pressure drop versus the rotor spin rate for four different liquid systems with density differences of 0.05, 0.15, and 0.4 g/cm^3. Flow rate: 1 ml/min; mobile phase viscosity: 1 or 10 cP (dotted line); Sanki LLN centrifuge loaded with six 250W cartridges (2400 channels).

to watch the pressure during the CPC "column" equilibration; if it is noted that the pressure will pass the 70 kg/cm^2 limit, then the rotation speed, and not the flow rate, should be decreased. The effect of the flow rate on pressure drop is less dramatic than the spin rate, except when the mobile phase viscosity is high (dotted line in Fig. 2).

Losses of liquid stationary phase is a problem in CCC [7]. Such losses can be due to bleeding, which is a small but continuous carryover of the stationary phase by the mobile phase [6,7]. Also, the pressure inside the apparatus can induce a stationary phase solubilization by the mobile phase saturated at atmospheric pressure [6]. Such losses can be monitored by pressure readings. When the stationary phase bleeds from the CPC machine, the height h in each channel decreases (or the number n of channels equilibrated with both phases decreases, or both of these occur) and the pressure drop decreases. The stationary phase volume can be measured by injecting a compound of known partition coefficient **P**. Its retention volume gives the stationary phase volume by

$$V_S = \frac{V_R - V_T}{P - 1} \tag{13}$$

III. PRACTICAL OPERATION OF THE CENTRIFUGAL PARTITION CHROMATOGRAPH

A. Liquid System Preparation

The volumes of solvents that make the biphasic liquid system are measured and poured together in a container. The mixture is shaken vigorously for several minutes. The settling time of the phase, that is, the time needed to obtained a meniscus between the two liquid phases, should be observed. As described by Ito [1], if this time is less than one minute, the system has a good interfacial tension and density difference between phases, and it should be easily retained inside the CCC machines. If the settling time is much longer, the system will likely not be retained in an HSCCC machine. It will still be possible to use this liquid system with a CPC machine by using a higher rotation speed. If the phases are not too viscous, they should be filtered through a 0.5-μm filter. We did not filter butanol- or octanol-rich phases due to their high viscosity. We did not observe pump problems when using chromatographic grade solvents.

The phases should not be perfectly separated. A small amount of one phase should be in contact with the other to ensure they remain well saturated. It is not even necessary to separate the phases; the inlet filter can be lifted into the upper liquid layer or plunged into the lower layer as needed during the "column" preparation.

B. CPC Apparatus Preparation

The clean apparatus is first filled with the liquid stationary phase at a high flow rate and without rotation. The operating mode is selected: *descending* (or head to tail in the CCC literature for HSCCC machines) if the mobile phase is more dense than the stationary phase, or *ascending* (tail to head) in the opposite case. The centrifuge can be started at a spin rate chosen on the basis of the phase density difference, as shown in Fig. 2. The mobile phase pump is turned on at the desired flow rate, which should be lower than the "flooding flow rate" [8]. The outgoing liquid is collected in a graduated cylinder. As long as the liquid stationary phase exits the CPC machine, the equilibrium is not reached, and *the pressure increases*. When the first drops of liquid mobile phase exit the apparatus, equilibrium is reached and the pressure stabilizes. The volume of the stationary phase displaced is measured; it corresponds to the volume of the mobile phase, V_M, inside the CPC "column." The pressure should be noted. If it decreases, it is a sign of stationary phase loss. If everything is stable, including the baseline of the recorder, the "column" is ready and the sample can be injected.

When the same liquid system is to be used for several analyses, several injections can be done on the same "column." It is even possible to use the same "column" for several days. To keep a "column," the mobile phase pump is first stopped, and then the rotor is stopped. One trick that can be used to stop any liquid motion is to turn the injection valve and the mode valve to an intermediate position. If this is done, do not forget to return the valves to their correct working positions before turning the pump on. If not, the liquids cannot circulate and the increasing pressure will cause leaks where the various Teflon tubes connect. To resume a separation with an existing "column," the rotor is first started, and, after an equilibration time of a few minutes, the mobile phase pump is started at a low flow rate (0.1 ml/min). This is progressively increased to the desired flow rate. The pressure is measured. If the pressure reading is comparable with that of previous experiments, the "column" is again ready for use.

IV. EXAMPLES OF CPC SOLUTIONS TO ACTUAL EXPERIMENTAL PROBLEMS

A. Total Sample Recovery of a ^{14}C-Labeled Herbicide and Metabolites

1. The Problem

Metolachlor (2-chloro-N-(2'-ethyl-6'-methylphenyl)-N-(2-methoxy-1-methylethyl) acetamide) is one of the α-chloroacetamide herbicides that is used extensively in the cultivation of corn and soybeans as a preemergence control for most annual grasses and many broad leaf weeds. It is one of the most widely employed herbicides, and frequently it is applied in combination with atrazine [9]. The structures of metolachlor and its metabolites are shown in Table 2. The pure standards were supplied by Ciba Geigy Corporation (CG). The most common methods used to measure metolachlor in water are gas chromatography and immunoassay [10–12]. The detection of metolachlor in aqueous media has been reported in the range of 0.1–0.9 ppb by enzyme-linked immunosorbent assay (ELISA) [12]. Several key metolachlor metabolites in corn have been identified, and a possible metabolic pathway has been proposed [13,14]. Two-dimensional thin-layer chromatography (TLC) and ion exchange chromatography (IEC) have been used to fractionate and characterize metolachlor metabolites contained in crude extracts from corn [15].

All the aforementioned studies made effective use of a variety of analytical methodologies. As is often the case when working with complex extracts of biological materials, however, larger, pure or partially pure sample fractions are needed for more extensive multiple analyses. Preparative high-performance liquid chromatography (HPLC) is useful in this respect, but irreversible retention is often a problem with biological samples. Complete sample recovery is desirable, particularly when working with radioactively labeled compounds. Incomplete recovery in HPLC converts costly stationary phase into radioactive solid waste. Recently, Chang et al. [16] demonstrated the utility of CPC in the fractionation and characterization of [^{14}C]metolachlor metabolites contained in crude extract of mature corn plants. Both UV detection and liquid scintillation counting detection were used.

Table 2 Structure of Metachlor and Its Metabolites

Structure	Reference code	Partition coefficient
Ethyl 2, methyl 6 phenyl	P	—
[Metolachlor structure]	Metolachlor	—
[structure]	CGA-37913	1.4
[structure]	C-25702	4.1
[structure]	CGA-46576	7.3
[structure]	CGA-50720	8.5
[structure]	CGA-37735	9.0
[structure]	CGA-40919	17
[structure]	CGA-51202	19
[structure]	CGA-13656	25
[structure]	CGA-40172	30
[structure]	CGA-110186	35
[structure]	CGA-41507	48
[structure]	CGA-46127	63
[structure]	CGA-118243	95

2. CPC Operation

The CPC experiments were performed using a two-phase ternary solvent system: 1-butanol/acetic acid/water in the ratio 20/3/20 v/v/v. Four liters of this solvent system were prepared by mixing 1860 ml 1-butanol, 280 ml acetic acid, and 1860 ml water by volume. The two phases (one butanol rich and one water rich) can then be separated from one another and used in the CPC. Fractionation was optimized on a nonradioactive corn extract control using UV (254 nm) detection. The denser aqueous phase was chosen as the mobile phase. So, the CPC machine was first filled by the butanol phase. Then the rotor was started at 900 rpm, and the aqueous mobile phase was pushed in the *descending* mode to equilibrate the "column."

It was not possible to elute all the injected solutes with the aqueous mobile phase, some of them being highly retained. This is not surprising, because it is known that metolachlor metabolites are quite difficult to separate by LC or GC [13,14]. This problem is easily solved by CPC, changing the mode (i.e., using the phase role reversal). The stationary phase can become the mobile phase, and vice versa. In this way, any solutes in the previous stationary phase are eluted from the apparatus in the "new" mobile phase. The total sample recovery is assured. A 0.5-ml aliquot of the [^{14}C]radioactive crude extract was injected directly into the CPC in the descending mode (i.e., predominantly aqueous mobile phase), switching to the ascending mode (i.e., predominantly butanol mobile phase) after two hours. The density difference between the aqueous and the butanol phase is close to 0.18 g/cm^3. A 900 rpm spin rate was sufficient to retain the butanol phase in the descending mode with 4 ml/min aqueous flow rate. The pressure was 55 kg/cm^3 at 32°C, and continuous UV detection was used. After 400 ml of the aqueous phase was pumped in the descending mode, the pump was stopped but the rotor was not. The incoming lines were flushed with the butanol phase, the mode valve was set in the ascending position, and the apparatus was cooled down to 23°C. The rotor spin rate was increased to 1100 rpm because the aqueous phase was bleeding at too high a rate at 900 rpm, though the butanol flow rate was only 1.2 ml/min. This was due to the high viscosity of the butanol water-saturated phase. The pressure was 60 bars at 1100 rpm in the ascending mode.

Fractions were collected, and 5 ml of scintillator cocktail was added to a 1-ml portion of each fraction that had been pipetted into a scintillation vial. A Beckman LS 7500 system was used to record the β-decay events. Blank solutions consisted of 5 ml scintillator cocktail and 1 ml water saturated with 1-butanol. All fractions and blanks were counted for 5 minutes. As can be seen in Table 2, two of the metolachlor metabolites are weak bases, a primary amine (C-25702) and an amino alcohol (CGA-37913). All the rest of the compounds are amides. Three of the amide compounds also contain carboxylic acid functional groups (CGA-118243, CGA-110186, and CGA-51202), however, while one compound contains an amino acid moiety (CGA-46576). Using UV detection, the CPC fractionation of the crude corn plant extract was optimized so that the total UV peak area in the descending mode was roughly equivalent to that in the ascending mode [16].

3. Results and Discussion

Figure 3 shows the CPC chromatograms obtained in both the ascending and descending modes. It appears that there are at least four distinct peaks in the descending mode (at 15, 19, 30, and 45 minutes). The first peak at 15 minutes also represents the dead time (t_0) for this separation. It is highly likely that each peak is composed of several closely or coeluting compounds. This is particularly apparent in the ascending mode (Fig. 3A), where many of the peaks are so close together (after the initial one at 117 minutes) that they are often difficult to distinguish one from another.

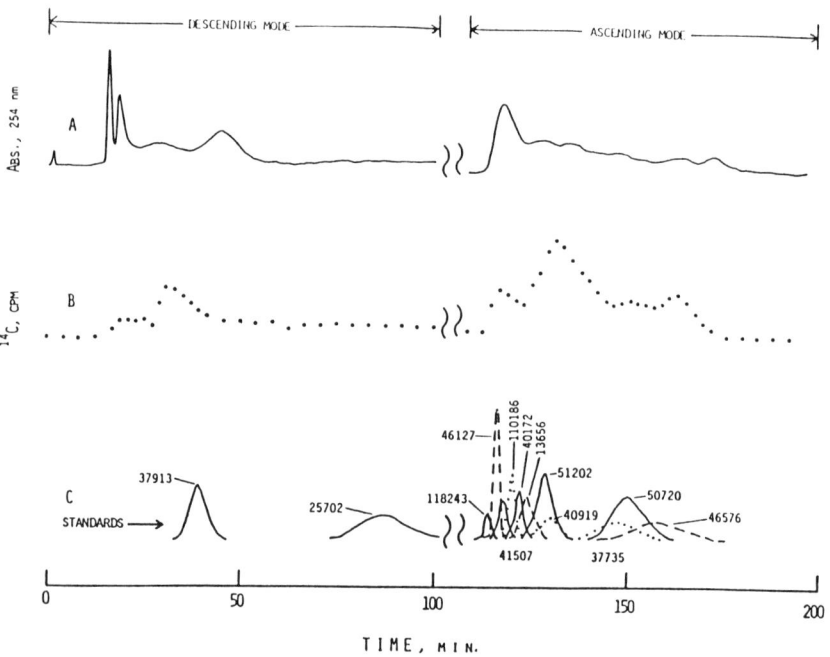

Figure 3 CPC fractionation of metolachlor metabolites. (A) Crude corn plant extract and UV 254 nm detection; (B) crude corn plant extract and ^{14}C liquid scintillation counting detection; (C) composite chromatogram obtained with the injection of the 13 pure standards of the metolachlor metabolites whose structures and code numbers are given in Table 2. All chromatograms were obtained with the following conditions: *Descending mode*: water mobile phase flow rate 4 ml/min, 900 rpm, 55 kg/cm², six cartridges, 32°C. *Ascending mode*: butanol mobile phase flow rate 1.2 ml/min, 1100 rpm, 60 kg/cm², 23°C.

Using exactly the same chromatographic conditions that were used in Fig. 3A, crude corn plant extracts previously sprayed with [^{14}C]metolachlor were fractionated. The resulting radiochromatogram is shown in Fig. 3B. It is apparent from this figure that more than two-thirds of the radioactive compounds are eluted in the ascending mode. When the radiochromatogram is directly compared with the UV chromatogram, it is apparent that the relative magnitudes of similarly retained peaks are very different. Also, peaks sometimes appear in one chromatogram but not in the other. Though essentially the same material was fractionated in both cases, this is not an unexpected result given the very different selectivities of the detection methods. Clearly, there are a number of UV-absorbing pigments and polar compounds that elute in the descending mode that are completely unassociated with [^{14}C]metolachlor or its degradation products. It seems there may be a higher correlation between the UV peaks and those of the radiochromatogram in the ascending mode (Figs. 3A and 3B). It appears, however, that the radiochromatogram, like the UV chromatogram, is made up of a large number of overlapping peaks [16].

Figure 3C shows the elution order of 13 known metolachlor metabolites when separated by CPC using conditions identical to those employed for the corn plant extract. Only the two amine compounds are eluted in the descending mode. Compound 25702 has no corresponding peak in either the UV chromatogram or the radiochromatogram. Interestingly, a major portion of both the UV absorbance and radioactivity in the descending mode chromatogram does not correspond to any of the tested metolachlor degradation products. Eleven of the thirteen metolachlor metabolite standards elute in the ascending mode. The retention times of the standards correspond to many of the peaks seen in the radiochromatogram and the UV chromatogram. The largest peak in the radiochromatogram has a maximum at about 130 minutes. It is broad and has shoulders indicating that it includes several compounds. Two metolachlor metabolites (51202 and 40919) elute at about this time. Both compounds result from the dehalogenation of metolachlor (Table 2), which seems to be an important early step in its biodegradation. Clearly, in the ascending mode there are similarities between the retention of the metabolite standards (Fig. 3C) and the various maxima in the radiochromatogram (Fig. 3B). This was not the case in the descending mode [16].

It is possible to estimate the partition coefficients of solutes eluted in the reversed mode if the lines were correctly flushed between the ascending and descending mode [17,18]. The partition coefficient of a given solute depends only on the ratio of the volume of aqueous phase pumped in the ascending mode divided by the retention volume (butanol phase) in the descending mode:

$$P = \frac{V_{aq}}{V_{but}} \tag{14}$$

Table 2 lists the estimated partition coefficients of the metolachlor metabolites. These partition coefficients indicate the solute affinity for the butanol phase; they should be closely related to their octanol–water partition coefficients. The 1.4 and 4.1 values obtained for metabolites CGA-37913 and C-25702, ionizable solutes, are not the molecular partition coefficients; see Chapter 7 in this book for more information on partition coefficients.

B. Enhanced Selectivity in the Fractionation of a Petroleum Product Using Gradient Elution

1. Problem Description

With the increased use of heavier crude petroleum reserves and the development of nonconventional fossil fuels such as shale and coal oil, and with the increased demand for light petroleum products, it is necessary to convert these heavy residues into light petroleum products in more economical, efficient, and environmentally friendly ways. This conversion is performed by either a thermal or catalytic cracking (cat-cracking) process that chemically breaks, or "cracks," higher molecular weight molecules into lighter fragments. Studies have shown that different classes of compounds typically found in heavy residues react differently in the cracking process [19]. In order to better understand and improve the cracking process, it is important to thoroughly characterize these residues. Typically, the analysis consists of physicochemical measurements such as density, viscosity, elemental analysis, and asphalt and ash content. Unfortunately, these basic measurements yield little information that can be used to optimize the refining process. Much more useful information may be elicited through spectroscopic analysis, but only after preparative separation of the complex petroleum materials [19–21].

Preparative high-performance liquid chromatography (HPLC) has been used to separate the components of heavy residues and crude oil samples into various classes. In a separation technique called "SARA" (for *s*aturate, *a*romatic, *r*esin, and *a*sphaltene), the asphaltenes are precipitated by hexane or heptane, and then the saturates, aromatics, and resins are separated by HPLC [22]. Other schemes separate components according to functionality (i.e., acid, base, and neutral) followed by chemical group-type fractionation (saturate, aromatic, and polar) or chromatographic separation by aromatic ring number (saturate, mono-, di-, tri-, and polyaromatic, and polar) [21]. Analyses are subject to many limitations with these preparative techniques, including low sample-mass capacity, irreversible retention, poor selectivity, tailing, and considerable sample preparation [21]. Much attention has been given to the detailed study of molecular structure and substituent effects on the retention characteristics of aromatic hydrocarbons on various stationary phases such as alumina, chemically bonded silica-NH_2 and silica-$R(NH_2)_2$. All of the stationary phases reported for this use had very low capacities. Consequently, it was necessary to use very large columns. Alumina stationary phases showed excellent selectivity for aromatic-ring containing molecules but not for alkanes or cycloalkanes [23]. Heteroatom-containing compounds (polar components) often were irreversibly adsorbed on the alumina stationary phase. Also, alumina stationary phases needed extensive equilibration or reactivation times and suffered from poor reproducibility due to varying levels of alumina activity. Silica stationary phases in conjunction with alumina were used for the separation of alkanes and cycloalkanes from aromatic hydrocarbons, but polar components still irreversibly adsorbed to the stationary phase [24]. Silica-NH_2 and dual-functional alkyl aminealkylnitrile–bonded stationary phases facilitated the elution of many of the polar components, retained roughly equivalent selectivity, and eliminated the extensive equilibration times [19]. Silica-NH_2 stationary phases are commercially available and well characterized; however, they still lack the high capacity necessary for preparative scale separations.

2. Use of CPC for Fractionation

CPC has a liquid stationary phase free of solid support. The large stationary phase to mobile phase volume ratio will be useful for accommodating large sample volumes. The liquid stationary phase eliminates irreversible retention and allows strongly retained samples to be recovered by flushing the system or using dual-mode elution. The cost of operating the CPC is less than other forms of preparative chromatography because a new stationary phase for the CPC (fresh solvent) is relatively inexpensive when compared with silica gel–based column packings.

The cat-cracker feed samples were prepared by adding approximately 0.5 ml of hexane to 1.2–1.6 g of cat-cracker feed (melting point ~35°C) to reduce the viscosity of the sample and to give a total sample volume of approximately 2 ml. The sample was shaken for 1 minute and then directly injected without filtration (sample loop volume was 1.9 ml). Several runs were done in isocratic conditions with several compositions in the methanol–water–hexane system. They did not give acceptable results, producing broad baseline drifts in both the ascending and the descending modes. The use of a gradient elution was investigated successfully. It is possible to make gradient separations with the methanol–water–hexane system because the solubility of water and methanol in hexane is very low [30]. Then, when the water content in the water–methanol phase changes, the hexane phase remains essentially pure hexane. This is not true with any ternary solvent system [25].

Hexane was the stationary phase. The initial mobile phase was 75% methanol and 25% water, and the final was 95% methanol and 5% water. Both the initial and final mobile phases were saturated with hexane by stirring, and the hexane stationary phase was saturated with the initial mobile phase in a similar manner. Nine cartridges, model 250W of the Sanki LLN machine, composed the column (a total column volume of 180 ml). To minimize the stationary phase changes, we chose to put less stationary phase volume than is optimal. The column was prepared by filling the empty cartridges with the hexane phase in the descending mode. Then, in the ascending mode, with the centrifuge spinning at only 700 rpm (density difference ~ 0.2 g/cm^3), approximately 95 ml of stationary phase was displaced by the initial mobile phase at a flow rate of 4.0 ml/min. The volume ratio was 1, but with most of the hexane stationary phase volume located in the head or top of the apparatus. The mode was then switched to descending, pushing the initial mobile phase (25% water) through the head of the apparatus. This pushed the known volume of hexane phase down in the rotor. The "column" was equilibrated for 45 minutes at a flow rate of 4.0 ml/min of 25% water phase. Sodium iodide was used as a dead volume marker. UV detection in the descending mode was set at 254 nm [25].

Once the "column" is equilibrated at 700 rpm with the 25% water mobile phase, the sample was injected (1.9 ml) and eluted with a mobile phase composition changing following a 400-minute gradient from 75% methanol–25% water to 95% methanol–5% water. As the experiment progressed, the mobile phase density decreased and so did the pressure; then the rotor spin rate was increased to ensure the stationary phase was not pushed out. The spin rate was increased by a 100-rpm step every 2 hours (800 rpm at 120 minutes, 900 rpm at 240 minutes, and 1000 rpm at 360 minutes, Fig. 4). After 400 minutes, the mobile phase flow in the descending mode was stopped. The mobile phase was then changed to hexane, and the retained components were eluted at a hexane flow rate of 1.0 ml/min in the ascending mode with 1000 rpm and UV detection at 320 nm (Fig. 5).

Figure 4, in the descending mode, illustrates the separation of the cat-cracker feed into 16 peaks with a shoulder clearly present on the first peak. The ninth and fifteenth peaks also show the presence of shoulders. Seventeen fractions, which correspond to these major peaks and shoulders, as shown by the arrows in Fig. 4, were collected for further analysis. Figure 5 depicts the ascending mode elution of the retained components. The ascending mode elution profile indicates there is somewhat less selectivity in this mode than in the descending mode, but there is some fractionation present [25]. This trend is to be expected, because strongly retained components in the descending mode partition mainly to the new mobile phase and interact very little with the new stationary phase. Five fractions were collected in the ascending mode, as depicted by the arrows in Fig. 5. Solvent remaining in the CPC chambers after separation was collected for the subsequent analysis as well. Mass spectroscopy, LC, GC, and nuclear magnetic resonance (NMR) were used to identify the separated components of the cat-cracker feed [25]. It is not possible to calculate the partition coefficient of the peak obtained because the liquid composition was changing due to the gradient analysis.

CPC can be used for the preparative scale separation of heavy distillates such as cat-cracker feed. It has several advantages over preparative HPLC in similar applications. This includes a sample capacity that is at least 10 times greater [25].

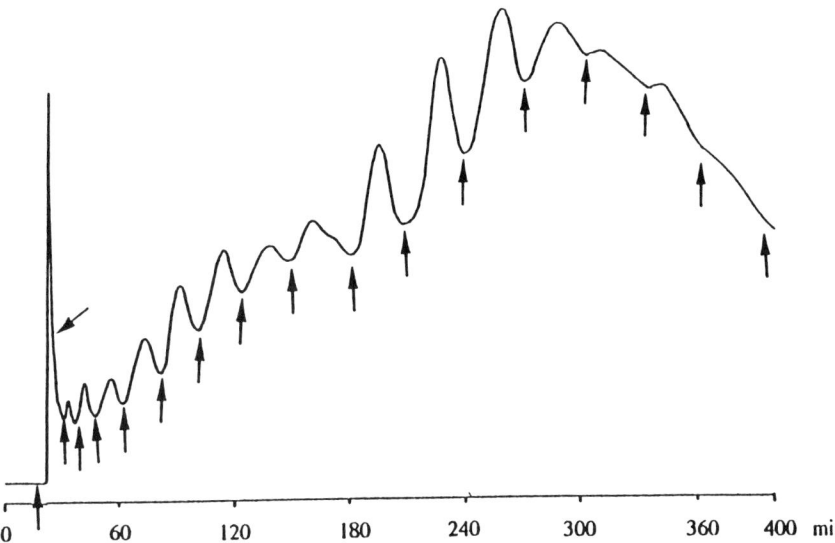

Figure 4 CPC chromatogram showing the descending fractionation of cat-cracker feed. Mobile phase composition is changing from 25% water–75% methanol to 5% water–95% methanol in a 400-minute gradient, flow rate 4 ml/min; stationary phase: hexane, nine cartridges, apparatus volume 185 ml, rotation speed increasing from 700 to 1000 rpm, variable pressure and density difference. The arrows indicate the fraction collection points.

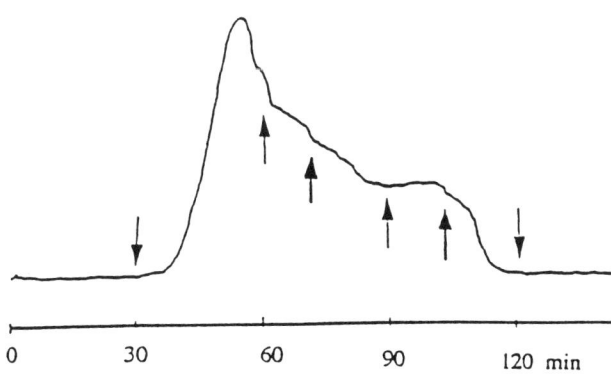

Figure 5 CPC chromatogram showing the ascending fractionation of cat-cracker feed after switching modes (see Fig. 4 for descending chromatogram). Hexane flow rate 1 ml/min, 1000 rpm, nine cartridges, 59 kg/cm^2. The arrows indicate the fraction collection points.

Also, there is no irreversible retention when using CPC. The cost of the stationary phase (a solvent in CPC) is only a fraction of that for HPLC. A greater number of distinct fractions (peaks) were found for CPC than has been reported for preparative HPLC on analogous samples.

C. Extraction of Trace Surfactant Contaminant from Waste Water

1. Problem Description

Nonionic surfactants, particularly those containing a polyoxyethylene chain as the hydrophilic moiety, are commonly used in detergents, cleaners, and process aids. The most important classes are linear alcohol ethoxylates (AEO) and alkylphenol ethoxylates (APEO). $RO(CH_2CH_2O)_nH$ and $RC_6H_4O(CH_2CH_2)_nH$ are the general chemical formulas for AEO and APEO, respectively. The R-group can be either linear or branched and may contain 8–16 carbon atoms, and n can be an average of 5–18 ethoxylate units, though commercial ethoxylated nonionic surfactants contain measurable amounts of ethoxymers that range from $n = 0$ to $n = 18+$. AEO and APEO compose a large percentage of the world's nonionic surfactant market. Approximately 40% of all household detergents that can be purchased in U.S. supermarkets contain nonionic surfactants as a major active constituent, and more than 60% of all surfactants used in industrial applications are nonionic. APEO have been used extensively as the main active component in institutional and industrial surface cleaners [26].

The extraction of APEO from water is usually done with ethyl acetate as the extractant with the addition of salts to the aqueous raffinate. The two liquids are placed in a sublator apparatus, and a stream of nitrogen is purged through the liquids to enhance the extraction of the surfactants [27]. The extraction of APEO from sewage sludge is performed by a modified soxhlet extraction with methylene chloride as the extractant [28]. The feasibility of using centrifugal partition chromatography (CPC) for extraction of nonionic surfactants from waste water has been reported [29]. CPC can be used to extract nonionic surfactants from waste water and for concentrating it in the extractant liquid used as the stationary phase of the CPC. The "waste water raffinate" is the mobile phase. This allows a continuous extraction of new raffinate into a relatively small volume of extractant. The extractant can easily be recovered from the CPC by reversing the direction of flow.

2. CPC Operation

The extraction from waste water of nonylphenol ethoxylate (NPEO), an easily UV-detectable surfactant (254 nm), was conducted in the descending mode. Of course, if the extractant liquid used has a density greater than the raffinate, the ascending mode should be used. The mobile phase and the stationary phase solvents must be mutually presaturated prior to the CPC extraction. The NPEO-spiked water was prepared by presaturating water or salt solutions with ethyl acetate and then adding approximately 10–15 mg/L of NPEO. The extraction was performed very simply: the desired amount of ethyl acetate, the extracting agent, was loaded into the CPC machine. The density difference was 0.1 g/cm^3, which allowed the use of moderate spin rates (600 rpm) to retain the ethyl acetate phase. Liters of aqueous phase were pumped through the ethyl acetate phase. The UV absorbance of the effluent was continuously monitored; it gave immediately the amount of unextracted NPEO. After extraction, the ethyl acetate phase was recovered. The mass of NPEO in the extractant phase and in the column effluent were measured, and mass balances were obtained [29].

3. Results and Discussion

Figure 6 shows the fraction of NPEO in the raffinate unextracted by the CPC as a function of the effluent volume for two CPC extraction experiments. Both extractions were performed under the same conditions except for the number of cartridges used for the column. Extraction (b) was performed using three cartridges in the CPC (column volume is 60 ml with 30 ml of the extractant ethyl acetate). Extraction (a) was performed using only one cartridge (column volume 20 ml, and 10 ml of extractant). The 0.5 line (dashed line, Fig. 6a) is crossed after 6.7 L of aqueous phase was pushed through the 10 ml ethyl acetate. The NPEO

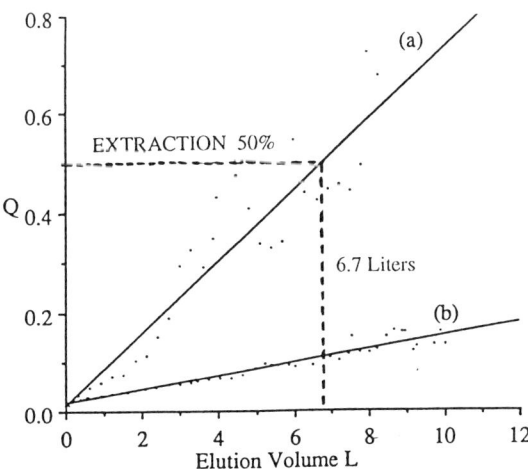

Figure 6 Extraction of NPEO with one cartridge with 10 ml of ethyl acetate (a) or three cartridges and 30 ml of ethyl acetate extractant (b). Water flow rate 5 ml/min, 600 rpm, 23°C. The inlet NPEO concentrations were 14.2 and 16.5 mg/L for (a) and (b), respectively. Q is the ratio [NPEO]$_{effluent}$/[NPEO]$_{initial}$.

partition coefficient can be estimated as 6700/10 = 670. After 5 L of raffinate was pumped through the CPC column, the CPC using three cartridges was still removing approximately 92% of the NPEO from the mobile phase; however, the extraction being done with one cartridge was removing only 62.5% of the NPEO from the mobile phase. It is obvious that increasing the volume of extractant and the number of extractions (i.e., increasing the number of cartridges) increases the percentage of NPEO extracted from the raffinate [29].

The quality of the extraction is closely related to the CPC efficiency. Figure 7 shows the fraction of NPEO unextracted from the raffinate as a function of eluent volume for two different flow rates. Curve (a) is for 5 ml/min and curve (b) is for 3 ml/min. After 2 L of raffinate was pumped through the CPC column, 95% of the NPEO was still being extracted from the stream for a flow rate of 5 ml/min, but if the flow rate was reduced to 3 ml/min, 97% could be extracted. Of course, decreasing the flow rate increases the total extraction time [29].

Figure 8 shows the eluant concentration of NPEO as a function of the effluent volume. The column consisted of three cartridges and contained only 30 ml of ethyl acetate extractant. The flow rate was 5 ml/min. Twenty liters of waste water containing 16.5 mg/L of NPEO was extracted. The total mass of NPEO was 20 L × 16.5 mg/L = 330 mg. At the end of the experiment, >75%

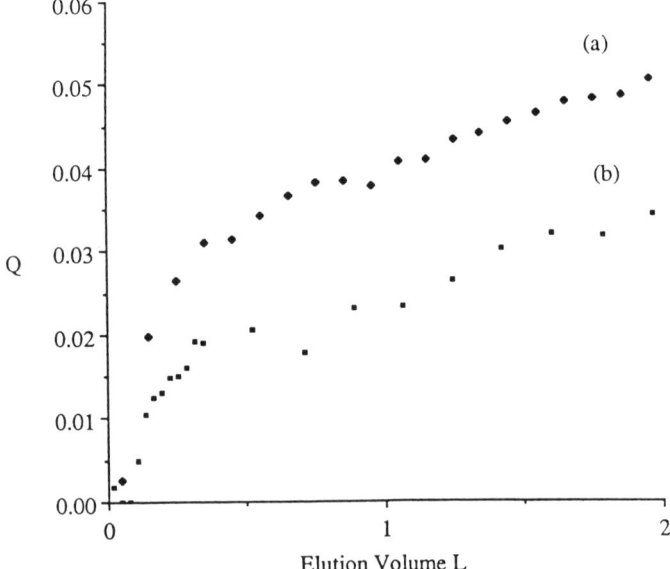

Figure 7 Effect of the flow rate on the extraction efficiency. A lower flow rate gives a higher efficiency. (a) 5 ml/min; (b) 3 ml/min. Both measurements were made with six cartridges, 600 rpm, 23°C, and 60 ml ethyl acetate extractant volume.

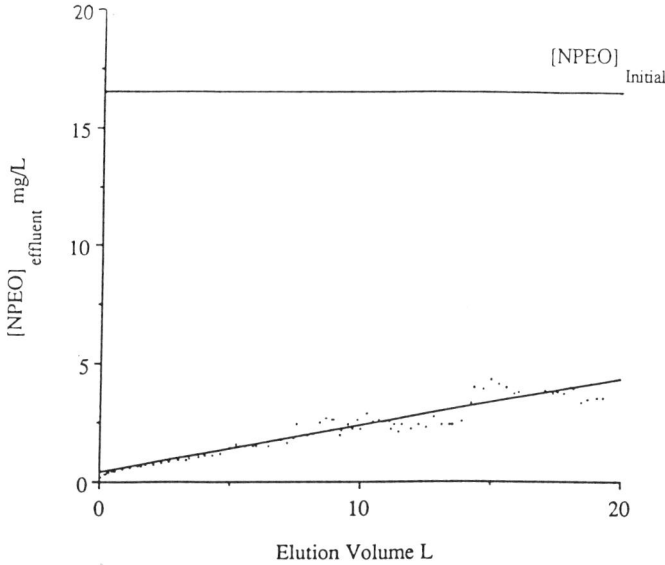

Figure 8 Effluent concentration profile for a 30-L CPC extraction experiment performed with only three cartridges, 30 ml of ethyl acetate extractant, flow rate 5 ml/min, 600 rpm, 23°C, $[NPEO]_{initial}$ = 16.5 mg/L.

of the NPEO was still being removed. Overall, >86% of the NPEO present in the raffinate was extracted; 285 mg of NPEO was concentrated into 30 ml of ethyl acetate extractant. For comparison, assuming the 30 ml of ethyl acetate were equilibrated with the 20 L of waste water (polluted by 300 mg NPEO), they would extract only 165 mg of NPEO (50%), leaving the other 165 mg in the 20 L of water. The water NPEO concentration would be half (8.25 mg/L), and the ethyl acetate concentration would be 5500 mg/L with the ratio 5500/8.25 = 670, the NPEO ethyl acetate–water partition coefficient. With CPC, there are only 45 mg of NPEO left in the aqueous phase instead of 165 mg, or 3.6 times less. The NPEO concentration in ethyl acetate is 9.5 g/L, 73% higher. Clearly, CPC offers the advantage of concentrating materials of interest into tens of milliliters of solvent rather than the hundreds of milliliters required by classical solvent sublation techniques.

This technique may be a viable approach for removing nonionic surfactants from waste water. Although this method was only shown to be effective for liter quantities of waste water, it serves as a good model for possible larger-scale industrial processes. In fact, much larger CPC apparatuses are available for larger countercurrent extractions. One area of caution one must be aware of is that of "column flooding." When a relatively high concentration of surfactant has been extracted into the stationary phase, the possibility of forming an emulsion increases. If an emulsion forms between the stationary phase and the mobile phase, the raffinate is pumped out of the column. When the emulsion band is eluted, a large peak is observed on the detector. Naturally, this leads to a large decrease in the volume of the stationary phase. It is difficult to predict when or if flooding will occur. It depends on the nature and concentration of the surfactant and the nature of the stationary and mobile phase solvents, as well as the instrumental parameters (e.g., spin rate, flow rate, cell geometry, etc.). If flooding occurs, the CPC column can be reloaded with the same amount of fresh extracting agent and the process can go on.

V. CONCLUSION

A CPC apparatus is no more difficult to operate than many other analytical instruments. The main problem is to retain an adequate amount of liquid stationary phase. The presence of a sufficient amount of stationary phase inside the CPC machine can be checked by monitoring the pressure or injecting a standard "marker" compound. The knowledge of the liquid systems, liquid–liquid equilibrium, solubility, density differences, interfacial tension, and especially partition coefficients is probably the most difficult part of CCC. The advantages of a liquid stationary phase for total sample recovery, high sample loading and enhanced selectivities, and solute extraction was outlined in this chapter. CCC also has unique capabilities in liquid–liquid partition coefficient determination and sample purification or preconcentration, as is discussed in other chapters. All these advantages make the CCC technique useful in the analytical chemistry laboratory. The development of the CCC instrumentation has improved tremendously. Now, modern CCC instruments are sufficiently reliable that they can be used on a routine basis.

REFERENCES

1. Y. Ito, in *Advances in Chromatography* (J. C. Giddings, E. Grushka, and J. Cazes, eds.), Vol. 24, Marcel Dekker, New York, Chapter 6, p. 181 (1984).
2. W. D. Conway, *Countercurrent Chromatography, Apparatuses, Theory and Applications*, VCH Publishers, Weinheim (1990).
3. N. B. Mandava and Y. Ito, *Countercurrent Chromatography*, Chromatographic Science Series, Vol. 44, Marcel Dekker, New York (1988).
4. L. R. Snyder and J. J. Kirkland, *Introduction to Modern High Performance Liquid Chromatography*, 2nd ed., John Wiley & Sons, New York, p. 36 (1979).
5. A. Berthod, *J. Chromatogr.*, 549:1 (1991).
6. A. Berthod and D. W. Armstrong, *J. Liq. Chromatogr.*, 11:547 (1988).
7. A. Berthod and N. Schmitt, *Talanta*, 40:1489 (1993).

8. A. Berthod and D. W. Armstrong, *J. Liq. Chromatogr.*, *11*:567 (1988).
9. H. R. Gerber, G. Muller, and E. Ebner, *Pro. Br. Weed Cont. Conf.*, *12*(2):787 (1974).
10. T. L. Potter, T. Carpenter, R. Putnam, K. Reddy, and J. M. Clark, *J. Agr. Food Chem.*, *39*(12):2184 (1991).
11. P. L. Wylie and R. Oguchi, *J. Chromatogr.*, *517*:131 (1990).
12. J. M. Schlaeppi, H. Moser, and K. Ramsteiner, *J. Agr. Food Chem.*, *39*(8):1533 (1991).
13. P. Blattmann, D. Gross, H. P. Kriemler, and K. Ramsteiner, *6th International Congress of Pesticide Chemistry (IUPAC)*, August 1986, Ottawa, Canada, Abstract 7A-02.
14. P. C. Kearney and D. D. Kaufman, *Herbicides: Chemistry, Degradation, and Mode of Action*, Vol. 3, Marcel Dekker, New York, p. 358 (1988).
15. T. J. Fleichmann, personal communication (Ciba-Geigy Corporation), December 1992.
16. C. D. Chang, D. W. Armstrong, and T. J. Fleichmann, *J. Liq. Chromatogr.*, *17*:19 (1994).
17. R. A. Menges, G. L. Bertrand, and D. W. Armstrong, *J. Liq. Chromatogr.*, *13*:3061 (1990).
18. S. J. Gluck and E. J. Martin, *J. Liq. Chromatogr.* 13:3559 (1990).
19. C. Bollet, J. C. Escalier, C. Souteyrand, M. Caude, and J. R. Rosset, *J. Chromatogr.*, *206*:289 (1981).
20. M. Radke, H. Willsch, and D. H. Welte, *Anal. Chem.*, *56*:2538 (1984).
21. P. L. Grizzle and D. M. Sablotny, *Anal. Chem.*, *58*:2389 (1986).
22. D. M. Jewell, E. W. Albaugh, B. E. Davis, and R. C. Ruberto, *Ind. Eng. Chem. Fundam.*, *13*:278 (1974).
23. A. Matsunga and M. Yagi, *Anal. Chem.*, *50*:753 (1978).
24. J. W. Vogh and J. S. Thomson, *Anal. Chem.*, *53*:1345 (1981).
25. R. A. Menges, L. A. Spino, and D. W. Armstrong, *Anal. Chem.*, *65*:2873 (1993).
26. P. L. Layman, *Chem. Eng. News*, Jan. 20, p. 21 (1986).
27. S. L. Boyer, K. F. Guin, R. M. Kelley, M. L. Mausner, H. F. Robinson, T. M. Schmitt, C. R. Stahl, and E. A. Setzbom, *Environ. Sci. Technol.*, *11*:1167 (1977).
28. E. Stephanon and W. Giger, *Environ. Sci. Technol.*, *16*(11):800 (1982).
29. R. A. Menges, T. S. Menges, G. L. Bertrand, D. W. Armstrong, and L. A. Spino, *J. Liq. Chromatogr.*, *15*:2909 (1992).
30. A. Foucault and K. Nakanishi, *J. Liq. Chromatogr.*, *12*:2587 (1989); and *13*:3583 (1990).

2
Theory of Centrifugal Partition Chromatography

Alain P. Foucault
Laboratoire de Bioorganique et Biotechnologies, Centre National de la Recherche Scientifique, Paris, France

I. INTRODUCTION

In this chapter, we will try to highlight some simple concepts that distinguish centrifugal partition chromatography (CPC) from its parent, high-performance liquid chromatography (HPLC), following from the liquid nature of its stationary phase.

At the present time, it is uncertain how the mobile phase really flows through the stationary phase in CPC. The mobile phase travels from channel to channel through the ducts, where it is a continuous phase, while once in the channel it may be continuous, like a film against one of the walls, flowing very quickly, then accumulating like in a pool, and waiting to be transferred in the next channel through the duct; or it may be discontinuous, like clouds or droplets going their own way, under the influence of the gravitational field, and being transferred when they reach the outlet of the channel.

Whatever the model, when the steady state is reached, with the channels having their volume shared by the stationary and the mobile phases, the governing law is always the same: What is coming in, comes out, directed by the flow rate.

Armstrong et al. [1] developed a model in which the mobile phase, once it comes out of the duct, flows very quickly to reach an intermediate emulsified layer and then decants in a third step before being transferred. We strongly suggest that the interested reader go back to the original publication for a complete description of this model.

We will describe here a simpler model, which may not be satisfying since it does not account for what is observed at very low flow rate in CPC, but is very useful for visualizing how the mobile phase flows into the stationary phase, since it

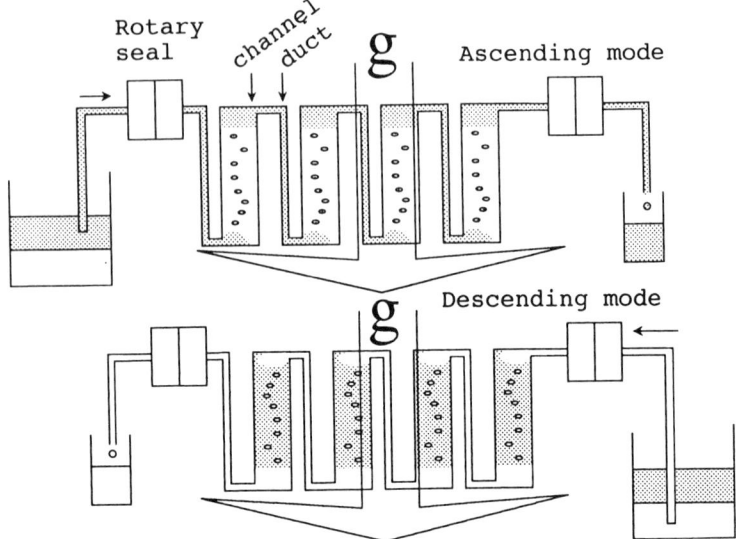

Figure 1 Simple scheme used by the pioneers of CPC to represent the arrangement between channels and ducts, with the droplets of mobile phase flowing in the stationary phase.

assumes this mobile phase consists of droplets flowing in the stationary phase with a linear velocity given by Stokes law (free fall of a rigid sphere in a continuous environment with a given viscosity); this simple model (which has always been that adopted by the pioneers of CPC, who represented their instrument with the simple scheme shown in Fig. 1) is consistent with all the experimental results we obtained in our laboratory and that are reported in several publications [2–4].

II. V_m^{min}: THE MINIMUM VOLUME OF MOBILE PHASE COMPATIBLE WITH A GIVEN FLOW RATE

The usual way of filling the CPC column is first to completely fill with stationary phase and then, when the experimental conditions have been set up (i.e., rotational speed, ascending or descending mode), to start pumping the mobile phase at a desired flow rate until it emerges from the outlet of the column. This steady state equilibrium, compatible with the experimental conditions, leads to a volume of mobile phase in the column that corresponds to the displaced volume of stationary phase that has been pushed out, and that is the minimum volume of mobile phase we can get for the actual flow rate. We will call this volume V_m^{min}.

THEORY OF CPC

V_m^{min} corresponds to the steady state equilibrium between all the parameters acting upon the hydrodynamics of the two phases in the column. This can be simply described as follows: The mobile phase is introduced at the inlet of the channel; then it travels freely in the stationary phase, "driven" by the gravitational field; then, as soon as it reaches the outlet, it is transferred to the next channel through the connecting duct. There is no real accumulation of mobile phase at the outlet of the channel.

We can, however, artificially fill the CPC with more mobile phase than allowed by the steady state equilibrium, and in this case, there will be a pool of mobile phase at the outlet of each channel, continually renewed at its surface while the bottom is transferred through the duct to the next channel. In other words, it is possible to fill the CPC column with more mobile phase than V_m^{min}, compatible with a given flow rate, and not with less (see Appendix I for the various ways to fill the CPC column).

III. $V_m^{min} = f(F)$, OR BETTER, $V_m^{min}/V_c = f(F/S)$

As already stated, let us imagine the mobile phase traveling in ducts, pushed by the pump at a given flow rate F; then, once having arrived in a channel, traveling freely into the stationary phase, "driven" by the gravitational field. The quantity of mobile phase arriving per time unit at the inlet of the channel will vary with F; the exact form of the mobile phase that will get in the channel and the velocity it will acquire will depend on the characteristics of the two phases, the geometry of the channel, and the gravitational field.

Figure 2 is a schematic of the situation in a channel under steady state

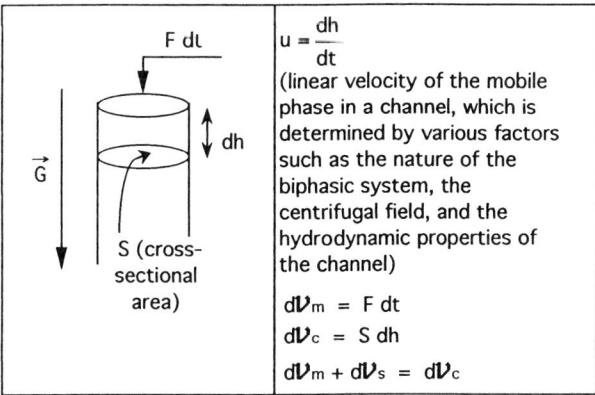

Figure 2 Steady state equilibrium in a channel at V_m^{min} conditions.

equilibrium, at V_m^{min} conditions: during the elapsed time dt, the volume introduced at the top of a channel will be $dv_m = F\,dt$. During this elapsed time, the mobile phase that was present near the inlet has been flowing down with a linear velocity u, which is related to the nature of the biphasic system (the density difference $\Delta\rho$, the viscosity of the stationary phase η_s, the interfacial tension γ), to the amplitude of the centrifugal field g, and to the design of the channel, so that it leaves a volume of the channel $dv_c = S\,dh$ ($dh = u\,dt$) to the incoming mobile phase, $F\,dt$, this volume, $S\,dh$, being shared with the stationary phase in the ratio dv_m/dv_s, corresponding to the steady state equilibrium. The volume $F\,dt$ is "sprayed out" as droplets, clouds, or other forms in the volume $Su\,dt$, so that $F\,dt/Su\,dt$ represents the relative volume of the mobile phase in the elementary volume of channel $Su\,dt$:

$$\frac{dv_m}{dv_c} = \frac{F}{S}\frac{1}{u} \tag{1}$$

If we summarize for a channel, then for all the channels it becomes

$$\frac{v_m^{min}}{v_c} = \frac{1}{\bar{u}}\frac{F}{S} + d' \tag{2}$$

where $\bar{u}$ is the average velocity in a channel (it may be not constant in the entire channel) and d' a constant from integration; limit conditions (for $F = 0$, the mobile phase is only in ducts and not in channels) lead to $d' = 0$. Adding the volume of the ducts, which are filled with the mobile phase only ($V_{ducts} = dV_c$; then $v_m^{min} = V_m^{min} - dV_c$, and $v_c = V_c(1 - d)$), it becomes

$$\frac{V_m^{min}}{V_c} = \frac{1-d}{\bar{u}}\frac{F}{S} + d \tag{3}$$

At this stage, we do not know if the linear velocity depends upon the flow rate or not. This is the first important experimental result: We found experimentally that there is a linear relationship between V_m^{min}/V_c and F/S over a broad range of values of F, which means that $\bar{u}$ is independent of F in centrifugal partition chromatography. Figure 3 shows the experimental results for two different CPC devices. The slope, $(1 - d)/\bar{u}$, is an estimate of the average velocity of the mobile phase in the channels, and the higher $\bar{u}$ is, the lower the variation of the volume of the stationary phase in the channels, for a variation of F/S, whatever the column is. This slope gives a good estimate of the stability of the stationary phase for various instruments, and it allows comparisons between them.

At this stage, let us note that $\bar{u}$ is the average velocity of the mobile phase in the channels, not in the ducts, where it is related to the flow rate and to the section of the ducts through a trivial relationship. The resulting average linear velocity in the entire instrument (channels + ducts), which has no physical

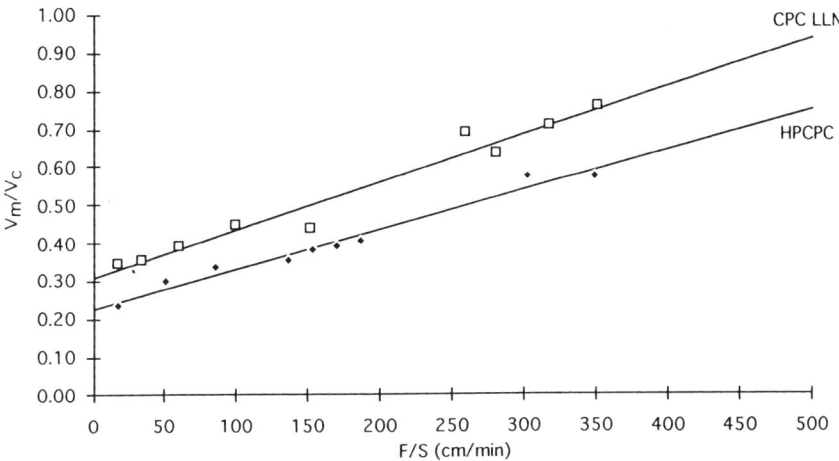

Figure 3 Linear relationship $\dfrac{V_m^{min}}{V_c} = \dfrac{1-d}{\bar{u}} \dfrac{F}{S} + d$ for the two CPC columns.

CPC column	V_c (ml)	S (cm^2)	$(1-d)/\bar{u}$ (min/cm)	d	$\bar{u}$ (cm/s)
CPC model LLN	108	0.026	0.00125 (± 0.00008)	0.31 (± 0.016)	9.2
HPCPC	240	0.059	0.00104 (± 0.00006)	0.23 (± 0.012)	12.3

meaning, is calculated from the retention time of a nonretained solute, as it is in HPLC.

This first important experimental result, "there is a linear relationship between the minimum volume we can get at a given flow rate and that flow rate," gives Eq. 3 great promise.

This has been verified for 12 solvent systems in both descending and ascending mode using the "high-performance CPC" (HPCPC). These 12 systems represent a wide range of physical and chemical properties (see Tables 1 and 2 for abbreviations and physical data). EtAOc/water and $CHCl_3$/water are not very useful for purification purposes, but they have a large interfacial tension; Oct/water is used for partition coefficient (hydrophobicity parameter) determination, and its upper phase is rather viscous; HEP/MeOH is widely used as a nonpolar system; WDT2, 4, and 5 are a new class of medium-polarity biphasic systems that were introduced recently [5], containing water, dimethylsulfoxide, and tetrahydrofuran, and they show very good solvating properties; MIBK/AcO/W and $CHCl_3$/MeOH/PrOH/W are medium polarity systems, too, while the three butanol-containing systems are polar systems, widely used for purification of polar

Table 1 Abbreviations Used in This Chapter

Abbreviations	Solvents and their volume ratio		
EtOAc/Water	Ethyl acetate/Water, 50/50		
CHCl$_3$/Water	Chloroform/Water, 50/50		
n-BuOH/Water	n-Butanol/Water, 50/50		
sec-BuOH/Water	sec-Butanol/Water, 50/50		
HEP/MeOH	Heptane/Methanol, 50/50		
Oct/Water	Octanol/Water, 50/50		
WDT group	Water	Dimethylsulfoxide	Tetrahydrofuran
WDT2 (see Ref. 5)	11.7	26.3	62
WDT4	21.5	21.2	57.3
WDT5	24.5	16.2	59.3
BAW	n-Butanol/Acetic acid/Water, 40/10/50		
MIBK/AcO/W	Methylisobutyl ketone/Acetone/Water, 25/50/25		
CHCl$_3$/MeOH/PrOH/W	Chloroform/Methanol/n-Propanol/Water, 29/38.7/6.5/25.8		

Table 2 Physical Data of the Biphasic Systems

System[a]	$\Delta\rho$ (g/cm^3)	γ (dyne/cm)	η (cP) Upper	η (cP) Lower
EtOAc/Water	0.097	13.2	0.47	1.10
CHCl$_3$/Water	0.478	32.8	0.98	0.55
HEP/MeOH	0.073	1.16	0.41	0.57
WDT2	0.102	0.59	0.72	2.37
WDT4	0.130	2.15	0.62	2.88
WDT5	0.115	1.26	0.72	2.60
MIBK/AcO/W	0.084	0.25	0.70	1.42
CHCl$_3$/MeOH/PrOH/W	0.213	0.42	1.78	0.97
Oct/Water	0.137	8.5[b]	7.10	0.80
n-BuOH/Water	0.144	2.91	3.05	1.33
BAW	0.101	1.21	2.89	1.51
sec-BuOH/Water	0.093	0.53	3.64	2.02

[a]The solvent systems are roughly sorted according to the Ito Classification in hydrophobic, intermediate, and hydrophilic groups; data: $\Delta\rho$ = density difference; γ = interfacial tension; η = viscosity.
[b]From *Handbook of Chemistry and Physics* [10].
Data are from Ref. 4.

compounds such as peptides. Regression analysis for the 12 systems in the descending and ascending modes (except for the Oct/Water system in the ascending mode) is shown in Table 3 for various rotational speeds of the HPCPC column; the calculated linear velocity $\bar{u}$ of the mobile phase in the channels is shown, too.

Three significant observations emerge from these results:

1. It is surprising to find that the slope of the regression line is not dependent upon the rotational speed of the HPCPC column, in the range ordinarily used by chromatographers (about 500–2000 rpm). The lower limit of this nondependence with the system WDT4, in the descending mode, was tested. It was still true at 400 rpm, while the stationary phase was no longer stable at 200 rpm. How the lower limit was system-dependent was not checked since, as we will show, it is always better for the chromatographer to work at higher rotational speeds.

2. Even if one can describe a system as "more stable" or "less stable," they are all stable enough to allow for chromatographic runs. For example, one can calculate from the data of Table 3 that the retention of the stationary phase ($1 - V_m^{min}/V_c$) will be 54% for $F = 5$ ml/min for the system $CHCl_3$/MeOH/PrOH/W in the descending mode, and 41% for the same flow rate, for the system WDT2 in the ascending mode. For the system BAW, the retention of the stationary phase will be 57% and 40% for $F = 10$ ml/min in the descending and ascending modes, respectively. We will define as *very stable* the systems where the retention of the stationary phase is >60% for $F = 10$ ml/min, as *less stable* those where the retention is <50% for the same flow rate, and as *stable* the intermediate systems; rules in Table 3 separate these categories.

3. The ratio d of the volume of ducts and connections to that of the column should be a constant since they are geometrically defined; but we see in Table 3 that d varies from 14% (WDT4 in the ascending mode) to 33% ($CHCl_3$/MeOH/PrOH/W in the ascending mode), with an average of 22% for the descending mode and 21% in the ascending mode. Using the dimensions given by the manufacturer for the HPCPC, we find that the ducts represent a total volume of ≈ 32 ml; adding 5 ml for the connections between disks and for the two rotary seals in the system leads to a minimum value of $d \approx 15\%$, which is approximately the minimal value we find for the system WDT4 in the ascending mode. Higher values for d must be understood as a consequence of the geometry of the channels and ducts and the way the mobile phase is injected into the channels, that is, the nature of the passage of a continuous mobile phase, in the ducts, to a discontinuous mobile phase, in the channels, and of the inverse phenomenon. We must then define a *dynamic duct*, which is the part of the column where there is no stationary phase and thus no chromatographic process. The best way to measure the dynamic ducts for a CPC column and a biphasic system is to calculate the slope and intercept of Eq. 3 by regression analysis. From our results, we find a dynamic duct of $\approx 22\%$ for most of the biphasic systems, the worst being 33% for the system $CHCl_3$/MeOH/PrOH/W

Table 3 Data for the Linear Regression Analysis of the Relationship: $\dfrac{V_m^{\min}}{V_c} = \dfrac{1-d}{\bar{u}} \dfrac{F}{S} + d$

Descending Mode: Lower Phase As Mobile Phase

System[a] (rot. speed, rpm)	$10^3 \dfrac{1-d}{\bar{u}}$ (min/cm) (10^3 std. error)	$10d$ (10 std. error)	r^2 (no. of exp.)	$\bar{u}$ (cm/sec)	
$CHCl_3$/Water (700)	0.46 (0.05)	2.02 (0.05)	0.988 (3)	28.8	very stable
Oct/Water (1000)	0.52 (0.06)	2.23 (0.06)	0.988 (3)	24.8	
EtOAc/Water (1200)	0.64 (0.04)	2.65 (0.05)	0.984 (6)	19.0	
n-BuOH/Water (800, 1200)	0.96 (0.08)	2.13 (0.09)	0.96 (8)	13.6	
WDT5 (800, 1200, 1400)	1.1 (0.06)	2.26 (0.07)	0.96 (13)	11.7	stable
WDT4 (400, 1200, 1400)	1.1 (0.05)	2.28 (0.06)	0.99 (6)	11.9	
HEP/MeOH (1200)	1.3 (0.09)	2.3 (0.07)	0.986 (5)	9.6	
BAW (1200)	1.3 (0.1)	2.1 (0.1)	0.985 (4)	9.8	
WDT2 (1000, 1400)	2.5 (0.2)	2.6 (0.1)	0.97 (9)	4.9	less stable
sec-BuOH/Water (1200)	2.7 (0.2)	1.6 (0.1)	0.97 (10)	5.1	
MIBK/AcO/W (1000)	2.9 (0.4)	2.5 (0.5)	0.96 (4)	4.3	
$CHCl_3$/MeOH/PrOH/W (700)	3.16 (0.09)	1.9 (0.1)	0.997 (5)	4.3	

THEORY OF CPC

Ascending Mode: Upper Phase As Mobile Phase

System[a] (rot. speed, rpm)	$10^3 \frac{1-d}{\bar{u}}$ (min/cm) (10^3 std. error)	$10d$ (10 std. error)	r^2 (no. of exp.)	$\bar{u}$ (cm/sec)	
CHCl$_3$/Water (700)	0.46 (0.01)	2.29 (0.01)	0.998 (3)	28.1	very stable
EtOAc/Water (1200)	1 (0.05)	1.48 (0.05)	0.990 (6)	14.2	
n-BuOH/Water (800, 1200)	1.2 (0.05)	1.54 (0.05)	0.990 (7)	11.8	
HEP/MeOH (1200)	1.5 (0.2)	1.9 (0.2)	0.988 (3)	9.0	stable
WDT5 (800, 1400)	1.8 (0.1)	1.8 (0.1)	0.97 (8)	7.53	
WDT4 (1200)	2.0 (0.3)	1.4 (0.4)	0.97 (3)	7.15	
sec-BuOH/Water (1200)	2.3 (0.1)	2.8 (0.2)	0.97 (8)	5.3	less stable
BAW (1200)	2.5 (0.2)	1.8 (0.2)	0.986 (5)	5.5	
CHCl$_3$/MeOH/PrOH/W (700)	2.6 (0.7)	3.3 (0.7)	0.81 (5)	4.3	
MIBK/AcO/W (1000)	2.8 (0.1)	2.9 (0.2)	0.988 (7)	4.3	
WDT2 (1400)	5.0 (0.6)	1.7 (0.5)	0.95 (5)	2.8	
Octanol/Water	deficient pumping				

[a]Systems are sorted in order of decreasing stabilities, and the rules mark the separations between very stable, stable, and less stable systems (see text).
Other data: $Vc = 240$ ml; $S = 0.059$ cm^2; F from 1 to 10 ml/min.

in the ascending mode. Twenty-two percent instead of fifteen percent corresponds to a layer of pure mobile phase of ≈ 8 μl in each channel, which is ≈ 88 μl.

IV. THE STOKES MODEL

As is customary in other fields (e.g., sedimentation of proteins in a centrifuge), we will adopt a very simple model to describe the hydrodynamics of the flow in a channel. We will imagine the mobile phase in a channel as droplets with an average radius a flowing in the stationary phase at constant linear velocity $\bar{u}$, previously defined, and corresponding to the radius of rigid spheres, which will follow the Stokes law:

$$\bar{u} = \frac{2a^2 \Delta\rho\, g}{9\eta_{SP}} \tag{4}$$

where $g = \omega^2 R$, R being the average radius of the centrifuge and ω the rotational speed, $\Delta\rho$ is the density difference between the two phases, and η_{SP} is the viscosity of the stationary phase, which can be the upper or the lower one, depending upon the chosen mode of operation.

In other words, instead of

> The mobile phase in a channel is something discontinuous flowing with an average linear velocity $\bar{u}$,

we say now

> The mobile phase is envisioned as regular droplets with average radius a, flowing with a constant linear velocity $\bar{u}$.

As a model it is not necessary to know if the mobile phase is actually flowing as droplets, but what we have to do is to verify that this model is compatible with experimental results.

If we adopt the Stokes model to describe the flow pattern, then the linear velocity $\bar{u}$ of the droplets in the channels does not depend upon the flow rate, in the range studied. The same result occurs using the CPC type LLN or the HPCPC; it seems to be an inherent feature of centrifugal partition chromatography. This means that once the mobile phase coming from the ducts is sprayed out as droplets into the channel, these droplets acquire their speed limit very quickly and travel in the channel with no "memory" of the way they were produced.

A second important experimental result is the following: For a given biphasic system, and at a given flow rate, the volume of the mobile phase in the column is not dependent upon the rotational speed, in the range ordinarily used by the chromatographer ($\omega > 700$ rpm). Two examples are given in Figs. 4 and 5, and other data can be found in Table 3.

THEORY OF CPC

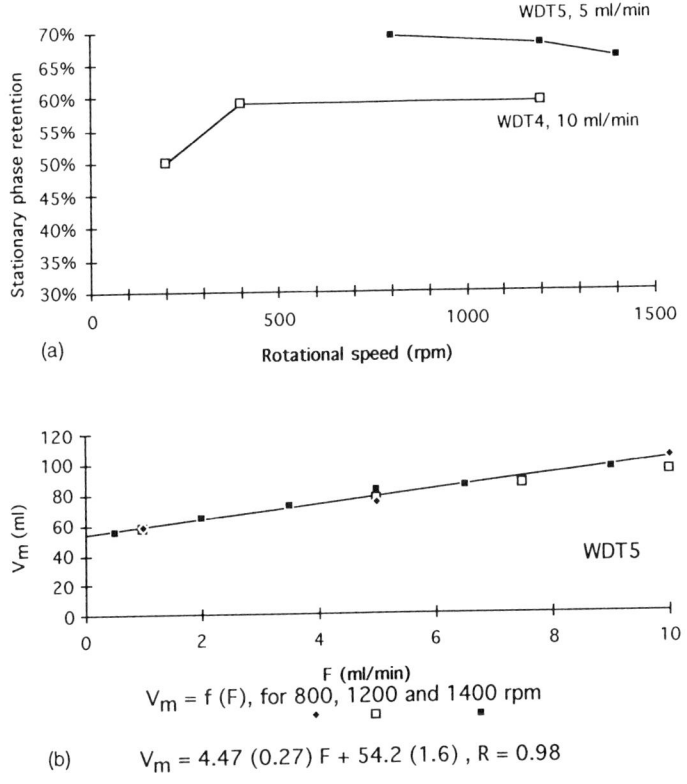

Figure 4 Independence of the stationary phase retention at higher rotational speed. Apparatus: HPCPC (2136 channels), V_c = 240 ml. Biphasic system: WDT5: water/DMSO/THF 25/16/59 v/v/v; WDT4: water/DMSO/THF 22/21/57 v/v/v; descending mode, F as calculated. (a) Stationary phase retention versus rotational speed; (b) mobile phase volume versus flow rate, for various rotational speeds.

With our simple model, this means that the linear velocity $\bar{u}$ is not dependent upon g, the centrifugal field, with the immediate consequence

$$a^2 g = A \tag{5}$$

A being a constant for a specific system and mode. This means that the higher the acceleration field, the smaller the radius of the "droplets" of mobile phase in a channel. Comparing various biphasic systems, A will characterize the dispersion of the mobile phase in the stationary phase, since a smaller value of A means that, for a given acceleration field, the continuous mobile phase coming from the duct will break into many smaller droplets, while a larger value of A means it will

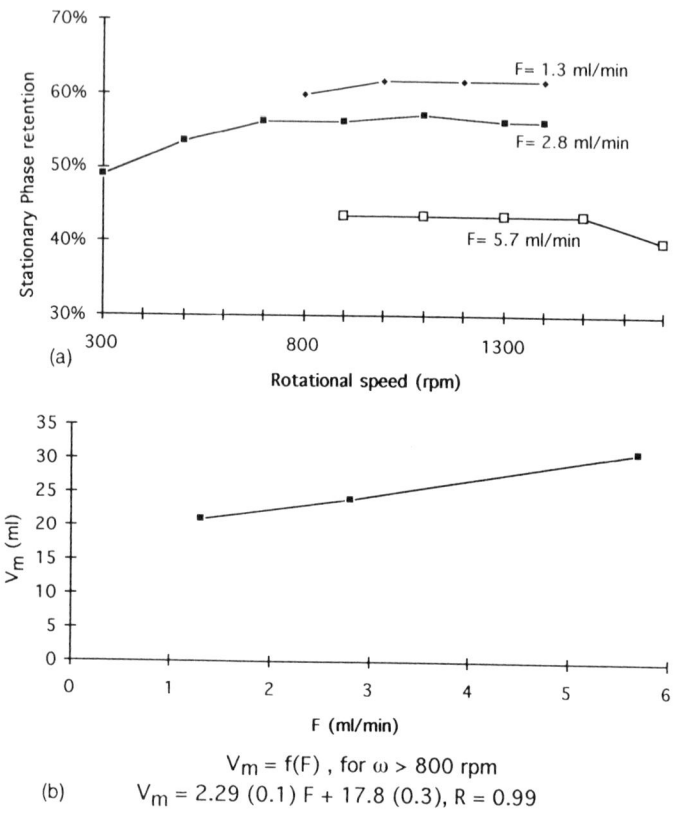

$V_m = f(F)$, for $\omega > 800$ rpm
(b) $V_m = 2.29 (0.1) F + 17.8 (0.3)$, R = 0.99

Figure 5 Independence of the stationary phase retention at higher rotational speed. Apparatus: CPC-LLN, with 3 cartridges (1200 channels), $V_c = 55$ ml. Biphasic system: $CHCl_3$/EtOAc/MeOH/Water 2/2/3/2 v/v/v/v; descending mode, F as calculated. (a) Stationary phase retention versus rotational speed; (b) mobile phase volume versus flow rate, for various rotational speeds.

break into few larger droplets. A may be called the *dispersion term*; its dimension is $L^3 T^{-2}$. From Eqs. 4 and 5 we get (for calculating A)

$$A = \frac{9}{2} \frac{\bar{u}\eta_{SP}}{\Delta\rho} \qquad (6)$$

As the dimension of A is the same as that of the ratio $\gamma/\Delta\rho$ (i.e., $L^3 T^{-2}$), where γ is the interfacial tension, we can then write

$$A = B\frac{\gamma}{\Delta\rho} \qquad (7)$$

where B is a dimensionless number. From Eqs. 5 and 7, this gives

$$B = \frac{a^2 g \, \Delta\rho}{\gamma} \tag{8}$$

Equation 8 can be used to compare B with the numerous dimensionless numbers we can find in the literature (there are some very useful tables at the end of the *Handbook of Chemistry and Physics*) and thus find out if it has already been described. B was defined in 1928 by W. N. Bond et al. [6] and is called the Bond number [7]. It characterizes the relative importance of gravitational to surface-tension forces, and accounts for the fragility of a droplet. "A droplet of liquid in motion through another liquid differs in its behavior from a solid sphere in that it may (a) be deformed, (b) have a circulation set up within itself by the shearing effect of the relative motion of the two fluids. These effects upset the stability of the drop, causing it to oscillate about the spherical shape and eventually to burst into fragments or, at least, into smaller drops" [7]. From Eqs. 6 and 7, we get (for calculating B)

$$B = \frac{9}{2} \frac{\bar{u}\eta_{SP}}{\Delta\rho} \tag{9}$$

Like A, B is not dependent upon the acceleration field. Table 4 gives the values of A and B calculated with Eqs. 6 and 9 for the 12 solvent systems in the descending and

Table 4 The Dispersion Term, A, and the Bond Number, B

System[a]	Descending mode		Ascending mode	
	A[b]	B[b]	A	B
$CHCl_3$/Water	2.66	0.04	1.46	0.02
Octanol/Water	57.87	0.93	—	—
EtOAc/Water	4.14	0.03	7.28	0.05
n-BuOH/Water	13.54	0.67	4.70	0.23
WDT5	3.67	0.33	7.62	0.69
WDT4	2.56	0.15	7.15	0.43
Heptane/MeOH	2.42	0.15	3.18	0.20
BAW	12.64	1.06	3.69	0.31
WDT2	1.58	0.27	2.90	0.50
sec-BuOH/Water	9.04	1.57	5.14	0.90
MIBK/AcO/W	1.61	0.53	3.27	1.08
$CHCl_3$/MeOH/PrOH/W	1.61	0.82	0.88	0.45

[a]The systems are sorted according to their decreasing stability in the descending mode.
[b]A in $cm^3 \, s^{-2}$; B dimensionless.

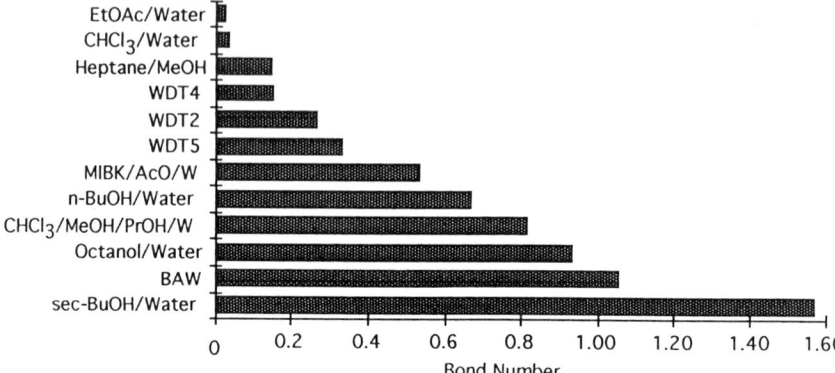

Figure 6 The 12 systems and their Bond numbers, corresponding to the descending mode.

ascending modes; systems are sorted in order of decreasing stability in the descending mode. Figure 6 shows the 12 systems with the Bond number corresponding to the descending mode, and sorted in order of increasing Bond number. From Table 4 and Fig. 6, we can conclude that B does not account for the stability of the systems in centrifugal partition chromatography, but it corresponds to the Ito classification of the so-called "hydrophobic and hydrophilic systems"; this

Table 5 Comparison of the Parameter $\gamma/\Delta\rho$ and the Experimental Average Linear Velocities of the Mobile Phase, in Descending and Ascending Models

System	$\gamma/\Delta\rho^a$	$\bar{u}$ DM[a]	$\bar{u}$ AM[a]
EtOAc/Water	136.51	19.0	14.2
$CHCl_3$/Water	68.62	28.8	28.1
Octanol/Water	62.04	24.8	—
n-BuOH/Water	20.21	14.2	11.3
WDT4	16.61	11.9	7.2
Heptane/MeOH	15.96	9.6	9.0
BAW	11.97	9.8	5.5
WDT5	10.97	13.1	7.5
WDT2	5.84	4.9	2.8
sec-BuOH/Water	5.74	5.1	5.3
MIBK/AcO/W	3.02	4.3	4.3
$CHCl_3$/MeOH/PrOH/W	1.97	4.3	4.3

[a] $\bar{u}$ in cm/s, $\gamma/\Delta\rho$ in cm^3 s^{-2}

THEORY OF CPC

accounts for the hydrodynamic behavior of solvent systems in the Ito CCC apparatus [8,9]. From Table 4, it appears that the stabilities of the solvent systems are related to both the dispersion of the mobile phase (A) and to the fragility of the droplets (B); higher values of A (large droplets) combined with lower values of B (less fragility) result in very stable systems (e.g., $CHCl_3$/water, EtOAc/water); lower values of A (small droplets) combined with higher values of B (more fragility) yield less stable systems (e.g., MIBK/AcO/W, $CHCl_3$/MeOH/PrOH/W).

Table 5 shows the 12 systems sorted according to $A/B = \gamma/\Delta\rho$, a parameter that accounts both for the dispersion of the mobile phase and for the fragility of the droplets; Fig. 7 shows the correlation between this parameter and the average linear velocities of the mobile phases in the channels. It can be seen from these data that $\gamma/\Delta\rho$ varies in the same way as the average linear velocities that were experimentally determined.

This correlation could be interpreted as follows:

1. $\dfrac{\gamma}{\Delta\rho}$ small $\equiv$ A small and B large

   Small and unstable droplets moving slowly, and easily broken into smaller ones, leading to some possible emulsification, dragging the stationary phase out of the column. These systems can be used at low flow rate, in order to minimize emulsification and keep a sufficient amount of stationary phase in the column.

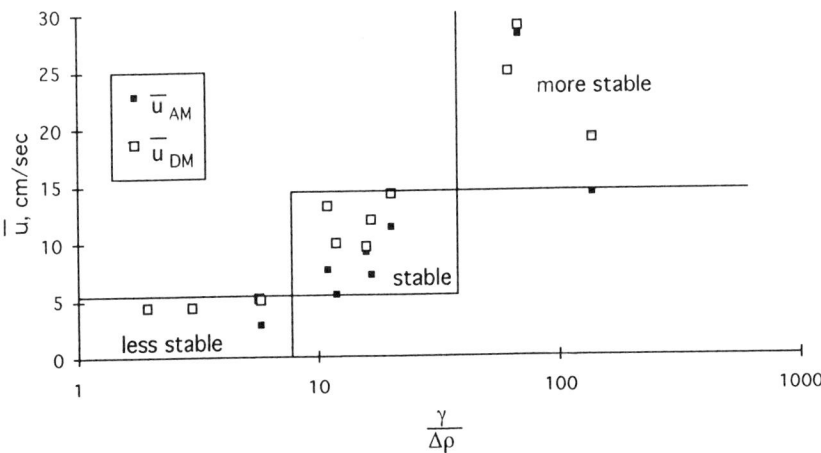

Figure 7 Correlation between the average linear velocity $\bar{u}$ of the mobile phase in the channels of the HPCPC and the parameter $\gamma/\Delta\rho$. $\bar{u}_{AM}$ = average linear velocity in the ascending mode; $\bar{u}_{DM}$ = average linear velocity in the descending mode.

2. $\dfrac{\gamma}{\Delta\rho}$ large ≡ A large and B small

   Large and very stable droplets moving fast; if so, rapid mass transfer between mobile and stationary phase is not favored for these systems, which should display poor chromatographic efficiencies.

3. $\dfrac{\gamma}{\Delta\rho}$ medium ≡ A and B medium

   This is the common case and is the best suited for chromatographic applications; the droplets are small enough to allow for a reasonable rate of mass transfer between the two phases, and fast enough to keep a large amount of stationary phase in the column. Some emulsified layer may be present in the channels, as described by Armstrong et al. [1], but this has no negative repercussions upon the stability of the stationary phase.

V. CONSEQUENCE OF THE STOKES MODEL

Using our very simple model to describe the flow pattern of the mobile phase, we can express the interfacial area between the two phases in the channels of the CPC column. From Eq. 3, the volume of mobile phase in the channel (not in ducts and connections) is

$$v_m^{min} = V_m^{min} - V_{ducts} = \frac{1-d}{\bar{u}} \frac{F}{S} V_c \qquad (10)$$

Since the droplets have surface area $s = 4\pi a^2$ and volume $v = (4/3)\pi a^3$, the number of droplets in the channels is $n = v_m^{min}/v$, and the interfacial area between the two phases in the channels is

$$I = ns = v_m^{min}\frac{3}{a} \qquad (11)$$

From Eqs. 5 and 11, we obtain

$$I = 3v_m^{min} A^{-1/2} g^{1/2} \qquad (12)$$

Combining this with Eqs. 6 and 11, we obtain

$$I = (1-d)V_c \frac{F}{S} \frac{27}{2} \frac{v_{SP}}{\Delta\rho} A^{-3/2} g^{1/2} \qquad (13)$$

where all terms are constant for a given system and mode, except for F (or v_m^{min} in Eq. 12) and g.

Assuming that the efficiency is linearly related to the interfacial area, verification of Eq. 12 or 13 with respect to g has been done using the HPCPC. We

THEORY OF CPC

used the system n-heptane/methanol in the descending mode, the rotational speed being varied between 700 and 2000 rpm. The flow rate was kept constant and equal to 7 ml/min, and the volume of mobile phase, V_m, was 120 ± 4 ml, 27 ml more than the calculated V_m^{min}, in order to keep this volume as constant as possible throughout the experiments—this parameter has a strong influence upon the efficiency, as we will show. V_m was determined using 4-hydroxybenzoic acid (nonretained solute), and diethyl phthalate was the analyte, with a partition coefficient of 0.27 ± 0.01. The number of plates has been estimated by using both the width at half height and at the base of the peak, and runs where the two values were not in close agreement were rejected. Runs where V_m was not constant were also rejected. Ultimately, 59 out of 77 injections were taken into consideration.

Figure 8 shows the results of this investigation. As predicted by Eqs. 12 or 13, based on the Stokes model, N is proportional to $g^{1/2}$, that is, to the rotational speed ω, in the range ordinarily used by the chromatographer (500–2000 rpm).

Regression analysis leads to the equation:

$$N = 0.98\ (0.03)\ \omega - 175\ (40)$$
$$n = 59, \quad r^2 = 0.95, \quad s = 87, \quad \mathcal{F} = 1126$$

where n is the number of experiments, r the correlation coefficient, s the standard deviation, and $\mathcal{F}$ the Fisher's test parameter (ω in rpm).

The dependence of the efficiency on the volume ratio of the mobile phase will be shown in the next section.

We conclude from these results that even if it has not been proven that the mobile phase actually flows as droplets through the stationary phase in the

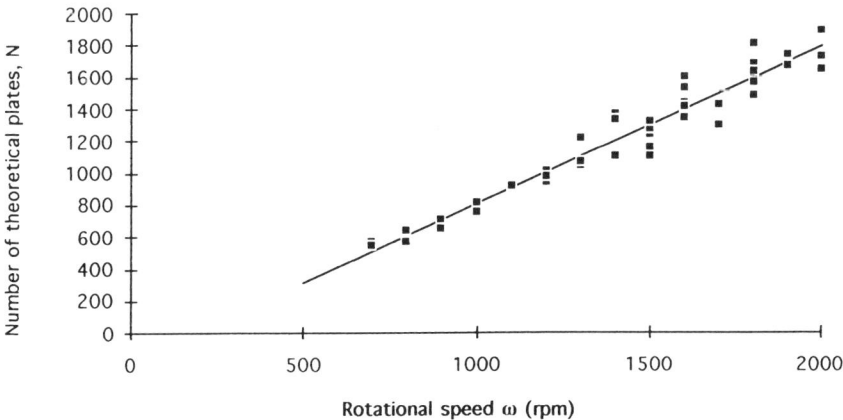

Figure 8 Relationship between N, the number of theoretical plates, and ω, the rotational speed of the HPCPC column, all other parameters being kept constant.

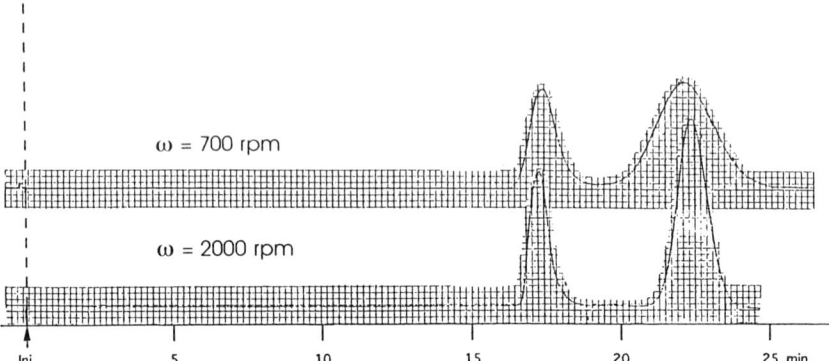

Figure 9 Comparison of HPCPC chromatograms obtained at 700 and 2000 rpm, all other parameters being kept constant. Solvent system: heptane/methanol, descending mode; unweighted but identical quantities injected; flow rate: 7 ml/min; $V_m/V_c = 0.5$.

channels of the HPCPC column, we can nevertheless use the simple Stokes model to account for the dependence of the efficiency upon the rotational speed. The upper limit of this relationship should be reached when N approximates the number of physical plates, that is, the number of channels in the instrument, but this has not been investigated yet. Figure 9 allows the comparison of two chromatograms obtained at 700 and 2000 rpm, all other parameters being kept constant.

VI. IMPORTANCE OF V_m/V_c AND F FOR THE VARIATION OF THE EFFICIENCY IN CENTRIFUGAL PARTITION CHROMATOGRAPHY

The relative influence of the two fundamental chromatographic parameters, that is, the flow rate F and the volume ratio of the mobile phase in the column, V_m/V_c, can be easily determined by independently varying these two parameters, which is possible using the various ways we can fill a CPC column (see Appendix I). This study, which has been fully developed (see Refs. 2 and 3) leads to some major conclusions:

  In the range 1–10 ml/min, the efficiency does not vary with the flow rate, but increases sharply when the ratio of the mobile phase in the column increases. This may be explained using our simple model, since the velocity of the mobile phase within a channel does not depend upon the flow rate. Put another way, the exchange between the two phases should be favored by a better balance in the channel, that is, around $V_m/V_c = 0.5$.

  Using a similar biphasic system (hexane/methanol with 4% water), and the same test solutes as in the previous section, we estimated the relative influence of F and V_m/V_c. In Fig. 10 (left side) the efficiency (calculated on a peak given by the

THEORY OF CPC

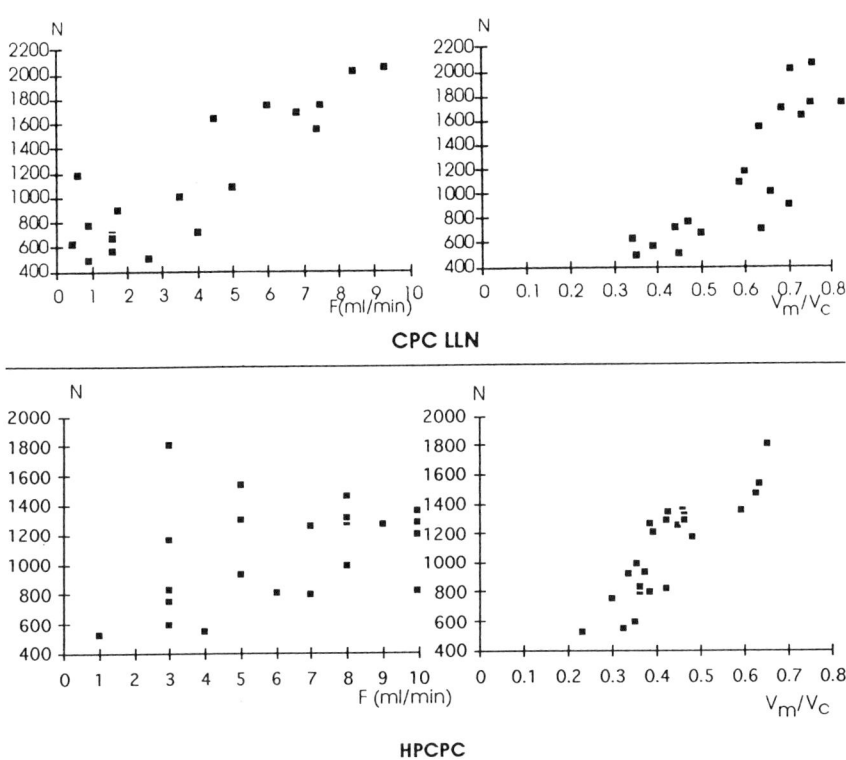

Figure 10 Evolution of the efficiency with the flow rate, whatever the ratio of the mobile phase in the column (i.e., $N = f(F, \forall V_m/V_c)$, left), and with the ratio of the mobile phase in the CCC column, whatever the flow rate (i.e., $N = f(V_m/V_c, \forall F)$, right), for the three CCC columns. Biphasic system: hexane/methanol, 4% water, descending mode; solute: diethyl phthalate.

diethyl phthalate, for two CPC columns) has been plotted against the flow rate, whatever the relative volume of the mobile phase in the column (i.e., $N = f(F, \forall V_m/V_c)$), and on Fig. 10 (right side) the same efficiency has been plotted against the relative volume of the mobile phase in the CPC columns, whatever the flow rate (i.e., $N = f(V_m/V_c, \forall F)$). ($\forall$ means *for any and all*.)

The same results can be observed for the two CPC columns: the efficiency is much more dependent upon the relative volume of the mobile phase in the column than upon the flow rate. There is no apparent correlation at all between N and F, while the relationship between N and V_m/V_c appears to be an increasing linear relationship for the CPC model LLN and the HPCPC.

Assuming that the efficiency is linearly related to the volume of mobile phase in the channels (Eq. 12) and that it is similar to V_m^{min}/V_c, verification of this equation with regard to V_m^{min}/V_c has been done using the HPCPC and the CPC model LLN. Figure 11 shows the efficiency plotted against V_m^{min}/V_c for the two instruments, the straight lines corresponding to the following regression analysis:

$$\text{CPC LLN:} \quad N = 4000\,(350)\,\frac{V_m^{min}}{V_c} - 1000\,(200)$$

$$n = 9, \quad r^2 = 0.94, \quad s = 166, \quad \mathcal{F} = 126$$

$$\text{HPCPC:} \quad N = 4600\,(450)\,\frac{V_m^{min}}{V_c} - 600\,(150)$$

$$n = 6, \quad r^2 = 0.96, \quad s = 61, \quad \mathcal{F} = 98$$

where n is the number of experiments, r the correlation coefficient, s the standard deviation, and $\mathcal{F}$ the Fisher's test parameter.

This correlation has to be further investigated in order to better define the range where the various parameters are acceptable.

Since the efficiency strongly correlates with the relative volume of the mobile phase, it is important to keep this parameter constant if the relationship between another parameter, such as the flow rate or the rotational speed, and the efficiency needs to be determined. Otherwise, the observed effects may not be due to the studied parameter, but to the uncontrolled variation of the relative volume of the mobile phase.

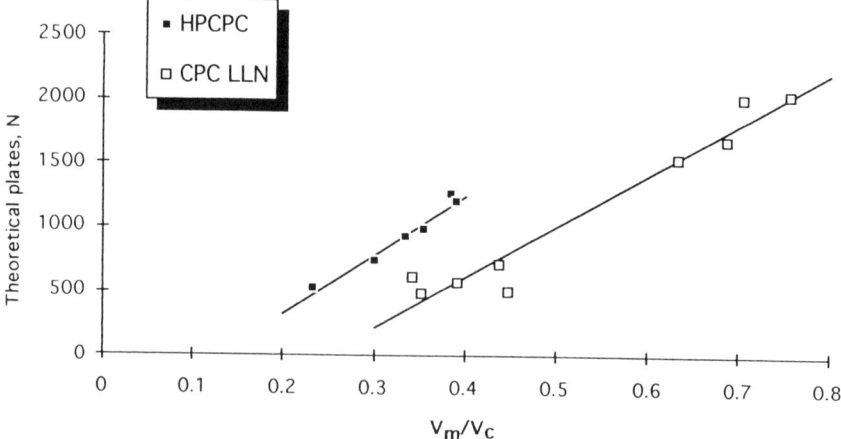

Figure 11 Correlation between the efficiency and the volume of mobile phase in the CPC column, at V_m^{min} conditions. Same system and solute as in Fig. 10.

THEORY OF CPC

When experiments are restricted to the V_m^{min} conditions, where V_m^{min} and F are linearly dependent (Eq. 3), one can either express the efficiency as a function of F or of V_m^{min}.

A. Importance of V_m/V_c for the Variation of the Resolution in Centrifugal Partition Chromatography

From the classical equations

$$\text{Rs} = 2\frac{V_{r2} - V_{r1}}{\omega_1 + \omega_2} \tag{14}$$

and

$$N = 16\left(\frac{V_r}{\omega}\right)^2 \tag{15}$$

where V_r is the retention volume for a peak, ω its base width, N the number of theoretical plates, and Rs the resolution between two peaks, and assuming that the number of theoretical plates is the same for two close peaks (and this is nearly true for CPC), we can rewrite Eq. 14 as

$$\text{Rs} = \tfrac{1}{2}\sqrt{N}(\alpha - 1)\frac{K_1}{K_1(\alpha + 1) + 2V_m/V_s} \tag{16}$$

where α is the selectivity factor (K_2/K_1), and K_1 is the partition coefficient of solute 1. The strong correlation between N and V_m/V_c, together with the supplementary term V_m/V_s, which appears in the denominator, should result in a very strong correlation between Rs and V_m/V_c, whatever the flow rate; that is, $\text{Rs} = f(V_m/V_c, \forall F)$.

Figure 12 shows the results obtained with the HPCPC for various flow rates in the range 1-10 ml/min. Plots linked together are those corresponding to the V_m^{min} conditions. As we can observe, there is a strong correlation between Rs and V_m/V_c, whatever the flow rate; and for $V_m/V_c > 0.2$, the resolution slowly decreases when V_m/V_c increases.

B. Resolution at V_m^{min} Conditions

Most CPC users work at V_m^{min} conditions, to find the best combination of a large volume of stationary phase with a high flow rate, and they often adjust the composition of their biphasic system to get partition coefficients around 1 for compounds of interest. Since, for that value ($K = 1$), the retention volume is exactly one column volume ($V_r = V_m + KV_s$), we found it interesting to calculate the resolution given by instruments for a given number of column volumes per hour, that is, the frequency of the experiment. In Fig. 13 is plotted the resolution given by the HPCPC, against F/V_c, in hour^{-1}.

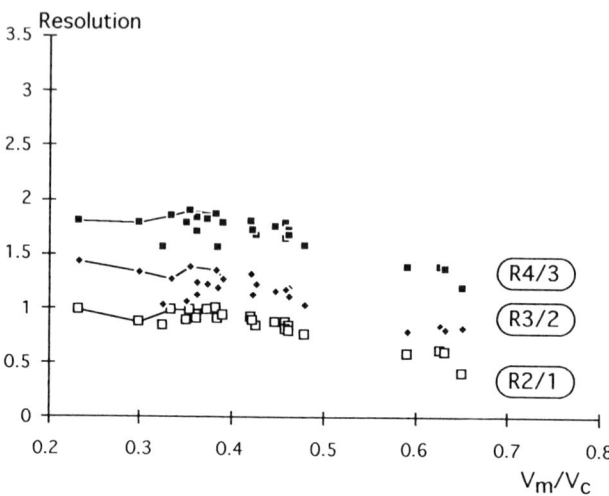

Figure 12 Variation of the resolution Rs with the ratio of the mobile phase in the column, V_m/V_c, whatever the flow rate, that is, Rs = $f(V_m/V_c, \forall F)$; linked plots are those corresponding to the V_m^{min} conditions. Same system as in Fig. 10; test solutes: 1: diethyl phthalate, 2: dipropyl phthalate, 3: ethyl phenylacetate, 4: butyl phenylacetate. R4/3, R3/2, R2/1 means resolution between peaks 4 and 3, 3 and 2, 2 and 1, respectively.

It clearly appears that when the elapsed time and throughput become significant (preparative and production scale), the HPCPC allows one to achieve a good resolution in a short time, because of its quasiindependence with the flow rate at V_m^{min} conditions. Figure 14 is an example of chromatograms given by the HPCPC column at low frequency (one run every 3 to 4 hours) and higher frequency (one run every 25 minutes). We must note that since the efficiency can be increased by increasing the rotational speed, which has no repercussion upon V_m^{min} and thus no effect on the third term of Eq. 16, we can easily increase the average resolution shown in Fig. 14 by increasing the rotational speed, provided the CPC instrument is used within the limit given by the manufacturer (particularly the pressure limit, which is a nonchromatographic parameter and appears to be the major obstacle for better performance).

VII. CONCLUSION

The Stokes model, which is very simple and simulates the mobile phase as droplets with an average radius calculated using the Stokes law and the average linear velocity deduced from the regression analysis of the dependence of the volume of the mobile phase in the CPC column with its flow rate, gave us a simple

THEORY OF CPC

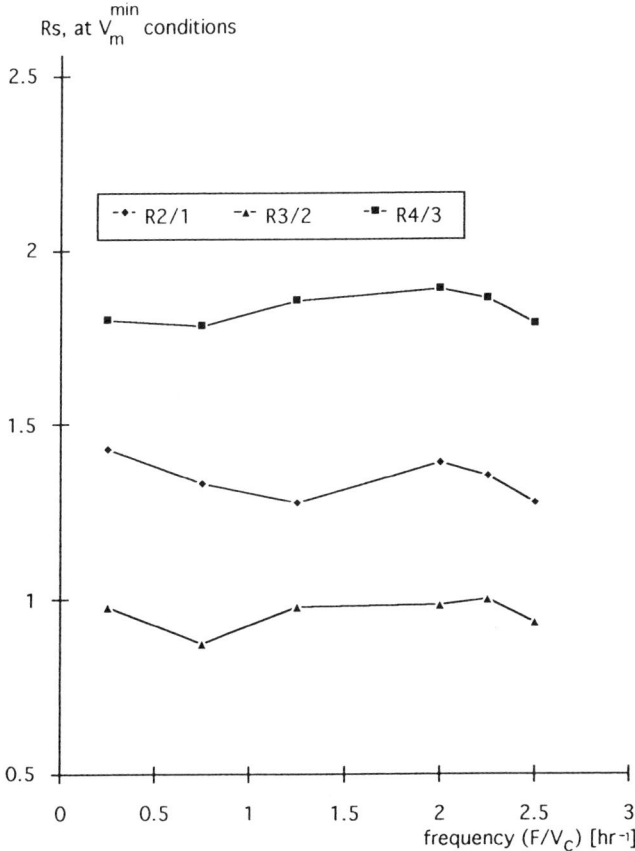

Figure 13 Resolution at V_m^{min} conditions, plotted against the frequency of the separation (number of column volumes per hour). Same system as in Fig. 10.

parameter ($\gamma/\Delta\rho$) to estimate the stability of a biphasic system in centrifugal partition chromatography. It is fully compatible with the evolution of the efficiency of the CPC column with the acceleration field and with the volume ratio of the mobile phase in the column. The best results will be obtained for higher rotational speed, whatever the system is, not because the retention of the stationary phase is higher, but because the dispersion of the mobile phase in the stationary phase becomes better. Exchange between the two phases is favored if droplets are numerous in the channels, and this is favored by higher flow rates, provided the stability of the stationary phase remains acceptable. For most of the systems

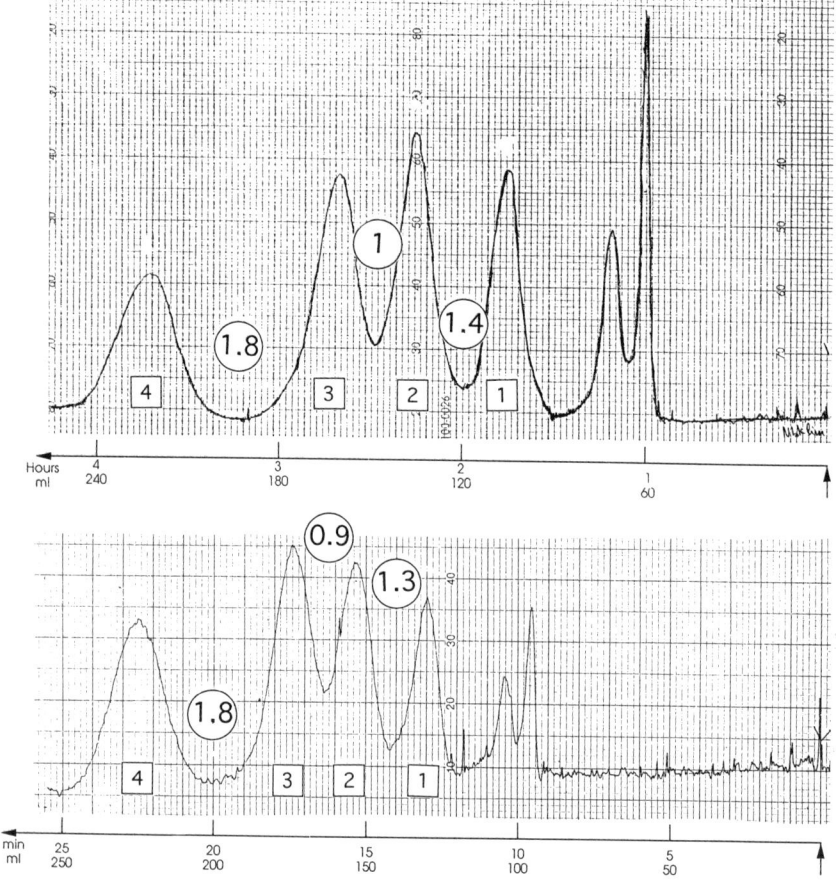

Figure 14 Similarity of two chromatograms which were run at V_m^{min} conditions for $F = 1$ and 10 ml/min with the HPCPC. Same system as in Fig. 10. *Solutes*: □, peak identification; see Fig. 12; ○, resolution between peaks. *Upper*: $V_m/V_c = 0.23$, $F = 1$ ml/min; *lower*: $V_m/V_c = 0.39$, $F = 10$ ml/min.

tested, which cover a wide range of physical and chemical properties, flow rates of 5–10 ml/min, or even higher, can be used with the HPCPC apparatus.

The back pressure, a nonchromatographic parameter mainly due to hydrostatic pressure ($\Delta P \approx \Delta \rho \, g$) [1], will be the major obstacle to achieving the best performance of a CPC apparatus. Instruments must be able to operate at high pressure in order to use them with any biphasic system at higher rotational speeds, to obtain as many theoretical plates as the column can yield.

REFERENCES

1. D. W. Armstrong, G. L. Bertrand, and A. Berthod, *Anal. Chem.*, *60*:2513 (1988).
2. A. P. Foucault, O. Bousquet, and F. Le Goffic, *J. Liq. Chromatogr.*, *15*:2691 (1992).
3. A. P. Foucault, O. Bousquet, F. Le Goffic, and J. Cazes, *J. Liq. Chromatogr.*, *15*:2721 (1992).
4. A. P. Foucault, E. Camacho-Frias, C. G. Bordier, and F. Le Goffic, *J. Liq. Chromatogr.*, *17*:1 (1994).
5. A. P. Foucault, P. Durand, E. Camacho Frias, and F. Le Goffic, *Anal. Chem.*, *65*:2150 (1993).
6. W. N. Bond and D. A. Newton, *Phil. Mag.*, *5*:794 (1928).
7. E. G. Richardson, *Flow Properties of Disperse Systems* (J. J. Hermans, ed.). Interscience, New York, Chapter VI (1953).
8. Y. Ito, *Countercurrent Chromatography* (N. B. Mandava and Y. Ito, eds.). Marcel Dekker, Inc., New York (1988).
9. W. D. Conway, *Countercurrent Chromatography* (W. D. Conway, ed.). VCH Publishers, New York (1990).
10. *Handbook of Chemistry and Physics*, 49th Edition, 1968–1969, CRC Press, Boca Raton (1969).

3
Pressure Drop in Centrifugal Partition Chromatography

M. J. van Buel, L. A. M. van der Wielen, and K. Ch. A. M. Luyben
Delft University of Technology, Delft, The Netherlands

I. INTRODUCTION

Centrifugal partition chromatography (CPC) is a novel form of the well-known countercurrent chromatographic technique that is based on the difference in partitioning behavior of components over two immiscible liquid phases. In CPC, one of these phases is kept stationary, whereas the mobile phase flows through. To retain the stationary phase in the chromatographic column, the column has a tortuous internal geometry and is operated in a centrifugal field. As shown in Fig. 1, a CPC column consists of channels engraved in plates of an inert polymer. The channels are connected by narrow ducts. Several plates are put together to form a cartridge. The cartridges are placed in the rotor of a centrifuge and are connected to form the chromatographic column. The exact geometry of the channels and ducts may deviate from the geometry shown in Fig. 1. The length of the column is variable and depends on the number of connected cartridges. The mobile phase enters and leaves the column via rotary seals.

Since two immiscible liquids are present in the channel, the denser liquid moves away from the axis because of the centrifugal force. The less dense liquid is pushed toward the axis. The mobile phase can be either the lighter or the denser phase. In the latter case, the mobile phase flows through the channels from the axis to the outside of the rotor. This is called the *descending mode* and is depicted in Fig. 1. The other case, the mobile phase flowing toward the axis, is called the *ascending mode*. Solutes with different partitioning behavior that are injected as a mixture distribute differently over the two phases, and consequently they develop different migration velocities and are separated in the column. Some applications for CPC are summarized by Foucault [1] and Meester et al. [2].

Berthod and Armstrong [3] have shown experimentally that a considerable

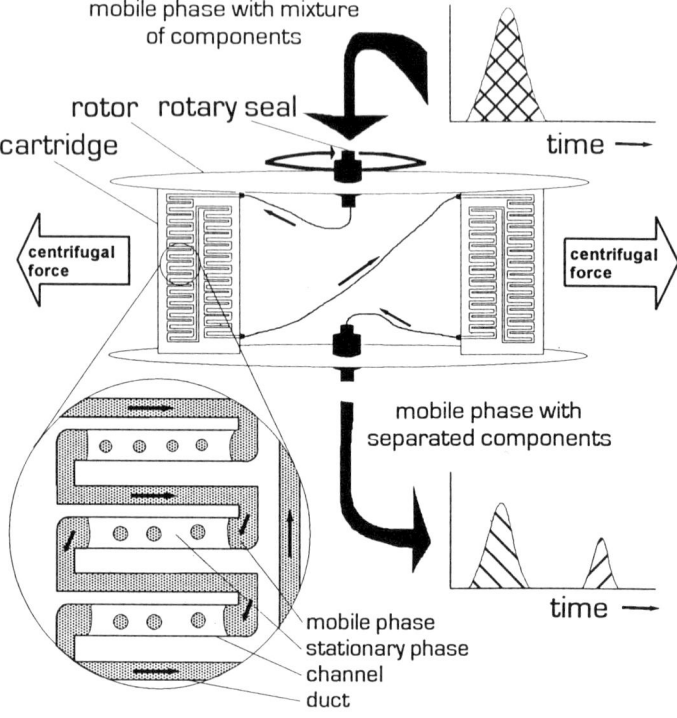

Figure 1 Schematic representation of the CPC apparatus.

pressure drop can arise over the column during CPC separation. The pressure drop is caused by the difference in density between the liquids in the ducts and in the channels (hydrostatic contribution) and by the friction of the mobile phase with the walls of the channels and ducts (hydrodynamic contribution). The overall pressure drop depends on the flow rate and rotational frequency (input variables), the physical properties of the two phases (system variables), the geometry of the channels and ducts and the number of channel–duct combinations (apparatus variables), and the hold-up of stationary phase in the channel. The maximum pressure is limited by the rotary seals, which start to leak above ±60 bar. The resolution and efficiency of a CPC separation also depend on the variables that determine the pressure drop [4,5]. Therefore, it is important to determine which combinations of input variables and liquid two-phase systems can be applied, with respect to the maximum pressure that can be exerted on the rotary seals. In the present chapter, a model that describes the overall pressure drop as a function of the aforementioned variables is presented. The model is tested for the prediction of pressure drops over an analytical-scale CPC column.

II. PRESSURE DROP MODEL

A. Overall Pressure Drop

In accordance with Berthod and Armstrong [3], the overall pressure drop has a hydrostatic and a hydrodynamic contribution. The contributions to the hydrodynamic pressure drop are the friction caused by the flow through the channels and the ducts and the pressure drop due to the bends in the ducts and the channels. The hydrodynamic pressure drop is equal to the sum of the pressure drops over the individual parts of the system, according to Bernoulli's law. Bernoulli's law, however, is only valid when the individual contributions do not influence each other. Hence, the overall pressure drop over the column, consisting of n channel–duct combinations, is calculated with

$$\Delta P_{\text{overall}} = \Delta P_{\text{stat}} + n(\Delta P_{\text{rect,c}} + \Delta P_{\text{rect,d}} + 2\, \Delta P_{\text{bend,c}} + 2\, \Delta P_{\text{bend,d}}) \quad (1)$$

in which ΔP_{stat} is the hydrostatic pressure drop, $\Delta P_{\text{rect,c}}$ and $\Delta P_{\text{rect,d}}$ are the viscous flow pressure drops over the rectangular channels and ducts, respectively, and $\Delta P_{\text{bend,c}}$ and $\Delta P_{\text{bend,d}}$ are the pressure drops over the bends in the ducts and the two bends that connect the ducts to the channels, respectively (see Fig. 2).

B. Hydrostatic Pressure Drop

Following Berthod and Armstrong [3], the hydrostatic contribution originates from the bulk density difference of the duct filled with mobile phase and the channel filled with the biphasic system:

$$\Delta P_{\text{stat}} = \Delta \rho\, \omega^2 R n h_s \quad (2)$$

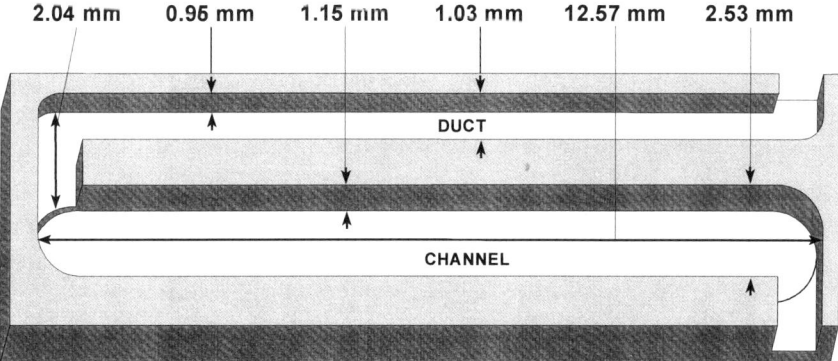

Figure 2 Schematic representation of a channel and duct engraved in plates of polymer in an analytical-scale 250W cartridge.

where $\Delta\rho$ is the density difference of the biphasic system, ω is the rotational frequency, R is the average rotational radius of the cartridge, and h_s is the height of the stationary phase in the channel. Assuming the ducts to be completely filled with mobile phase, and therefore the stationary phase being only present in the channels, the following substitution is valid:

$$nh_s = \frac{\epsilon V_t}{A_c} \tag{3}$$

where ϵ is the stationary phase hold-up in the column (the volume of stationary phase divided by the total volume of the column, V_t) and A_c is the cross-sectional area of a channel. In this way, the hydrostatic pressure drop is expressed as a function of parameters that can be obtained experimentally. Furthermore, the equation can be applied for every flow regime. Thus, the hydrostatic pressure becomes

$$\Delta P_{stat} = \Delta\rho\, \omega^2 R \frac{\epsilon V_t}{A_c} \tag{4}$$

C. Hydrodynamic Pressure Drop in Rectangular Channels and Ducts (ΔP_{rect})

For modeling the hydrodynamic pressure drop over channels and ducts, it is assumed that the single-phase flow is laminar and that the channels and ducts are rectangular. The pressure drop over a straight rectangular tube is calculated by multiplying the pressure drop over a straight cylinder by a correction factor that depends on the ratio of the width and depth of the tube.

The pressure drop over a straight cylinder (ΔP_{cil}) is given by [6]

$$\Delta P_{cil} = \psi_{cil} \tfrac{1}{2} \rho_m v^2 \frac{L_{cil}}{d} \tag{5}$$

where ψ_{cil} is the friction factor of a straight cylinder, ρ_m is the density of the mobile phase, v is the linear velocity of the mobile phase, L_{cil} is the length of the cylinder, and d is the internal diameter of the cylinder.

The linear velocity is obtained by relating the flow rate, ϕ_v, to the surface of the cross-sectional area of the channel or duct and then correcting for the hold-up of the stationary phase:

$$v = \frac{\phi_v}{ab(1-\epsilon)} \tag{6}$$

where a and b are the width and the depth of the channel or duct, respectively. Note that in a duct ϵ will be equal to zero.

The friction factor of a straight cylinder is a function of the Reynolds number, and equals, for the laminar flow regime [6],

PRESSURE DROP IN CPC

$$\psi_{cil} = \frac{64}{Re} \tag{7}$$

where the Reynolds number is given by

$$Re = \frac{\rho_m v d}{\eta} \tag{8}$$

where η is the dynamic viscosity of the mobile phase. The equivalent diameter (d) that must be used for calculating the pressure drop over a straight cylinder is the hydraulic diameter (d_h) of the rectangular tube, given by

$$d_h = \frac{2ab}{a+b} \tag{9}$$

The pressure drop over a rectangular tube is calculated by multiplying the pressure drop over a straight cylinder by a correction factor [6]:

$$\Delta P_{rect} = \Delta P_{cil} 2\zeta \frac{1 + (a/b)^2}{\pi a/b} \tag{10}$$

in which ζ depends on the ratio of a and b. Cornish [7] gives

$$\zeta = 0.878 + 0.566\alpha + 0.758\alpha^2 - 0.193\alpha^3 \tag{11}$$

where α is

$$\alpha = \frac{1 - b/a}{1 + b/a}, \qquad a > b \tag{12}$$

Equations 10–12 are valid for all ratios of $a/b > 1$.

D. Hydrodynamic Pressure Drop in Rectangular Bends (ΔP_{bend})

The pressure drop over a bend is calculated from [6]

$$\Delta P_{bend} = K_w \rho_m \frac{v^2}{2} \tag{13}$$

in which K_w is the friction loss factor for the bend. Two types of bends are present in the CPC cartridge, which are the two bends in the duct with equal inlet and outlet diameters, and the bends in the entrance and exit from the channel to the duct (see Fig. 2). For a circular 90° bend with equal inlet and outlet diameters, the friction loss factor (K_w) can be calculated from [6]

$$K_w = 0.785 \phi \psi_{bend} \frac{D}{d_h} \tag{14}$$

where ψ_{bend} is the friction factor for the bend, D is the curve diameter of the bend, and φ is a correction factor that is a function of the bend angle and the ratio of the curve diameter and the diameter of the bend. An empirical relation for the ratio of the friction factors of a circular bend and a straight cylinder in the case of laminar flow is given by Hausen [8]:

$$\frac{\psi_{bend}}{\psi_{cil}} = 0.805 + 0.0448 \left(Re \sqrt{\frac{d_h}{D}} \right)^{0.6} \quad (15)$$

and is valid for $11.6 < Re \sqrt{(d_h/D)} < 1000$. The correction factor ϕ for a 90° bend equals [6]

$$\phi = 0.95 + 17.2 \left(\frac{d_h}{D}\right)^{1.96} \quad (16)$$

and is valid for $d_h/D > 0.051$. Since the bends are rectangular, the same correction has been applied as for flow through straight cylinders (Eq. 10).

No information has been found in the literature on friction factors for bends with unequal inlet and outlet diameters for laminar flow. Therefore, the interconnecting bends are treated as if they had equal inlet and outlet diameters. Since the difference in inlet and outlet diameter is not taken into account, the model will underestimate the pressure drop. The hydraulic diameter of the channel is also used for the interconnecting bends.

It should be noted that the model for the hydrodynamic pressure drop is derived for a single-phase system without stationary phase being present in the channels. Nevertheless, the model is used for the hydrodynamic pressure drop when both stationary and mobile phases are present in the channels. The presence of a stationary phase might reduce the accuracy of the model

III. MATERIALS AND METHODS

A. Equipment for Pressure Drop Measurements

The pressure drop experiments were performed with a CPC-type LLN purchased from Sanki Engineering Ltd. The pump was a Shimadzu Model LC 6A. The column consisted of six cartridges, type 250W. Each cartridge was composed of four plates of polychlorotrifluoroethylene (PCTFE). In each side of a plate of PCTFE, 50 channels and ducts in two rows of 25 are engraved. This makes a total of 400 channels and 400 ducts per cartridge, and a total of 2400 channels and ducts for a six-cartridge column. Between each plate of PCTFE, a sheet of Teflon is placed to seal the cartridge. The exact dimensions of the channels and ducts are given in Fig. 2. The dimensions could be measured with an accuracy of ±0.05 mm. The total length of a duct is 14.59 ± 0.05 mm. The depths of the channels and the ducts were measured with a micrometer. Since Teflon exhibits deformation due to cold flow, part of the Teflon flows into the channels and ducts. Therefore, the depths of the channels and ducts could not be measured precisely, but were estimated at 1.15 and 0.95 mm, respectively. The total volume of the six cartridges was determined experimentally to be 122.25 ± 0.25 ml. The distance from the center of the cartridge to the rotor axis was 0.116 ± 0.001 m.

B. Chemicals

1-Butanol, octanol, and hexane have been purchased from J. T. Baker (Deventer, The Netherlands) and were all p.a. grade. Water was demineralized. All densities of the single-phase liquids, and the viscosities of water, octanol, and hexane have been taken from Weast [9]. The viscosity of butanol was determined experimentally by measuring the time for flow through a glass capillary. The densities and viscosities of the single-phase liquids are given in Table 1, at the temperature at which the pressure drop experiments were performed. The densities of the individual phases of the hexane–water system have been measured with a Paar DMA 48 density meter. The density difference of the biphasic hexane–water system is 337.3 kg/m^3.

C. Equipment for Visual Observation of Flow in Channels and Ducts

To obtain a visual impression of the phenomena that take place in the channels and ducts, a transparent CPC apparatus was constructed. Since the 250W cartridges supplied by Sanki are not transparent, new cartridges were made from Teflon with the same channel and duct dimensions. By using a system consisting of two mirrors, a stroboscope, and an image analyzer, the cartridges could be "freezed" during operation and pictures could be taken (see Fig. 3). The average rotational radius of the cartridge is larger in the transparent CPC apparatus (0.199 m ± 0.001). By using a lower rotational frequency compared with the one used in the Sanki CPC, however, the centrifugal force exerted on the phases is the same. A dye was added to the mobile phase to increase the contrast of the photographs.

D. Methods

To distinguish between the hydrodynamic and the hydrostatic contributions to the overall pressure drop, both single- and biphase experiments have been performed. The single-phase experiments have been performed by filling the column with the

Table 1 Density and Viscosity of the Liquids Used in the Pressure Drop Experiments

Liquid	Density (kg/m^3)	Viscosity (10^{-3} Pa s)
Water	998	1.0 (20°C)
Hexane	660	0.33 (22°C)
1-Butanol	810	2.8 (20°C)
Octanol	827	6.9 (21.5°C)

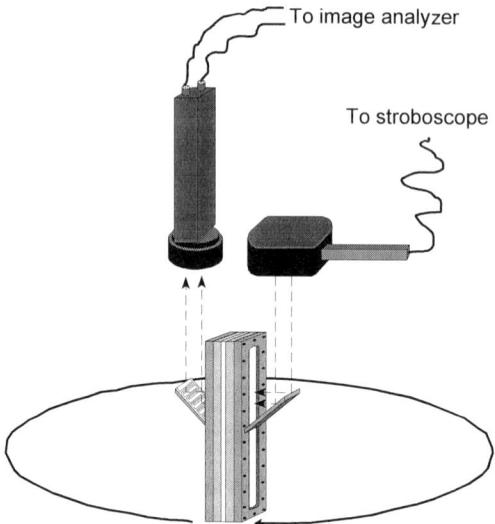

Figure 3 Schematic representation of the transparent CPC cartridge and the system used for visual observations.

liquid, adjusting the flow rate within the range of 0.5–19 ml/min, and measuring the pressure drop. When the single-phase liquid was replaced, the column was first rinsed with ethanol and then washed with 2.5 column volumes of the new single-phase liquid to remove the ethanol. The liquid was then recycled over the column. The temperature of the column was equilibrated by recycling the liquid through the column at the maximum flow rate until the temperature of the inlet was equal to that of the outlet.

The single-phase experiments have been performed in the descending mode at 0 rpm. It was checked, however, whether or not the mode and rotational frequency influenced the pressure drop for single-phase experiments. The volume of the pump head, chamber, and the connecting tubes were measured separately.

To perform the biphasic experiments, the column was filled with the stationary phase, and the required rotational frequency and flow rate were set. Subsequently, the inflow of stationary phase was replaced by an inflow of mobile phase. The volume of stationary phase pushed out of the column was measured. The temperature of the column was equilibrated by recycling the mobile phase. The pressure drop was measured as a function of either the rotational frequency or the flow rate, while keeping the other constant. The rotational frequency was varied between 200 and 1300 rpm, constrained by flooding of the column at low rotational frequencies and leaking of the rotary seals at high rotational frequencies. The flow rate has been varied between 0.5 and 19 ml/min. The flow rates and rotational frequencies used during the experiments were lower and higher, respectively, than those used during the filling of the column with the mobile phase. This was done to prevent leaking of stationary phase from the column during the experiments. Between experiments, the column was rinsed with ethanol for a thorough removal of the previous phase system. The mobile and stationary phase have been prepared by mixing the liquids thoroughly and allowing the two phases to settle.

IV. RESULTS AND DISCUSSION

A. Single-Phase System

During the single-phase experiments, only one phase is present in the channels and ducts. Therefore, the hydrostatic contribution of Eq. 1 is omitted, and only the hydrodynamic contributions are considered. Except for the curve diameter of the two types of bends, all dimensions were determined experimentally, and they are shown in Fig. 2. The curve diameters of the two types of bends cannot be measured, because the inside of the bend is rectangular whereas the outside is circular. This means that the inside of the bends has a curve diameter of 0 mm, while the outside of the bends has a curve diameter of 0.9 mm [3]. The same holds for the bends at the transition from the ducts to the channels. Therefore, one overall curve diameter is determined by fitting the pressure drop profiles of various single-phase liquids with the model. In Fig. 4, experimental pressure drop data are shown as a function of the flow rate of the mobile phase for single-phase hexane, water, 1-butanol, and octanol systems. The resulting pressure drop at a given flow rate is higher at higher viscosity of the corresponding liquid. This is in agreement with the theory. The model predictions are also shown in Fig. 4. By fitting the experimental data of the four monophasic systems with the model, a curve diameter of 1.9 mm was found, which is outside the range given by Berthod and Armstrong [3]. The model, however, is rather insensitive for the fitted curve diameter. A change of 25% in curve diameter changes the deviation from the model by only 2%. The average deviation from the model is 12%.

The model tends to slightly overestimate the pressure drop at higher viscosities (>2 Pa s), while it tends to underestimate the pressure drop at lower viscosities. The reason might be that at higher flow rates the flow becomes more turbulent, and therefore the assumption of complete laminar flow may not hold for the complete experimental range. Since the pressure drop at equal flow rate is lower for turbulent flow than for laminar flow, the overall pressure drop will be lower. This is supported by Schügerl et al. [10], who found that the transition from laminar to turbulent flow for irregular channels occurs at a Reynolds number of approximately 35, instead of 2000 for straight cylinders. The maximum Reynolds number for the ducts is 30 for octanol and 600 for hexane at maximum flow rate.

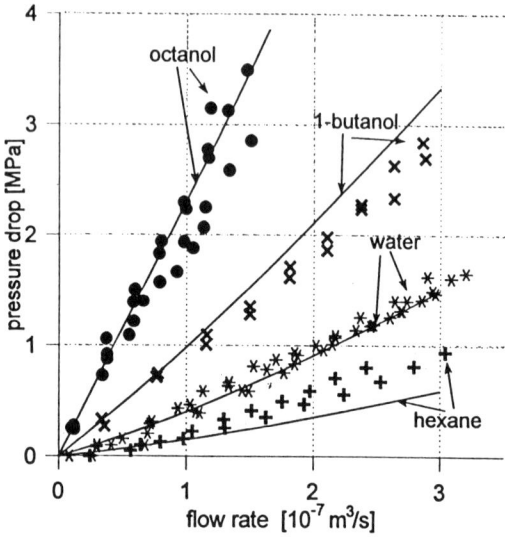

Figure 4 Hydrodynamic pressure drop as a function of the flow rate for various single phases. Markers are experimental data; lines are model calculations.

The transition of laminar to turbulent flow in the case of rectangular tubes might be at even lower Reynolds numbers than for cylindrical tubes. Furthermore, the hydrodynamic pressure drop for turbulent flow depends less on viscosity than for laminar flow. This means that the pressure drop (as calculated by the model) for the more viscous liquids (octanol, 1-butanol) decreases relatively more than for the less viscous liquids.

The various terms in Eq. 1 do not contribute equally to the hydrodynamic pressure drop. The terms that contribute the most are the viscous flow through the duct ($\Delta P_{rect,d}$) and the pressure drop over the two bends in the duct ($2\,\Delta P_{bend,d}$). With the liquids commonly used in CPC separations, such as water, hydrocarbons, and lower alcohols [1], these terms contribute at least 75% to the total hydrodynamic pressure drop. Their contribution is almost equal. The pressure drop over the bends is more important at lower viscosities and higher flow rates. The same is true for the hydrodynamic pressure drop over the channels and the bends in the channels.

Of the physicochemical parameters, viscosity is the most influential on the hydrodynamic pressure drop. A change of 5% in viscosity changes the hydrodynamic pressure drop by 4.5%. The density has hardly any effect on the hydrodynamic pressure drop. Of the geometry parameters, the dimensions of the ducts have the largest influence. A change of 5% in duct height and width, equal to the accuracy with which the dimensions could be measured, changes the hydrodynamic pressure drop by 15%.

B. Biphasic System

Pressure drop experiments have been performed with a water (mobile phase, m)–hexane (stationary phase, s) system. Figure 5 shows the total pressure drop as a function of the rotational frequency for various stationary phase hold-ups and flow rates. Figure 6 shows the total pressure drop as a function of the flow rate for various hold-ups and rotational frequencies. Experimental data are shown together with the model predictions for comparison. When both rotational frequency and stationary phase hold-up are low, model predictions and experimental results match closely. At elevated rotational frequencies, the model tends to overestimate the total pressure drop. The deviation is largest for systems with low stationary phase hold-up.

The ordinate intercept of Fig. 5 corresponds to the hydrodynamic contribution to the total pressure drop, which is assumed independent of the rotational frequency. These values are predicted rather well by the model. Likewise, the ordinate intercept in Fig. 6 corresponds to the flow rate–independent hydrostatic contribution. Here, the model predictions deviate from the experimental data

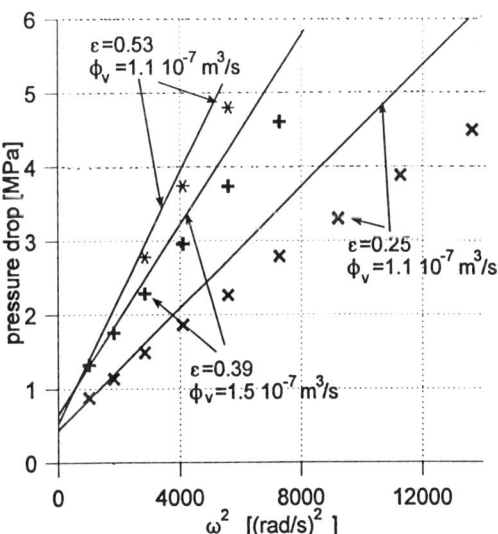

Figure 5 Pressure drop as function of the rotational frequency for the water (m)–hexane (s) system, for various hold-ups and flow rates. Markers are experimental data; lines are model calculations.

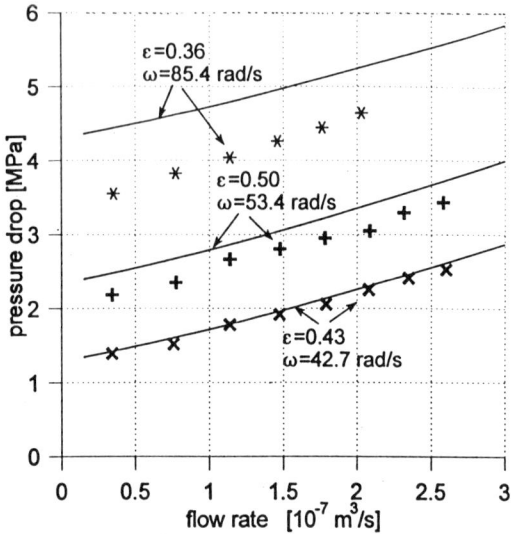

Figure 6 Pressure drop as a function of the flow rate for the water (m)–hexane (s) system, for various rotational frequencies and stationary phase hold-ups. Markers are experimental data; lines are model calculations.

obtained at a high rotational frequency, as can be seen in Fig. 5 also. Figure 6 shows that for high rotational frequencies, the hydrodynamic pressure drop increases more than the model predicts. Since the hydrodynamic pressure drop predictions for low and moderate rotational frequencies match closely with the experimental data, this may indicate that the hydrostatic pressure drop is a function of the flow rate, which is not predicted by the model.

The model for the hydrodynamic contribution is based on the assumption of single-phase flow. The biphasic system in the channels, however, could influence the hydrodynamic pressure drop. To find out whether this is true, the hydrodynamic pressure drop contributions of a single-phase experiment with water and five biphasic experiments with the water (m)–hexane (s) system are compared in Fig. 7. Figure 7 was constructed by subtracting the hydrostatic contribution, equal to the intercept with the ordinate, from the overall pressure drop. Although the variation in experimental points for the biphasic system is larger than for the single-phase system, they agree with the single-phase experiments. This indicates that the model for the hydrodynamic pressure drop is valid also for biphasic systems.

The hydrostatic pressure drop at high rotational frequency and the total pressure drop for a biphasic system at high rotational frequency and at high flow

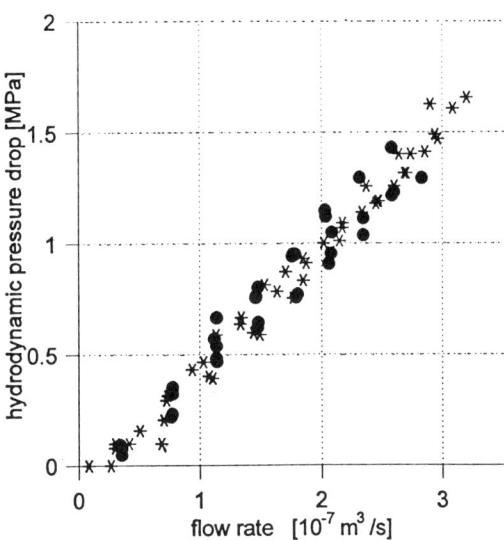

Figure 7 Hydrodynamic pressure drop as a function of the flow rate for single-phase water (∗) and the biphasic water (m)–hexane (s) systems (●), for various rotational frequencies (400–800 rpm) and stationary phase hold-ups (0.36–0.50).

rate are not predicted well by the model. A possible reason is that the stationary phase is not only present in the channel, as assumed, but also partially in the ducts. This can be evaluated by visual observations.

C. Visual Observations

The experimental set-up shown in Fig. 3 and described in Section III.C has been operated with the water (m)–hexane (s) system. Figure 8 shows the stationary phase distribution at increasing centrifugal force from 253 to 2597 rad^2 m/s^2 for a constant flow rate of 1.1×10^{-7} m^3/s. These preliminary results clearly demonstrate that the stationary phase is partly forced in the duct as a result of the large centrifugal force acting on it. Thus, experimental hydrostatic pressure drops appear to be lower than predicted by the model. The amount of stationary phase forced into the ducts increases with increasing centrifugal force. This corresponds to a lower increase of the hydrostatic pressure drop than the proportionality with the squared rotational frequency predicts. Thus, the pressure drop is not a linear function of the squared rotational frequency, but deviates increasingly from Eq. 4 at higher rotational frequencies. This is shown in the pressure drop data of Fig. 5. Since the relative effect of stationary phase accumulation in the duct is larger at

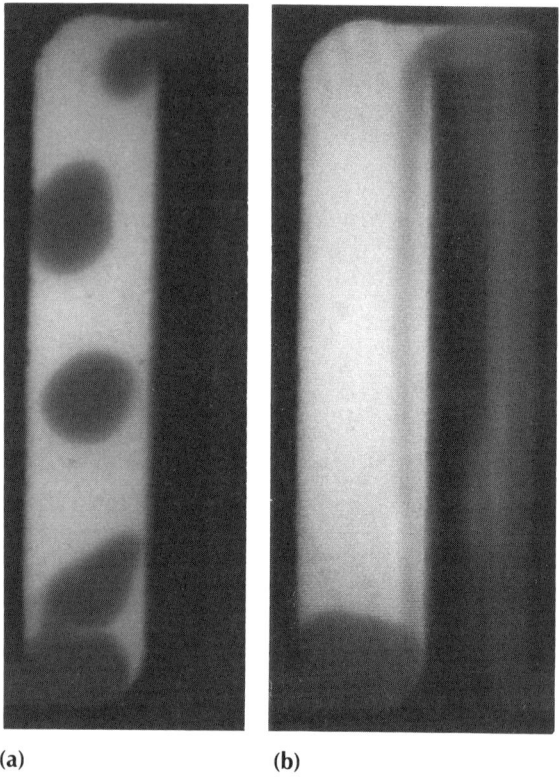

Figure 8 Picture taken using the transparent CPC cartridges. Water (m)–hexane (s) system, $\phi_v = 1.1 \times 10^{-7}$ m^3/s. (a) $\omega^2 R = 253$ rad^2 m/s^2; (b) $\omega^2 R = 555$ rad^2 m/s^2; (c) $\omega^2 R = 1044$ rad^2 m/s^2; (d) $\omega^2 R = 1874$ rad^2 m/s^2; (e) $\omega^2 R = 2597$ rad^2 m/s^2.

smaller stationary phase hold-up, deviations of model predictions from experimental data are larger at smaller stationary phase hold-up.

Figure 9 shows the stationary phase distribution at 1 and 9 ml/min for a centrifugal force of 1970 rad^2 m/s^2. The results clearly show that at increased flow rates, part of the stationary phase, which was forced into the duct because of the high centrifugal force, is pushed out of the column again. Therefore, the hydrostatic pressure drop becomes a function of the flow rate when part of the stationary phase is forced into the ducts at higher rotational frequencies, which is not predicted by the model. This can also be seen in Fig. 6 for the highest rotational frequency.

The visual observations also show different flow regimes for the mobile

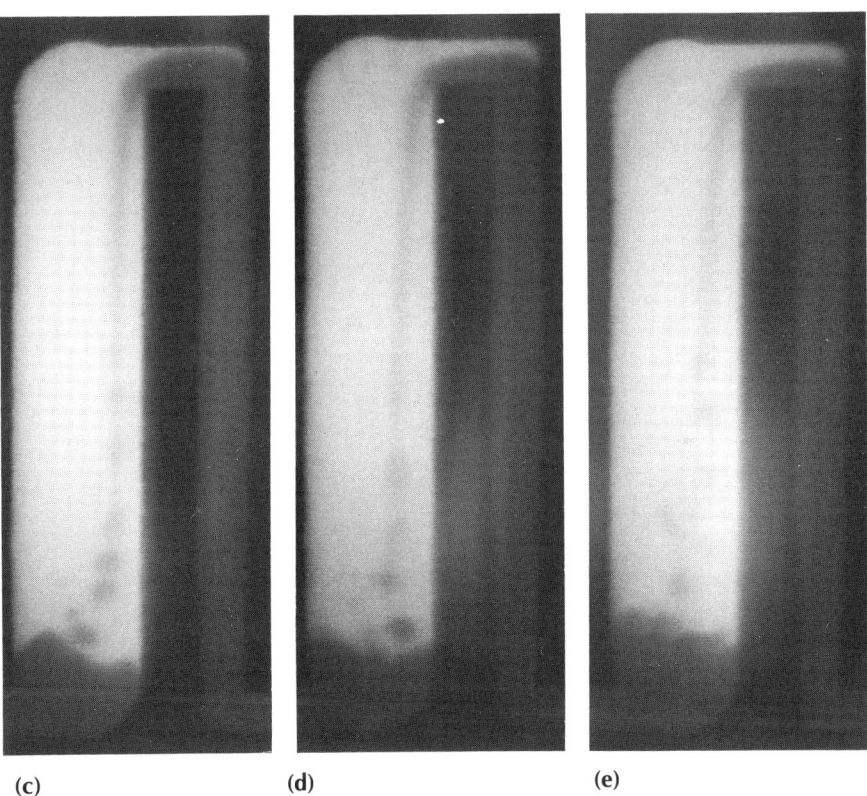

(c)   (d)   (e)

phase in the channel. Depending on the flow rate, rotational frequency, and physicochemical parameters (including interfacial tension), different types of flow regimes occur, such as droplets, jets, breaking jets, and sprays (not shown). Since the hydrodynamic pressure drop over the channel contributes only to a small extent to the total pressure drop, the influence of the flow regime on the hydrodynamic pressure drop will most probably be small. This is also indicated in Fig. 6, where the hydrodynamic contribution to the total pressure drop at different rotational frequencies, and therefore at different flow regimes, is described equally well.

The extent to which the hydrostatic pressure drop and the hydrodynamic pressure drop contribute to the overall pressure drop depends on the flow rate and rotational frequency, the geometrical parameters, and the physicochemical parameters. If the density difference between the two phases is large and the viscosity of the mobile phase is low, the contribution of the hydrostatic pressure drop may be as high as 95%. On the other hand, the contribution of the hydrostatic pressure

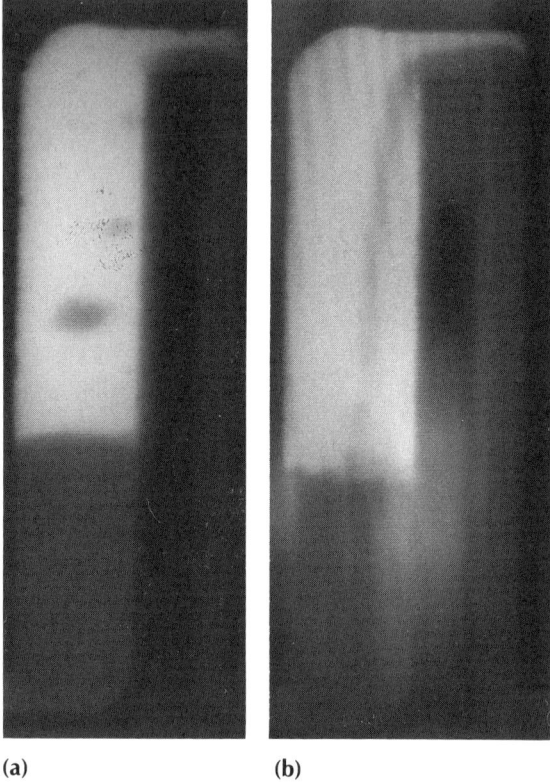

(a) (b)

Figure 9 Picture taken using the transparent CPC cartridges. Water (m)–hexane (s) system, $\omega^2 R = 1970$ rad^2 m/s^2. (a) $\phi_v = 0.16 \times 10^{-7}$ m^3/s; (b) $\phi_v = 1.5 \times 10^{-7}$ m^3/s.

drop might be negligible when systems with small density differences of the phases are used, such as aqueous two-phase systems. For channels and ducts with increasing dimensions, the hydrodynamic pressure drop decreases and the hydrostatic pressure drop becomes the most important.

V. CONCLUSIONS

A predictive model for the pressure drop over a CPC column was presented. The model contains two pressure drop contributions: hydrostatic and hydrodynamic. The model can be used to predict the pressure drop as a function of the flow rate of the mobile phase, the rotational frequency, the stationary phase hold-up, and the physical properties of the phases. The model contains one parameter and predicts the single-phase hydrodynamic pressure drop with an average deviation of 12%. The hydrodynamic pressure drop is due mainly to the viscous flow through the ducts and the bends in the ducts. The two important parameters that influence the hydrodynamic pressure drop as a function of the flow rate are the dimensions of the ducts and the viscosity of the liquids. The lack of information about the exact nature of the type of flow (laminar or turbulent) might be the cause for the deviation between the model and the single-phase experimental data. The model for the hydrodynamic pressure drop, which is based on the assumption of single-phase flow, also applies for biphasic flow. The hydrodynamic pressure drop is independent of the rotational frequency, as is predicted by the model.

It has been shown that Eq. 4 gives a good prediction of the hydrostatic contribution at lower rotational frequencies. As the distribution of the stationary phase deviates increasingly at increasing rotational frequencies from its assumed channel hold-up, pressure drop predictions deviate increasingly from the model predictions. Furthermore, the hydrostatic pressure drop becomes a function of the flow rate at a higher rotational frequency. At the moment, it is not clear how the model should correct for the stationary phase forced into the ducts at higher rotational frequency. Still, the model can be used to estimate the overall pressure drop from the properties of the phases and the input parameters. Hence, rapid evaluation and optimization of the pressure drop with regard to the various parameters is possible, which will prove to be beneficial for design and scale up.

VI. FUTURE DEVELOPMENTS

This chapter is a first study into parameters that might influence pressure drop over CPC columns. Future work will include an extension to other phase systems, column geometries, and modes of operation (ascending or descending). Hold-up of the stationary phase and actual flow regime of the mobile phase in relation to these factors will also be important items in hydrodynamic research. The setup that allows visual observation will prove to be a powerful tool in these investigations. Our aim is to relate the aforementioned aspects to efficiency, capacity, and resolution of CPC separations to make better design, operation, and control possible.

ACKNOWLEDGMENTS

The Center of Bio-Pharmaceutical Sciences, Division of Pharmacognosy, of the Leiden University is gratefully acknowledged for providing the CPC apparatus. The authors wish to thank A. Hermans and R. Verpoorte (Leiden University) and R. Meester for the stimulating discussions. Great thanks go to the people from the workshop and electronics department of the Kluyver Laboratory for Biotechnology, who constructed the CPC apparatus that was used for the visual observations. Finally, S. van Hateren is thanked for his support with the Image Analyzer.

NOTATION

a	width of a channel or duct [m]
A	cross-sectional area [m²]
b	depth of a channel or duct ($a > b$) [m]
D	curve diameter of a bend [m]
d	diameter [m]
d_h	hydraulic diameter according to Eq. 9 [m]
h_s	stationary phase height in a channel [m]
K_w	friction loss factor
L	length [m]
n	total number of channel–duct combinations
R	average rotation radius of the cartridge [m]
Re	Reynolds number
v	linear velocity of the mobile phase [m/s]
V_t	total volume of the column [m³]
ρ_m	density of the mobile phase [kg/m³]
ΔP	pressure drop [Pa]
$\Delta \rho$	density difference between the phases [kg/m³]
ϵ	stationary phase hold-up relative to the total column volume [m³/m³]
ζ	correction factor depending on a and b
η	dynamic viscosity of the mobile phase [Pa s]
ϕ	correction factor
ϕ_v	flow rate [m³/s]
ψ	friction factor
ω	rotational frequency [rad/s]

Subscripts

bend	bend in channel or duct
c	channel
cil	cylinder
d	duct
dyn	hydrodynamic
m	mobile phase
overall	overall
rect	rectangular
s	stationary phase
stat	hydrostatic

REFERENCES

1. A. P. Foucault, Countercurrent chromatography, *Analytical chemistry*, *63*:569A–579A (1991).
2. R. Meester, M. J. van Buel, L. A. M. van der Wielen, R. Verpoorte, and K. Ch. A. M. Luyben, Hydrodynamics of centrifugal partition chromatography, *Proceedings Precision Process Technology* (M. P. C. Weijnen and A. A. H. Drinkenburg, eds.), Delft, pp. 295–303 (1993).
3. A. Berthod and D. W. Armstrong, Centrifugal partition chromatography. I. General features, *J. Liq. Chromatogr.*, *11*:546–583 (1988).
4. W. Murayama, T. Kobayashi, Y. Kosuge, H. Yano, Y. Nunogaki, and K. Nunogaki, A new centrifugal counter-current chromatograph and its application, *J. Chromatogr.*, *239*:643–649 (1982).
5. A. P. Foucault, O. Bousquet, and F. Le Goffic, Importance of the parameter Vm/Vc in countercurrent chromatography: Tentative comparison between instrument designs, *J. Liq. Chromatogr.*, *15*:2691–2706 (1992).
6. H. Brauer, *Grundlagen der Einphasen- und Mehrphasenströmungen*, Sauerländer, Aarau/Frankfurt am Main (1971).
7. R. J. Cornish, Flow in a pipe of rectangular cross-section, *Proc. Roy. Soc. London (A)*, *120*:691–700 (1928).
8. H. Hausen, *Wärmeübergang im Gegenstrom, Gleichstrom und Kreuzstrom*, Springer, Berlin/Göttingen/Heidelberg (1950).
9. R. C. Weast, *CRC Handbook of Chemistry and Physics*, 60th ed., CRC Press, Inc., Boca Raton, Florida (1980).
10. K. Schügerl, H. G. Blaschke, and R. Streicher, *Recent Developments in Separation Science*, Vol. III, part A, p. 71 (1977).

4
Solvent Systems in Centrifugal Partition Chromatography

Alain P. Foucault
Laboratoire de Bioorganique et Biotechnologies, Centre National de la Recherche Scientifique, Paris, France

I. INTRODUCTION

Selection of a two-phase solvent system for centrifugal partition chromatography (CPC) is similar to choosing a column and an eluant for high-performance liquid chromatography (HPLC). Important criteria are the polarity of the sample and its solubility, charge state, and ability to form complexes, etc. The purpose of solvent optimization for CPC separations is first to find a solvent combination for which the sample is freely soluble, since the goal of CPC is preparative rather than analytical; second, to adjust this solvent combination to make the partition coefficients of the species that are to be separated different from each other. A tremendous variety of biphasic solvent systems can be found in the literature, most of them consisting of mixtures of three solvents, some with four or more solvents.

Most often, with CPC, optimization of a separation involves optimization of chromatographic selectivity, for it is here that CPC has the most to offer. Both phases are directly accessible; their compositions can be fine-tuned to achieve the desired resolution. Phase diagrams may be used, when available, to intelligently understand the effects of varying the composition of one phase upon the composition of the other. Ternary diagrams for many solvent systems have been compiled by Sørensen and Arlt [1]. They often consist of two immiscible solvents plus a third solvent that is soluble in the two primary solvents. These systems are called type 1 systems, and their ternary diagrams are similar to Fig. 1a. The binodal is the line separating the monophasic and biphasic zones. Any point along a tie-line will give the same composition for the left and right phases, but with different volume ratios. A point very close to the binodal and in the biphasic zone will produce roughly one phase, saturated with a few droplets of the other. When approaching

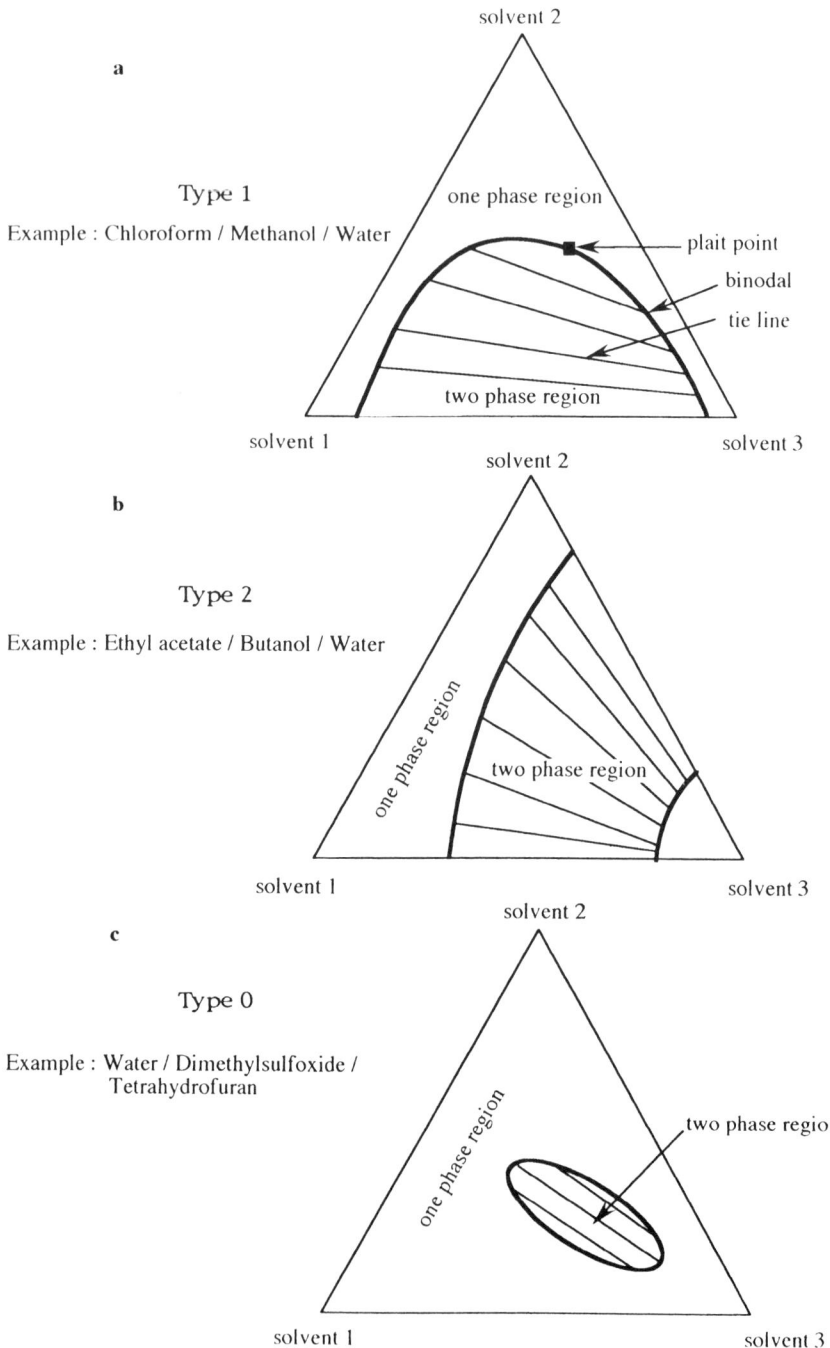

SOLVENT SYSTEMS IN CPC 73

the plait point, the composition of the left and the right phases become identical and all partition coefficients converge to 1. The position of the binodal is temperature sensitive, and a mixture that has a composition close to it can be monophasic or biphasic, depending upon the temperature; it is therefore not recommended to work too close to the plait point.

A few systems consist of two miscible solvents plus a third one that is insoluble in the two primary solvents. These systems are called type 2 systems, and their ternary diagrams look like the one in Fig. 1b. A very well known example is the ternary system ethyl acetate/butanol/water. The binodal is here in two parts, and there is no plait point.

Few systems consist of three solvents that are fully miscible among each other in pairs, but for which a zone exists in the ternary diagram where a biphasic system occurs when mixing them in a suitable ratio. These systems are called type 0 systems, and the ternary diagram looks like in Fig. 1c. The unique example recently introduced in CPC is the system water/dimethyl sulfoxide/tetrahydrofuran [2].

Figure 2 shows two ways to represent the composition of a mixture of three solvents, 1, 2, and 3, by mixing the volume S_1, S_2, and S_3, respectively. The axes are expressed in percentage by volume. The representation in Fig. 2a is widely used, but is not readily compatible with spreadsheets and computers, which prefer to deal with orthogonal representations for a given set of data; moreover, it is redundant, as $S_1 + S_2 + S_3 = 100$, whatever the point B, giving the two phases A and C.

It is easier to use the representation in Fig. 2b, where only the percentages of two of the solvents, S_2 and S_3 (or S_1 and S_2, or S_1 and S_3), are displayed. In this way, it is very easy to use computers to draw the tie-lines and the binodal, and to plot the volume percentage graph from mole percentage or mass percentage data.

Hundreds of ternary diagrams have been described in the literature; a large compilation was published by Sørensen and Arlt in *Liquid–Liquid Equilibrium Data Collection* [1], which gives the data in mole percentage and in the orthogonal representation.

Ternary diagrams are very useful for understanding how to vary the

---

Figure 1 Ternary diagrams encountered in centrifugal partition chromatography: (a) Most systems conform to type 1, made of one solvent miscible with two other immiscible solvents. A typical example is the ternary system chloroform, methanol, and water. (b) A few systems are like type 2, made of one solvent immiscible with two other miscible solvents. A very well known example is the ternary mixture ethyl acetate, butanol, and water. (c) Very few systems are like type 0, made with three solvents fully miscible by pairs, but for which a zone exists in the ternary diagram where a biphasic system occurs when mixing them in a suitable ratio. A recently introduced system is the ternary mixture of water, dimethyl sulfoxide, and tetrahydrofuran.

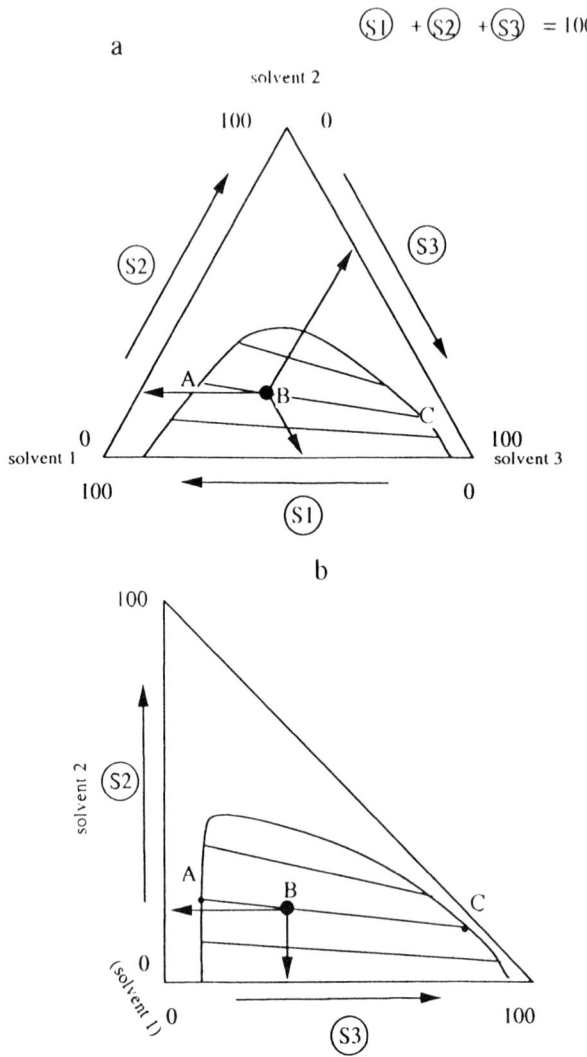

Figure 2 Composition of a ternary solvent system. (a) Ternary representation—laborious to manipulate with computers and spreadsheets. (b) Orthogonal representation—gives the same amount of information and is very easy to manipulate with computers.

composition of a ternary mixture to significantly change the properties of a biphasic system, and for predicting the consequences of varying the composition of one phase upon the composition of the other.

Figure 3 compares three ternary diagrams of chloroform/alcohol/water, with the alcohol as ordinate and water as abscissa. Tie-line orientation clearly indicates where an increase in the alcohol content of a mixture will lead:

1. For methanol (Fig. 3a), the composition of the chloroform-rich phase will vary only slightly when the alcohol content of the mixture is increased from M to M', while the composition of the water-rich phase will vary from ≈ 10 to $\approx 40\%$ methanol (v/v).
2. In the case of 2-propanol (Fig. 3b), the composition of the two phases varies greatly with the level of alcohol, which distributes evenly in water and chloroform.
3. For 1-butanol (Fig. 3c), the composition of the chloroform-rich phase will vary extensively (≈ 10 to $\approx 70\%$ 1-butanol (v/v)), while the water-rich phase will remain approximately constant.

Since most of the systems used in CPC are type 1 systems, the primary role of the solvent that is soluble in the two others has to be highlighted. This solvent, generally the ordinate on ternary diagrams, partitions in the two others exactly as we would like our sample to partition. It must then be a good solvent for the sample with which we are working, and therefore it is allowed to go into the two phases of the selected biphasic system. In other words, when a sample has to be fractionated by CPC, the first thing to do is to find the best solvents for the sample—the solvents where it is most soluble—since the goal is often preparative chromatography. Then, we have to partition this best solvent into two other solvents, in order to build a biphasic system; that is, we have to look for ternary diagrams where the best solvent plays the role of solvent 2 in Fig. 1a or Fig. 2. We will keep the name of *best solvent* through this chapter, since it characterizes the strategy for looking to a biphasic system when trying to solve a problem: first, the best solvent; then, the way to partition it (and thus the sample) into two other solvents to obtain a biphasic system. This solvent is also called the blending solvent, the consolute, or the bridge solvent.

Looking at the best solvent is the best way to begin a CPC run. From the numerous ternary diagrams found in the collection compiled by Sørensen and Arlt [1], we extracted some selected systems that have been used, or that could be useful, for CPC applications. The mole percentage data have been transformed into volume percentage diagrams, which we know to be not theoretically exact because we are neglecting contractions normally observed on mixing different solvents; this can be observed, for example, when making up a mixture corresponding to the midpoint of a tie-line, which generally leads to two phases that do not have exactly the same volume. These small discrepancies have no conse-

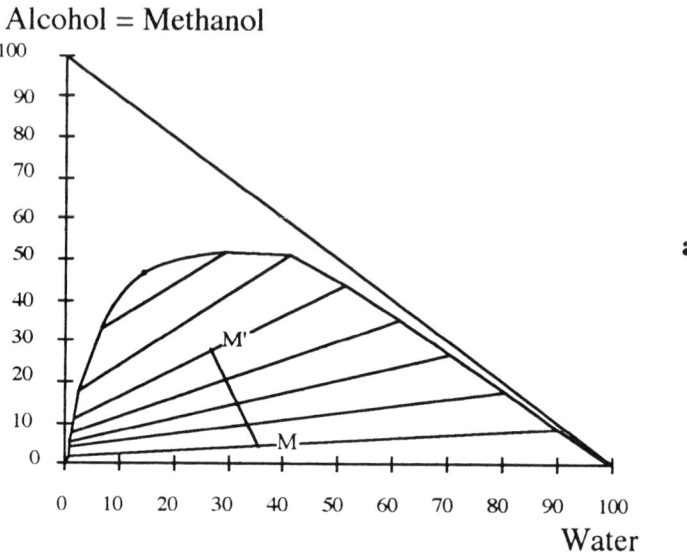

Figure 3 Ternary diagrams for three chloroform/alcohol/water solvent systems. Tie-line orientation clearly indicates the preferred direction of an increase in the alcohol content of a mixture (e.g., from M to M').

quences for CPC applications, however, and they are insignificant when compared to the advantages of using these diagrams. The ternary diagrams are located at the end of this book.

Following is the list of best solvents (Table 1), found by browsing through the ternary diagrams; they are presented in a column, facing a list of less polar solvents and one of more polar ones that will lead to biphasic systems when mixed together. It must be noted that water plays a special role, since it is a polar solvent for many ternary diagrams.

Chlorinated solvents, which have been widely used in the past, are often prohibited for industrial applications; they are still useful for laboratory-scale purifications.

Benzene is not used as much as the diagrams may indicate; the ternary diagrams involving benzene have been indicated because benzene can generally be replaced with toluene, which will result in a slightly different, but comparable ternary diagram.

In the same way, heptane can replace hexane in ternary diagrams involving hexane, but keeping in mind that the binodal will shift in a way that will reflect the difference in the solubility of heptane and hexane in the two other solvents.

SOLVENT SYSTEMS IN CPC

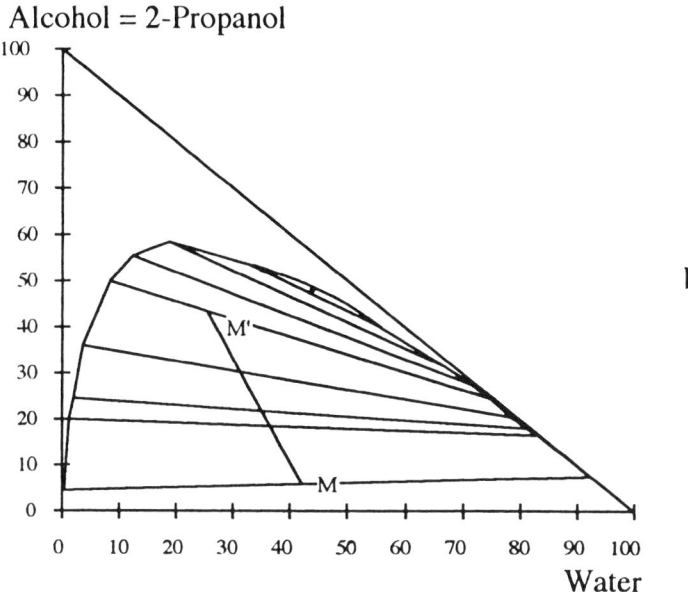

b

Alcohol = 2-Propanol
Water

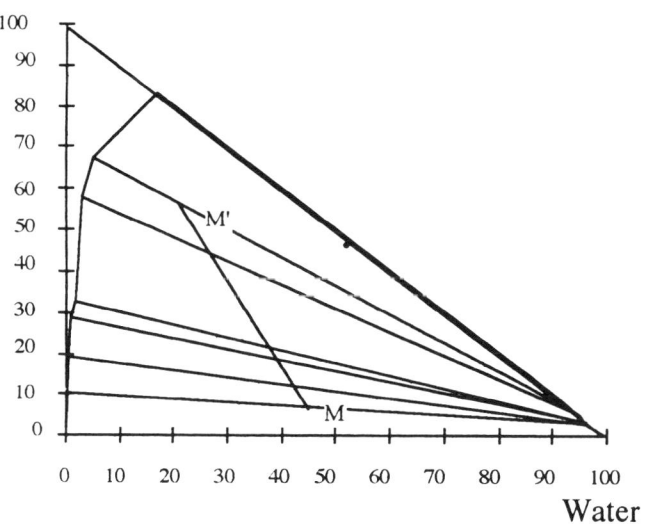

c

Alcohol = 1-Butanol
Water

Table 1 *The Best Solvent Table*: Starting with a New Sample, First Find the Best Solvent, and Then Try to Put It Between Two Other Solvents, As Shown

Less polar solvent	Best solvent	More polar solvent
BENZ $C_2H_4Cl_2$, $CHCl_3$	HCOOH	Water
cycloHEX, HEX TOL Et_2O $CHCl_3$, $C_2H_4Cl_2$ isoAmOH, 1-BuOH MIBK EtOAc	HOAc	Water
HEX HEP cycloHEX	Benzene Toluene	EG DMF Sulfolane DMSO MeCN
cycloHEX	$CHCl_3$	MeCN
cycloHEX	Et_2O	MeOH
TOL, ClBENZ MtBE, MIBK EtOAc	MeCN	Water
TOL Et_2O $CHCl_3$ MIBK cycloHEX, HEX, HEP EtOAc	AcO	Water EG
cycloHEX, HEX, HEP	MEK	Water
HEX, HEP CCl_4, $CHCl_3$ EtOAc BENZ, TOL	1-BuOH 2-BuOH	Water MeCN EG

Table 1 Continued

Less polar solvent	Best solvent	More polar solvent
BENZ, TOL HEP $CHCl_3$ isoPropyl Ether EtOAc 1-BuOH	1-PrOH 2-PrOH	Water
HEX, HEP $CHCl_3$ Et_2O EtOAc	EtOH	Water MeCN
HEX, HEP BENZ, TOL $CHCl_3$, $C_2H_4Cl_2$ EtOAc 1-BuOH	MeOH	Water MeCN
EtOAc	Dimethylacetamide	Water
TOL $CHCl_3$	DMF	Water
THF	DMSO	Water
TOL	Pyridine	Water
Perfluorohexane	HEX	BENZ
cycloHEX, HEP CCl_4, $CHCl_3$, CH_2Cl_2 $C_2H_4Cl_2$	THF	MeOH Water

Abbreviations: AcO, acetone; BENZ, benzene; BuOH, butanol; $C_2H_4Cl_2$, dichloroethane; CCl_4, carbon tetrachloride; $CHCl_3$, chloroform; ClBENZ, chlorobenzene; cycloHEX, cyclohexane; DMF, dimethyl formamide; DMSO, dimethyl sulfoxide; EG, ethylene glycol; Et_2O, diethyl ether; EtOAc, ethyl acetate; HCOOH, formic acid; HEP, heptane; HEX, hexane; HOAc, acetic acid; isoAmOH, isoamyl alcohol; MeCN, acetonitrile; MEK, methyl ethyl ketone; MeOH, methanol; MIBK, methyl isobutyl ketone; MtBE, methyl tertiobutyl ether; PrOH, propanol; Pyr, pyridine; THF, tetrahydrofuran; TOL, toluene.

II. STRATEGY FOR SOLVENT SELECTION

A. The Ternary Diagram Approach

1. Select the solvent(s) where the sample is the most soluble. CPC has a preparative goal, and so the final mixture of solvents must be able to dissolve large amounts of sample, and thus it should contain at least one of the "best" solvents that makes this sample freely soluble.

2. Aided by the polarities of the solvents (numerous polarity scales may be found in the literature), put a solvent on each side of the selected best solvent, that is, one less polar and one more polar, in order to obtain a biphasic system where the best solvent will partition into the two other solvents. This can be done by looking at Table 1 and then at the ternary diagrams.

3. Adjust the ratio of the best solvent to disperse the sample into the two phases; that is, the less polar fraction of the sample will go preferentially in the less polar phase, and the more polar fraction will go preferentially in the more polar phase, so that the average partition coefficient will stay around 1.

At this stage, the ternary diagram can be useful; it indicates where the best

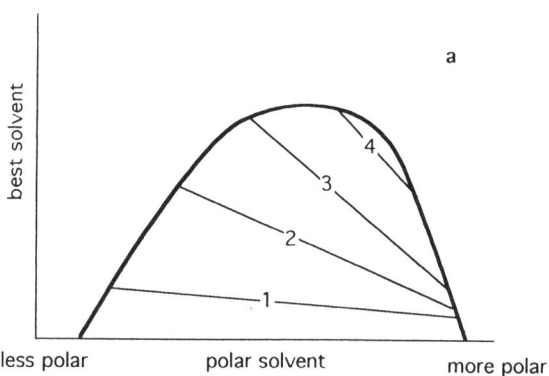

Figure 4 If in (a) the sample is mainly in the left phase, it will never partition well unless you reach tie-line 4, which is too close to the plait point for CPC applications; while if in (b) the sample is mainly in the left phase, it should easily partition when going to tie-line 2 or 3, since the best solvent prefers to go in the right phase, thus inducing part of the sample to go with it. (c) The best solvent evenly distributes in the left and right phases, and so the sample should do so; or, if on tie-line 3 the sample is still mainly in one phase, the left (right) one, for example, it is often better to add a fourth solvent than to go to tie-line 4, which is too close to the plait point; adding a solvent less (more) polar than the less (more) polar solvent of this ternary diagram will usually result in a variation of the partitioning of the best solvent, and thus of the sample.

solvent will preferably go, and depending where the sample is, it will reveal the possibility of a good partitioning or not. Consider the ternary diagrams of Figs. 4a–c.

Figure 4a: Working with the tie-line 1, if the sample is in the left phase, that is, the less polar phase, then it will stay in the left phase even if the volume ratio of the best solvent is increased (working with tie-lines 2 or 3), since the best solvent prefers the left phase, too. If, however, the sample is in the right phase, that is, the more polar phase, then increasing the volume ratio of the best solvent should allow a good partitioning, since it will favor the sample going in the left phase, where it is mainly going.

Figure 4b: The opposite rule applies: If the sample is in the left phase, then increasing the volume ratio of the best solvent (going from tie-line 1 to 3) will favor a good partitioning. If, however, the sample was in the right phase, it will not favor a good partitioning unless you reach tie-line 4, which is so close to the plait point that all the components of the sample will have the same partition coefficient (around 1).

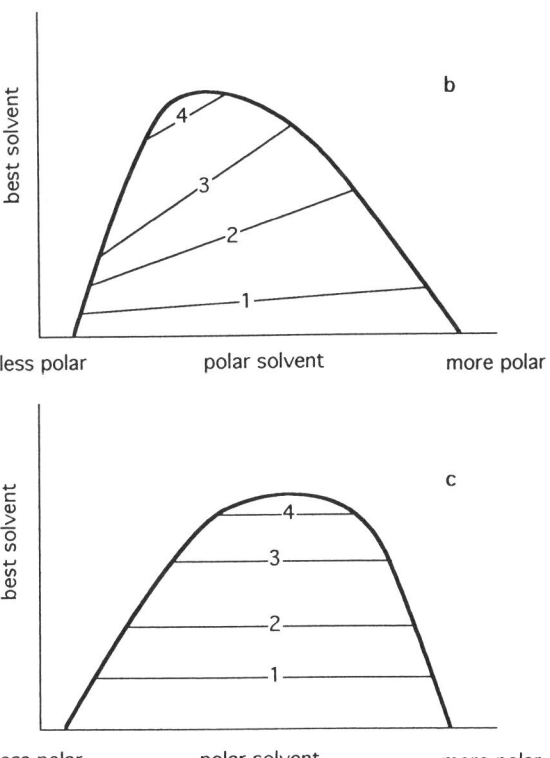

Figure 4c: This kind of ternary diagram should favor a good partitioning of the sample, since the best solvent is evenly distributed in the left and the right phases. If on tie-line 3 the sample is still mainly in one phase, the left (right) one, for example, it is often better to add a fourth solvent than to go to tie-line 4, which is too close to the plait point; adding a solvent less (more) polar than the less (more) polar solvent of this ternary diagram will usually result in a variation of the partitioning of the best solvent, and thus of the sample. The only disadvantage is that you will not be able to use the ternary diagram any more to make the two phases separately, since the introduction of a fourth solvent greatly modifies the distribution of the best solvent.

4. When step 3 has been optimized, and if the separation still has to be improved, then all the classical parameters of analytical chemistry applicable in conventional column chromatography, can be used, namely, the pH, ionic strength, nature of the added salts, and complexation.

B. The Four- and Five-Solvent Approaches

When routinely working with new but similar samples, such as plant extracts or pharmaceuticals, one may wish to have on his or her bench a collection of selected biphasic systems that will allow the quick finding of the one suitable for a successful fractionation. This is why some multisolvent systems have been developed; these cover a broad range of polarities by varying the ratio of each solvent. We now discuss four of these approaches.

1. The Oka Approach

Oka et al. [3] use 16 mixtures of n-hexane (HEX), ethyl acetate (EtOAc), n-butanol (n-BuOH), methanol (MeOH), and water (W) to go from the HEX/MeOH/W, 2/1/1 to the n-BuOH/W, 1/1 systems (Fig. 5). This solvent series covers a wide range of hydrophobicity continuously from the nonpolar n-hexane–methanol–water system to the polar n-butanol–water system. All these solvent systems are volatile and yield a desirable two-phase volume ratio of about 1.

Partition coefficients of various test samples, including dinitrophenyl (DNP) amino acids, herbicides, and various indole auxins have been determined; these show the 16 mixtures provide a simple way to vary the average hydrophobicity of a biphasic system and to find the one that is suitable to have a partition coefficient close to 1 for the desired compound. Dipeptides like Tyr-Gly, Tyr-Val, and Tyr-Leu stay in the aqueous phase even with the more polar system n-butanol–water, and more polar solvent systems are required for the separation of these compounds.

2. The Expanded Margraff Approach

Margraff et al. [4] use a mixture of n-heptane, ethyl acetate, methanol, and water to go from a 1/1/1/1 mixture to a 1/1 mixture of EtOAc/W; they define seven

SOLVENT SYSTEMS IN CPC

N°	n-Hexane	EtOAc	n-BuOH	MeOH	Water	U/L	Settling time(sec)
1	10	0	0	5	5	1.05	5
2	9	1	0	5	5	0.96	8
3	8	2	0	5	5	0.88	14
4	7	3	0	5	5	0.82	20
5	6	4	0	5	5	0.77	22
6	5	5	0	5	5	0.74	26
7	4	5	0	4	5	0.80	28
8	3	5	0	3	5	0.86	30
9	2	5	0	2	5	0.93	30
10	1	5	0	1	5	0.92	30
11	0	5	0	0	5	0.88	32
12	0	4	1	0	5	0.91	20
13	0	3	2	0	5	0.99	15
14	0	2	3	0	5	1.09	12
15	0	1	4	0	5	1.16	14
16	0	0	5	0	5	1.22	17

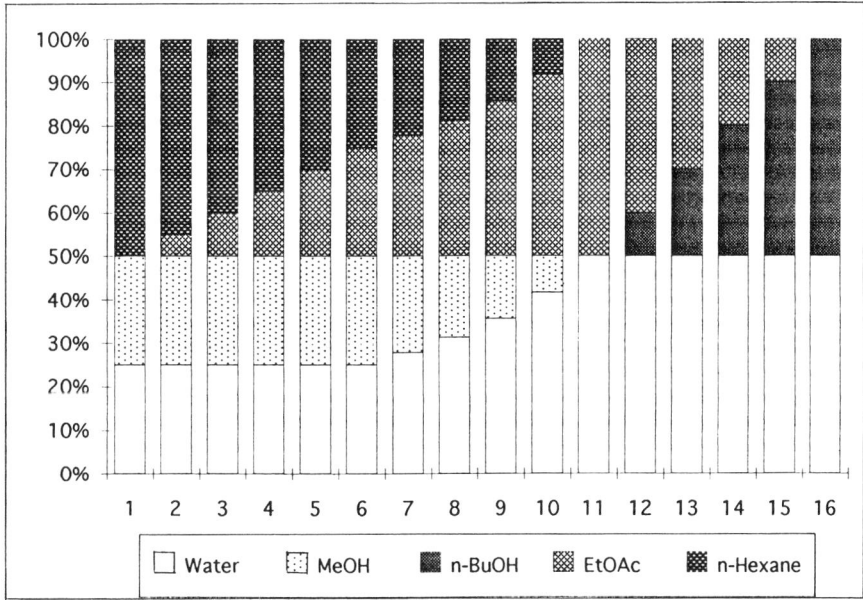

Figure 5 The Oka et al. solvent systems. U/L indicates the volume of the upper phase divided by that of the lower phase.

solvent systems, A, B, C, D, F, G, and H (they skip the character E for a mysterious reason). With this series, and for the families of compounds they are studying, they always find a system where the compound of interest has a partition coefficient near 1. We added the opposite series, named $-B$, $-C$, $-D$, $-F$, $-G$, and $-H$, in order to go to a 1/1 mixture of HEP/MeOH, which is useful for less polar samples (Fig. 6).

The relative polarity of each phase seems to vary smoothly, as shown by their EtN values [5] (or maybe these values are not, in these quaternary mixtures, a faithful witness of the polarity).

Using these systems assumes the samples are freely soluble in EtOAc and/or MeOH.

Starting with the solvent system A, if the sample is well partitioned, it will be used for CPC fractionation, and then adjustment will be made by using B or $-B$ if one wants the sample to favor the less polar phase or the more polar one, respectively.

If with A the sample is mainly in the more polar phase, one will test B, C, . . ., H to adjust the average polarity of the solvent system to the average polarity of the sample.

If with A the sample is mainly in the less polar phase, then one will test $-B$, $-C$, . . ., $-H$ for the same reason.

3. The HBMW Approach

Mixtures of heptane, *n*-butanol, acetonitrile, and water (HBMW) have been tested in our laboratory (they are UV transparent), and seven mixtures have been defined and named -3 to $+3$, going from a 1/1 HEP/MeCN system to a 1/1 *n*-BuOH/W system (Fig. 7).

The EtN values for these systems show they cover a broad range of polarities, as with the expanded Margraff systems, but some variations are sharper, and so it may be necessary to interpolate between two systems to achieve a proper fractionation. But it must be noticed that some mixtures of HEP/*n*-BuOH/MeCN/W lead to triphasic systems, the system 2/3/2/3 being an example. These systems will be useful if the sample is freely soluble in BuOH and/or MeCN (MeCN can be replaced by MeOH as well). As in the previous approach, the system 0 will be tested first, and if the sample is mainly in the more polar phase, then systems 1 to 3 will be tested; otherwise, if the sample was mainly in the less polar phase of the system 0, systems -1 to -3 will be tested.

4. The Multisolvent System Approach

Abbott and Kleiman [6] evaluated 13 biphasic systems for relative polarity using Reichardt's dye and the solubility index of compounds (Table 2). The relative polarity of each phase and some solvents are shown in Fig. 8, where they are classified from least polar (top of Fig. 8) to most polar (bottom of Fig. 8). They

Name	Heptane	EtOAc	MeOH	Water	EtN Lower	EtN Upper	
-H	6	0	6	0	0.73	0.23	(0.23 is an
-G	6	1	6	1	0.76	0.23	estimate)
-F	6	2	6	2	0.77	0.51	
-D	6	3	6	3	0.77	0.51	
-C	6	4	6	4	0.80	0.50	
-B	6	5	6	5	0.77	0.54	(EtN is 0 for
A	6	6	6	6	0.76	0.53	non polar
B	5	6	5	6	0.77	0.54	solvent, and 1
C	4	6	4	6	0.78	0.55	for water)
D	3	6	3	6	0.79	0.55	
F	2	6	2	6	0.82	0.53	
G	1	6	1	6	0.85	0.53	
H	0	6	0	6	1	0.50	

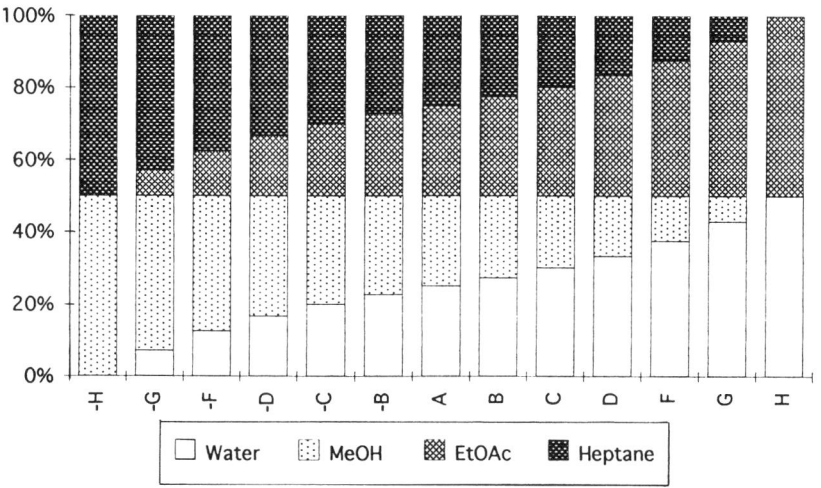

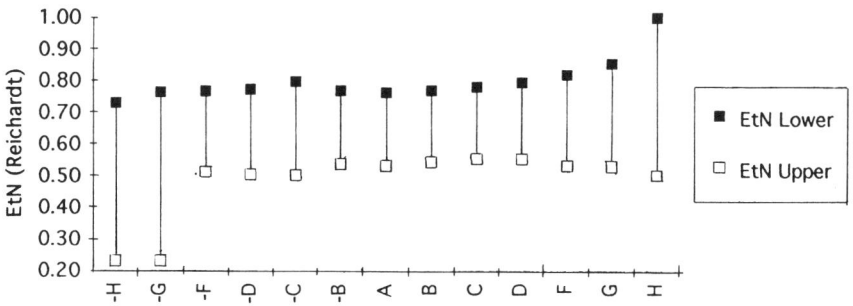

Figure 6 The expanded Margraff solvent systems.

85

N°	Heptane	n-BuOH	MeCN	Water	EtN Lower	EtN upper
-3	5	0	5	0	0.55	0.39
-2	5	2	3	0	0.60	0.56
-1	4	2.6	2.4	1	0.70	0.55
0	4	3.2	1.8	2	0.72	0.57
1	2	3.8	1.2	4	0.79	0.65
2	1	4.4	0.6	4	0.77	0.65
3	0	5	0	5	0.90	0.65

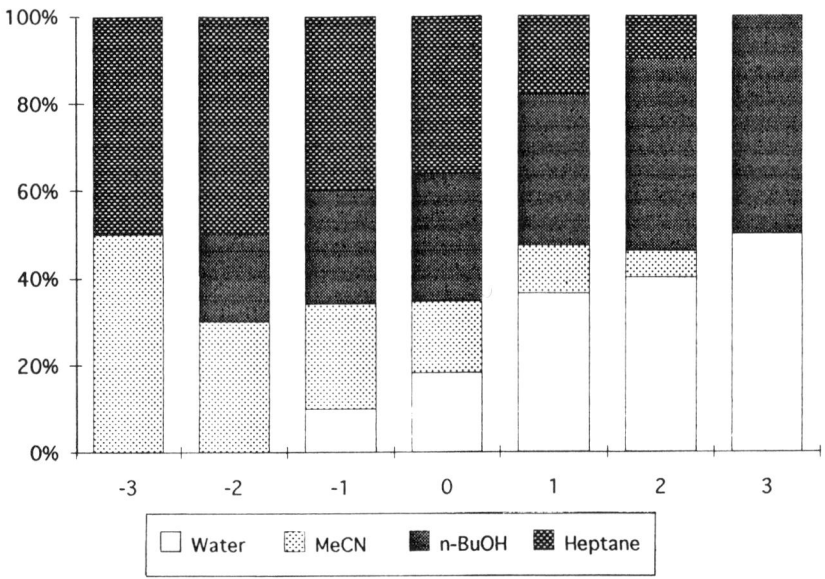

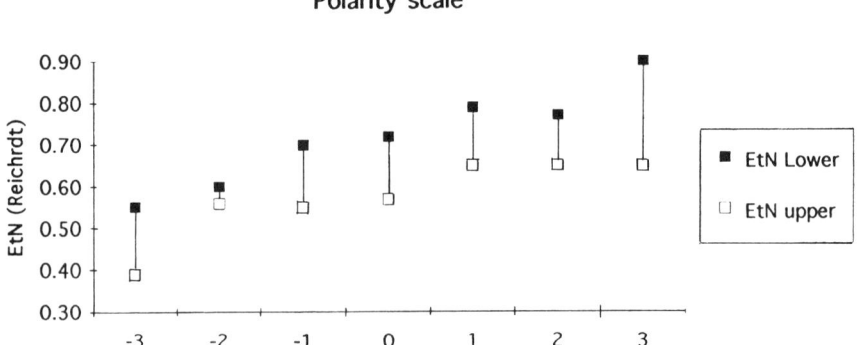

Figure 7 The HBMW solvent systems.

Table 2 Two-Phase Solvent Systems for Countercurrent Chromatography

System	Solvent and volume ratio
1	HEX/MeCN (1/1)
2	HEX/MeCN/CHCl$_3$ (5/5/1)
3	HEX/EtOH/W (6/5/1)
4	HEX/EtOAc/MeCN/MeOH (5/2/5/4)
5	HEX/EtOAc/MeOH/W (1/1/1/1)
6	CHCl$_3$/MeOH/W (13/7/2)
7	CHCl3/MeOH/W (1/1/1)
8	CHCl$_3$/MeOH/W (7/13/8)
9	TOL/MeCN/W/EtOH (3/4/3/2)
10	CHCl$_3$/MeOH/0.2 M HOAc (1/1/1)
11	EtOAc/EtOH/W (2/1/2)
12	n-BuOH/HOAc/W (4/1/5)
13	n-BuOHG/EtOAc/W (4/1/4)

Abbreviations: BuOH, butanol; CHCl$_3$, chloroform; EtOAc, ethyl acetate; HEX, hexane; HOAc, acetic acid; MeCN, acetonitrile; MeOH, methanol; TOL, toluene; W, water.
Source: Ref. 6.

defined three groups of systems—the lipophilic (systems 1–3), the polar (systems 9–13), and the intermediate (systems 4–8).

III. ELUTION MODE IN CPC

When a suitable biphasic system has been selected to perform a CPC run, the chromatographer has still to decide which elution mode he will choose, since the liquid nature of the two phases does not restrict one of them to be stationary, as it is in HPLC.

Moreover, one can plan to use alternatively each phase as mobile phase, in the so-called dual-mode elution.

Or, taking advantage of some favorable situations revealed by the shape of the ternary diagram, one may run a polarity gradient by varying the composition of the mobile phase while keeping that of the stationary phase roughly constant.

A. The Dual-Mode Operation

A useful feature of CPC is its ability to be used in either normal or reversed phase elution with the same two-phase partition solvent system, as illustrated in Fig. 9. The hypothetical sample contains five components a, b, c, d, and e, their affinity

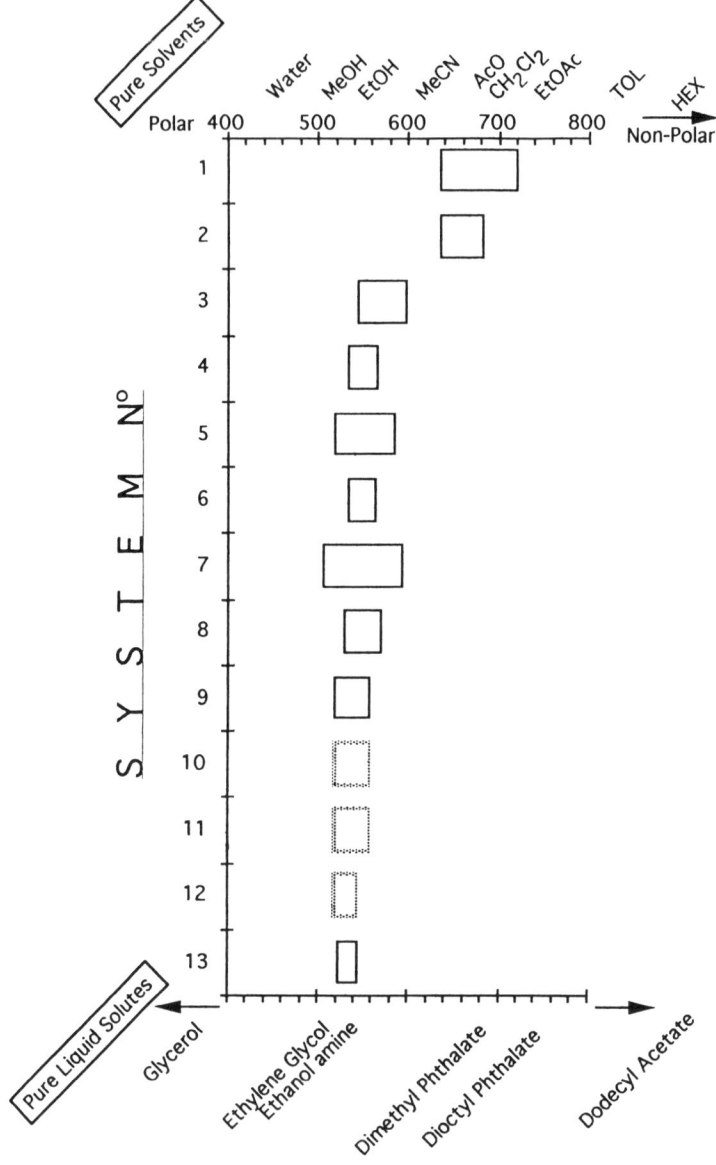

Figure 8 The multisolvent systems. Abscissae show the absorbance maxima of Reichardt's dye (in nm); dashed boxes were classified by solubility of index compounds, not Reichardt's dye. Abbreviations are the same as in Table 2, where solvent systems 1–13 can be found; additional abbreviations can be found in Table 1.

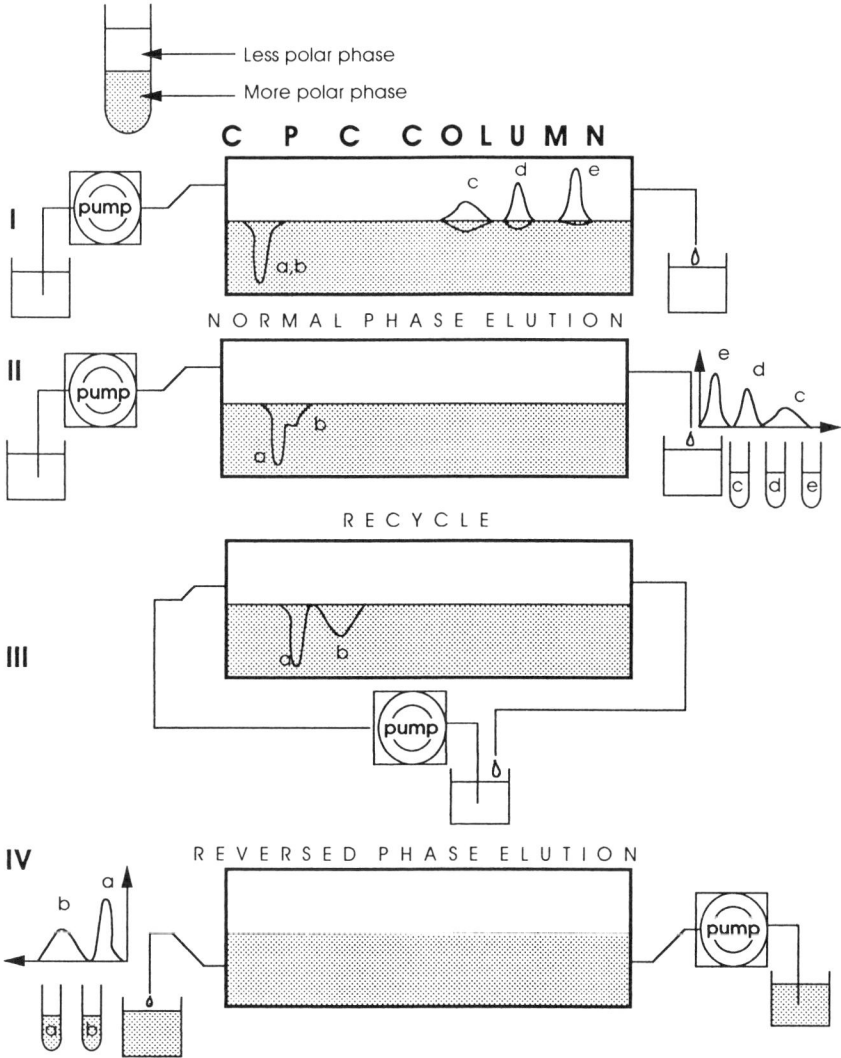

Figure 9 Dual-mode elution in CCC. See text for comments.

for the nonpolar upper phase increasing from a to e. The less polar components, e, d, and c, separate from each other via elution with the upper phase (normal phase chromatography), while the more polar components, a and b, remain in the stationary lower phase (I and II). During upper phase elution, components a and b slowly migrate and separate from each other in the stationary phase (III); after a sufficient amount of upper phase has passed through the CPC column (III), mobile phase and flow direction are reversed (IV), and components a and b are eluted. This feature makes CPC popular with natural product chemists for isolating active molecules from a very rich and diversified starting material, as they are sure to retrieve both polar and nonpolar compounds in a single run. CPC becomes also popular when biological products are involved, since we do not have to worry about irreversible adsorption to the stationary phase.

B. Gradient Elution

Just as with HPLC, gradient elution provides to CPC an easy way to fractionate solutes of widely differing polarities and partition coefficients, and to reduce run times. Some favorable situations may occur where the composition of one phase may be systematically varied while the composition of the other remains relatively constant. The most direct way to predict this condition is to refer to ternary phase diagrams.

Let us consider the ternary diagram shown in Fig. 10. It describes how water, ethanol, and hexane mix together. Tie-line orientation clearly indicates that ethanol prefers to go into water rather than into hexane. If we prepare the two biphasic systems corresponding to S_1 and S_2, (Fig. 10a), we will get approximately equal volumes of the lower polar phase (left) and the upper nonpolar phase (right); we can see, by following the tie-lines corresponding to S_1 and S_2, that the respective nonpolar phases will be roughly the same, while the respective polar phases will differ mainly by the EtOH content.

In this case, if we fill the CPC apparatus with the hexane-rich phase (stationary phase), we should be able to vary the polarity of the other phase (mobile phase) by increasing the amount of ethanol in water, provided we stay close to the binodal for the composition of the mixture used as the mobile phase. We can then go from ≈0 to 70% ethanol in water (v/v), saturated with hexane, to elute from a "hexane column," and run, in this case, a gradient of decreasing polarity on a "reversed phase column."

It can sometimes be much more convenient to prepare the stationary phase and the mobile phases separately; this can be with reference to the ternary diagram (Fig. 10b):

IMP (initial mobile phase) corresponds to water saturated with hexane, and to which approximately 1% ethanol can be added.

FMP (final mobile phase) corresponds to a mixture EtOH/water/HEX 66/31/3, but to prepare it it is very convenient to make a mixture EtOH/water 70/30 and add 3 to 4% hexane, until you see a small layer of nonpolar phase covering the lower phase; you are then sure "to be on the binodal."

SP (stationary phase) corresponds to hexane saturated with water and containing a small percentage of ethanol; similarly to FMP, it is better to have a small layer of lower phase at bottom of the solvent container, in order "to be on the binodal."

IMP and FMP can then be the reservoirs *A* and *B* of a gradient pump, and any combination of IMP and FMP will result in a phase the corresponding plot of which will be on the dotted line of Fig. 10b, that is, mainly left phase (polar, for this example), saturated with right phase.

Many ternary diagrams indicate this kind of behavior, most with water as

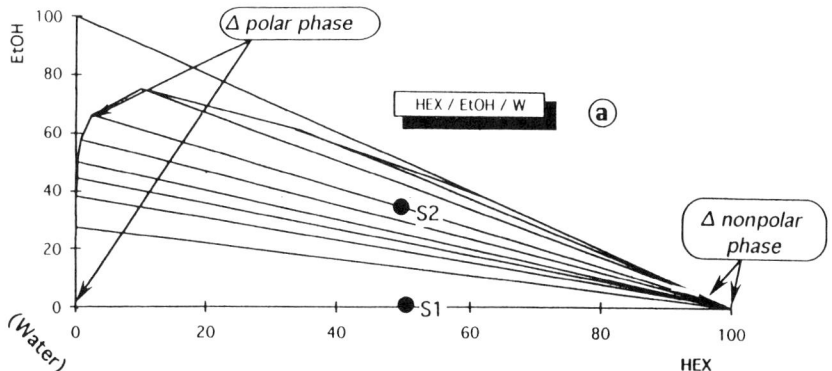

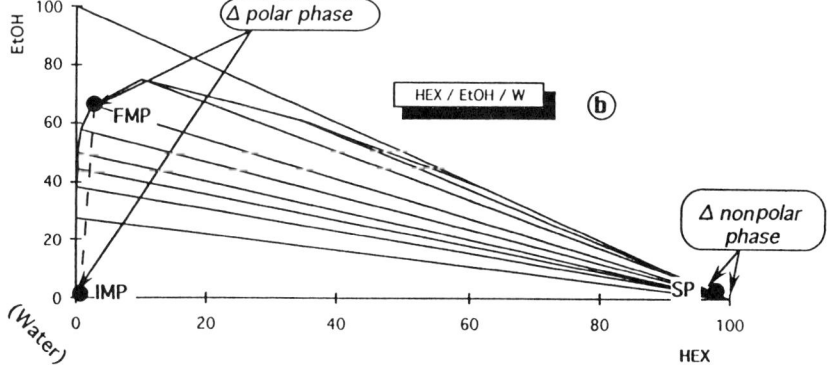

Figure 10 Ternary system of hexane/ethanol/water. (a) Compositions of the liquid phases used to generate a gradient of ethanol in water on a hexane-rich liquid stationary phase: S_1 corresponds to the initial system, and S_2 to the final system; (b) but it is much easier to use the ternary diagram in order to prepare the corresponding liquid phases separately, that is, IMP (initial mobile phase), FMP (final mobile phase), and SP (stationary phase).

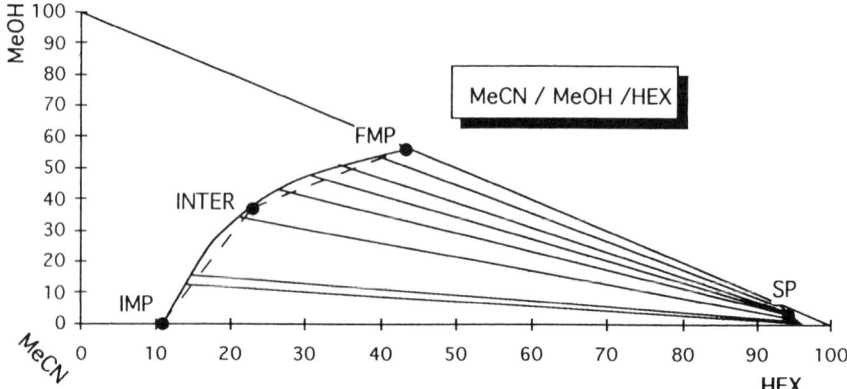

Figure 11 Ternary diagram for the solvent system acetonitrile/methanol/hexane, an example of a nonaqueous solvent system that can be used to build up gradients.

one of the constituents; some of them, however, are nonaqueous, for example, the system acetonitrile/methanol/hexane, the ternary diagram of which is shown on Fig. 11. The hexane-rich phase (SP) can be used as the stationary phase; and the other (mobile) phase can be varied from IMP ($\approx$90% acetonitrile, 10% hexane, ϵ methanol) to FMP ($\approx$55% methanol, 45% hexane, ϵ acetonitrile), via the intermediate phase called INTER ($\approx$35% methanol, 25% hexane, 40% acetonitrile), for approximately following the binodal (a three-reservoir gradient pump is then needed).

1. The Ethyl Acetate/Butanol/Water Ternary Diagram and Gradient

The two systems EtOAc/1-BuOH/W and EtOAc/2-BuOH/W are type 2 systems (see Fig. 1); that is, they consist of two solvents (EtOAc and BuOH), plus a third that is insoluble in the two primary solvents (water). The corresponding ternary diagrams are shown in Fig. 12.

These two ternary systems are very useful for both isocratic and gradient separation of rather polar compounds. Potentially, all compounds that prefer water to ethyl acetate but prefer butanol to water can be purified using these two systems. Their nature (type 2) makes analytical preexperiment extremely easy, since the less polar system (EtOAc/W) and the more polar one (BuOH/W) can be tested for partitioning, and then any mixture of intermediate polarity can be adjusted since there is no problem of plait point for these systems.

As the composition of the polar, water-rich, phase does not vary so much when adding butanol (since butanol preferentially goes into the ethyl acetate–rich phase (the left phases in Fig. 12)), these ternary diagrams show favorable conditions for a gradient run in the normal phase mode; the right part of the

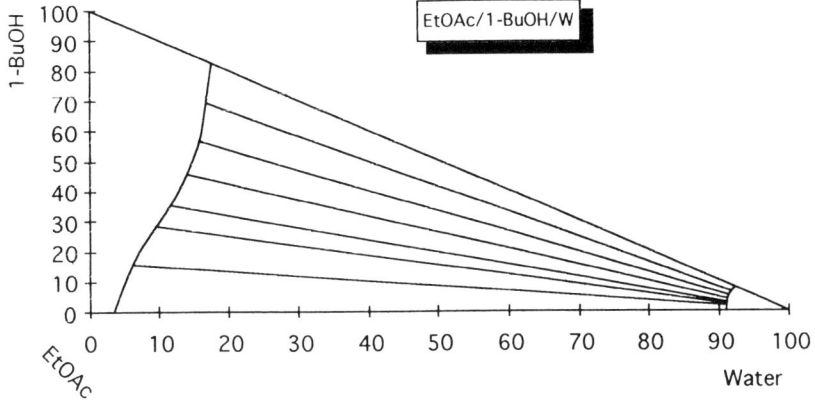

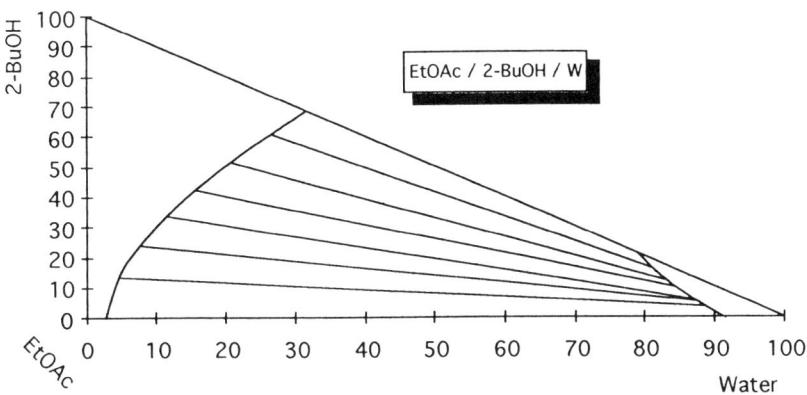

Figure 12 The two ternary diagrams ethyl acetate/1-butanol/water and ethyl acetate/2-butanol/water. The advantage of their being systems 2 is that one can go from the less polar system (EtOAc/water) to the more polar one (BuOH/water) without any problem, since they have no plait point.

binodal, which presents a small curvature, is favorable for running gradients by mixing the two mobile phases, IMP and FMP (like in Fig. 10b).

2. An Example of Application

A very promising field of application for CPC is peptide purification. It would be useful to find some ternary systems for which gradients can be run.

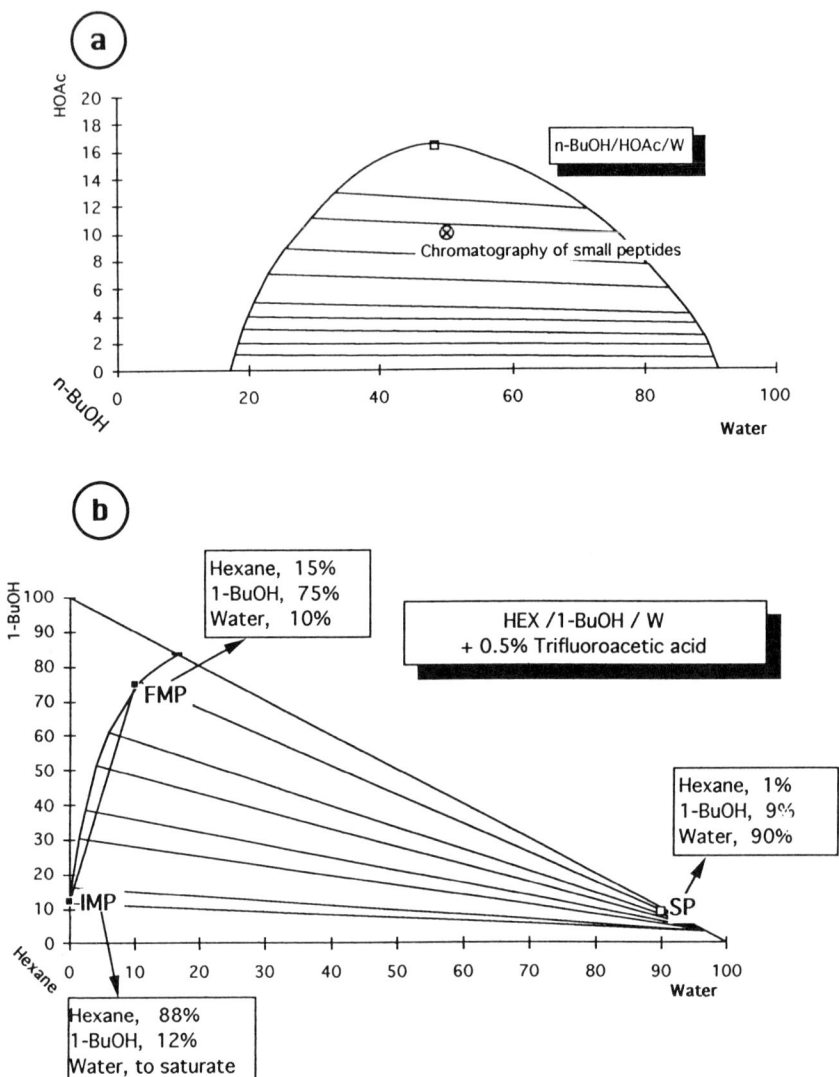

Figure 13 (a) The system *n*-butanol/acetic acid/water is very useful for fractionation of peptides that are rather polar. The experimental point shown is known as the BAW 4/1/5 system. (b) A less polar system, used to build up gradients for peptide fractionation.

SOLVENT SYSTEMS IN CPC

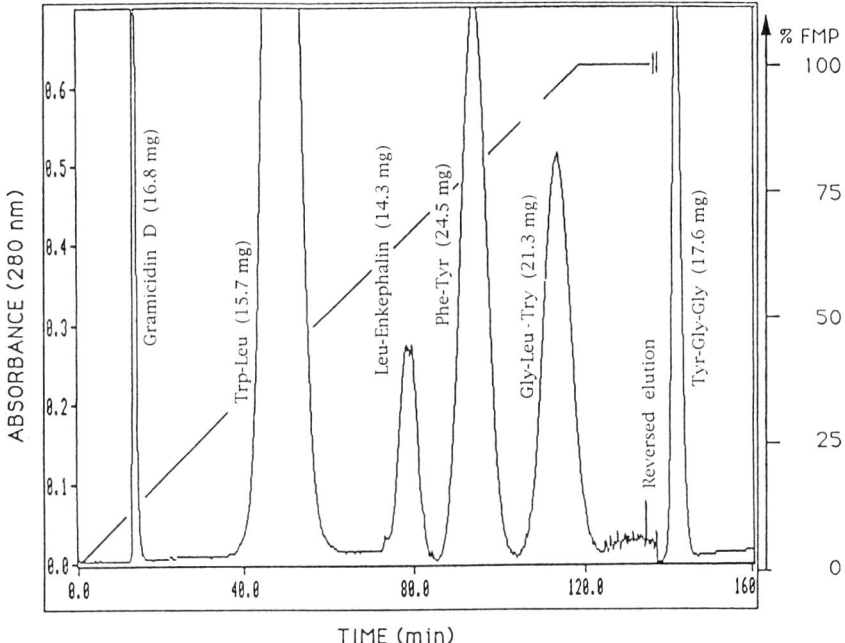

Figure 14 Gradient elution with the hexane/1-butanol/water system. Compositions of the initial (IMP) and final (FMP) mobile phases and of the stationary (SP) phase are calculated from the diagram of Fig. 13.

	Hexane	1-Butanol	H$_2$O
IMP	88	12	sat.
FMP	15	75	10
SP	1	9	90

(+0.5% TFA everywhere)

Chart Column: CPC LLN with six cartridges of type 250W; V_0 (mobile phase volume): 54 ml; mobile phase: hexane/butanol-rich phase, ascending mode; inverted elution at $T \approx 135$ min; rotational speed: 700–800 rpm; flow rate: 4 ml/min; pressure drop: 47 to 40 bars, room temperature; detection UV at 280 nm; sample in 4 ml: Gramicidin D, 16.8 mg; Trp-Leu, 15.7 mg; Leu-Enkephalin (Tyr-Gly-Gly-Phe-Leu), 14.3 mg; Phe-Tyr, 24.5 mg; Gly-Leu-Tyr, 21.3 mg; Tyr-Gly-Gly, 17.6 mg.

Most of the examples found in the literature have been reported by Knight et al. [7] using the ternary system butanol/acetic acid/water shown in Fig. 13a.

This solvent system is very useful for rather polar peptide fractionation, but this is an example of a system we cannot use to design a gradient, because the tie-lines are parallel (i.e., they do not converge), and no phase can be kept constant.

In Fig. 13b is shown the ternary system hexane/butanol/water (with a constant addition of 0.5% trifluoroacetic acid, TFA, (v/v)), which looks very favorable for running gradients in the normal phase mode, with the water-rich phase as the stationary phase. We have tested this gradient, from IMP to FMP, and we found it was extremely stable and reproducible, performing more than 30 repetitive runs with or without inverting the elution mode at the end of a gradient to elute the strongly retained solutes, and with the pressure being always the same when at initial conditions, reflecting the stability of the stationary phase.

Figure 14 is an example of separation of small peptides using a two-hour gradient, followed by dual-mode elution to elute the hydrophilic solutes. UV detection was at 254 and 280 nm, but we did some runs with UV monitoring at 220 nm with no problems.

Figure 15 shows the "pseudonumber of theoretical plates" and the retention times of the corresponding peaks. Due to the compression effect of the gradient for the late eluted peaks, we have more than 1000 plates for all the peaks eluting during the second hour of the two-hour gradient.

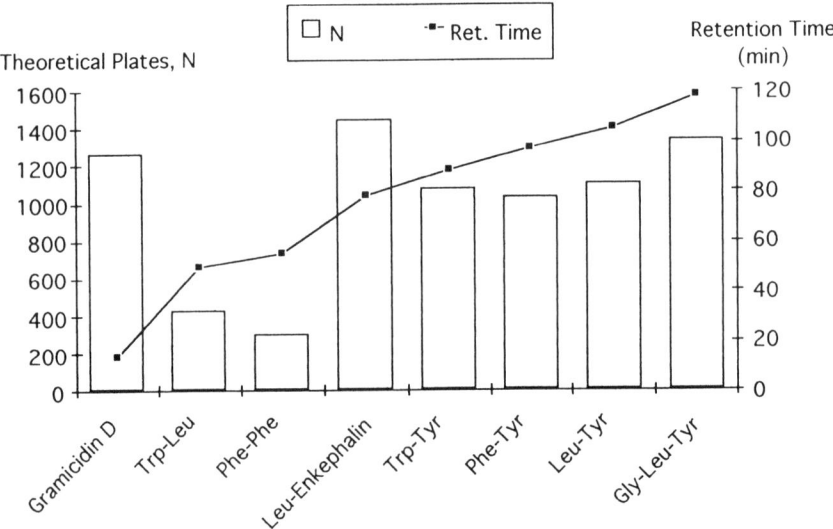

Figure 15 Effect of the gradient on the pseudonumber of theoretical plates estimated for some peptide peaks in a hexane/1-butanol/water system (+0.5% TFA).

IV. CONCLUSION

If I had to conclude this chapter with one idea, I would say, first find the best solvent for your sample, and then think chromatography!

REFERENCES

1. J. M. Sørenson and W. Arlt, in *Liquid–Liquid Equilibrium Data Collection*, (D. Behrens and R. Eckermann, eds.), Dechema, Frankfurt/Main (GFR) (distributed by Scholium International, Great Neck, New York) (1980).
2. A. P. Foucault, P. Durand, E. Camacho Frias, and F. Le Goffic, *Anal. Chem.*, *65*:2154 (1993).
3. F. Oka, H. Oka, and Y. Ito, *J. Chromatogr.*, *538*:99 (1991).
4. R. Margraff and A. Durand, personal communication (1991).
5. Solvent polarity can be estimated using Reichardt's dye (Aldrich, catalogue number 27,244-2), following the procedure described by S. J. Gluck and M. P. Wingeier in *J. Chromatogr.*, *547*:69 (1991).
6. T. P. Abbott and R. Kleiman, *J. Chromatogr.*, *538*:109 (1991).
7. M. Knight, in *Countercurrent Chromatography*, Marcel Dekker, New York, p. 583 and references therein (1988).

5
Fractionation of Plant Polyphenols

Takuo Okuda, Takashi Yoshida, and Tsutomu Hatano
Okayama University, Tsushima, Okayama, Japan

I. ADVANTAGES OF CENTRIFUGAL PARTITION CHROMATOGRAPHY IN FRACTIONATION OF PLANT POLYPHENOLS

A. Polyphenols in Human Life

1. Correlation of Plant Polyphenol and Tannin

The name *plant polyphenols* fundamentally means various types of compounds having phenolic hydroxyl groups, comprising tannins, flavonoids, lignans, coumarins, etc. Among them, tannins and flavonoids are the most widely distributed in the plant kingdom, often in useful plants. In recent years, the name *plant polyphenol* is sometimes employed in place of *tannin*, in spite of the confusion of excluding other types of polyphenols [1]. This is due to the recent revision of the concept of tannin, from a mixture of unidentified compounds to a compound of established chemical structure, which has followed from major advances in tannin chemistry in a few decades. The *plant polyphenols* in the present chapter are mostly those belonging to tannins, but they also include some flavonoidal compounds having tanninlike properties, such as the phenolic compounds in licorice. In fact, some tannins cannot be strictly discriminated from flavonoids since condensed tannins, one of the two main classes of tannins, are condensates of flavans (catechin and its analogs), which belong to the flavonoids. The main component of green-tea tannin, (−)-epigallocatechin gallate, is a gallate of a flavan analog.

One of the well-known uses of plant polyphenols in human life is leathering with tannin-rich plant extracts. The utilization of plant polyphenols in medicine and food, however, is far more significant for human life because of their wide distribution and structural variation [2,3]. There are also several other ways of utilizing tannins [4].

2. Polyphenols in Food and Medicine

The polyphenols, contained in large amounts in tea, coffee, red wine, and several other beverages, are examples of the polyphenols taken in everyday life by an enormous number of people worldwide. The polyphenols in tea are catechin analogs that are called tea tannin, and those in coffee are chlorogenic acid and its analogs called caffetannin. Among these compounds, (−)-epigallocatechin gallate, the main polyphenol in green tea, has been found to be a potent inhibitor of tumor promotion [5,6]. Some flavonoids and their glycosides are contained in practically all fruits and vegetables. There are also many medicinal plant species that are rich in polyphenols of the tannin class with various chemical structures.

3. Ambiguity in the Concept of Health Effects of Polyphenols, and Their Recently Found Medicinal Effects

Due to lack of attention to the differences of the properties of polyphenols induced by their structural varieties, the toxicity of phenolics is often overestimated. It is true that phenol (monohydroxybenzene), the simplest phenolic compound, and some other small phenolic compounds are quite toxic, but many multifunctional phenolic compounds are practically nontoxic. For instance, α-tocopherol (vitamin E), which is widely contained in food, is a phenolic compound, and gallic acid, having three phenolic hydroxyl groups and a carboxyl group on a benzene ring, merely has a very low systemic toxicity [7]. Many tannin molecules are composed of this gallic acid and/or its analogs that esterify the hydroxyl groups of the sugar or cyclitol core in a hydrolyzable tannin molecule [2].

It must be said that the effects on health of natural polyphenolic compounds contained in various species of plants have mostly been unknown, because it is only recently several hundred pure polyphenolic compounds (tannins) have been isolated from various plants and their chemical structures elucidated [8]. The old, vague concept of tannins prevailed when tannins were undefined mixtures of unknown phenolic compounds, and it should now be replaced by a new concept that differentiates each polyphenolic compound based on the recently developed polyphenol chemistry [8].

Among the pharmacological activities of these recently found polyphenols are the inhibition of tumor promotion and mutagenic and carcinogenic activities, the inhibition of lipid peroxidation and related injurious effects of active oxygen, and several others [9,10]. The host-mediated antitumor activity by the oligomeric hydrolyzable tannins are also one of the remarkable activities of tannins [11]. The potency of activity of each polyphenolic compound is often markedly different. Since the polyphenolic compounds contained in each plant extract are different, it is essential to isolate the constituent polyphenols from the mixture in the plant extract and determine their chemical structures, before the investigation of their biological activities, in order to know their effects on the human life.

B. Properties of Plant Polyphenols, Including Difficulties in Fractionation

Among the polyphenolic compounds biosynthesized in plants, those belonging to flavonoids, except for anthocyanins and catechin condensates (condensed tannins; proanthocyanidins), can be isolated from the plant extracts rather easily because of their stability and the absence of strong binding activity. The most difficult problems often encountered upon the fractionation of plant polyphenols are those involving the polyphenols of the tannin class [12].

Between the two main groups of polyphenolic compounds of the tannin class, the compounds of hydrolyzable tannins are apt to be hydrolyzed in aqueous solutions, particularly in the presence of acid, alkali, or enzyme. Condensed tannins, another main group of polyphenolic compounds, condense with each other in a similar environment to produce larger molecules, sometimes leading to the formation of high polymers, phlobaphenes. Air oxidation of polyphenols often occurs to various extents depending on the chemical structure of each compound, although many polyphenolic compounds of the hydrolyzable tannin class are fairly stable in air [12].

Binding of these polyphenolic compounds with other substances, particularly proteins, peptides, basic compounds such as alkaloids, metallic ions, and compounds of large molecular weight, etc., is another property causing difficulty in the isolation of polyphenols, because of their firm binding with the stationary phases in chromatography [12].

C. Problems in Chromatographic Fractionation of Polyphenols

Open column chromatography with silica gel and alumina is not applicable to the fractionation of tannins because of their firm binding to these adsorbents, which induces extensive (often complete) loss of tannins. High-performance liquid chromatography (HPLC) of polyphenolic compounds on a silica gel column, when applied in small scale for analytical purpose, can be performed without much difficulty by developing with acidic solvents [12]. The open column chromatography or medium-pressure column chromatography on several types of gels, such as Sephadex, DIAION, MCI-gel, and Toyopearl (TSK-gel), particularly when combined with fractionation with centrifugal partition chromatography (CPC), often gives good separation of the polyphenolic components in plant extract [12]. Loss of some components in the polyphenol mixtures, however, due to adsorption on the solid support, is inevitable.

Countercurrent chromatography that does not require a solid support inducing such losses of polyphenolic compounds has advantages over other methods of chromatography, particularly in preparative-scale chromatography. Among several techniques of countercurrent chromatography, droplet countercurrent chromatography (DCCC) is more favorable for the fractionation of complex mixtures than the classical Craig's method because of the higher efficiency obtainable by simple operation of a small instrument of DCCC [12].

The separation of hydrolyzable tannins has been facilitated by the combination of DCCC with column chromatography on some of the aforementioned gels, resulting in isolation of many new polyphenolic compounds from plant extracts [13]. The development in DCCC for the fractionation of tannins in a plant extract, however, usually takes a few days or longer. This long development often causes hydrolysis and diffusion of tannins in the columns.

Application of the CPC with rotary seal joints, which enables the solvent to be pumped fairly quickly into the rotating separation columns, to the separation of tannin mixtures markedly shortened the separation time and improved the separation [13,14].

II. APPARATUSES AND METHODS FOR CENTRIFUGAL PARTITION CHROMATOGRAPHIC SEPARATION OF PLANT POLYPHENOLS

A. Fractionation in Small Scale

1. Apparatus

Centrifugal partition chromatographs (1) Model L-90, comprising a centrifuge with 12 column cartridges, each containing a polyfluoroethylene resin block (Type 250W; 150 × 40 × 40 mm, total volume 180 ml), and (2) Model B-92-N equipped with 12 partition cell cartridges (Type 1000E, 900 ml in total) (both Sanki Engineering, Nagaokakyo, Kyoto), were used. The solvents were pumped into the rotating columns with a pump, Model CPC-LBP-II (Sanki), and collected with a fraction collector, Model SF-160K or Model SF-100 (both Advantec, Tokyo). Eluates were monitored with UV absorbance at 280 nm, and every 10 or 5 fractions were analyzed by high-performance liquid chromatography (HPLC). A Model CPC-UVM-I UV absorbance monitor, equipped with a cell of light path 0.2 mm (Sanki), was occasionally used at 254 nm.

For DCCC, which was performed for comparison with CPC, an apparatus consisting of 100 Pyrex tubes (1200 × 3.2 mm), mutually connected by fine resin tubes, was used. The internal diameter of the glass tube is the smallest one enabling the formation of droplets of the solvent system in which n-butanol is the main component of the upper layer [12].

2. Solvents

The solvent systems (1) chloroform–methanol–water (7:13:8, for licorice polyphenols); (2) chloroform–methanol–n-propanol–water (9:12:2:8, for tea polyphenols); (3) n-butanol–n-propanol–water (4:1:5, v/v/v); (4) n-butanol–n-propanol–water (2:1:3) were used for CPC, and the solvent system (3) was used for both CPC

FRACTIONATION OF PLANT POLYPHENOLS

and droplet countercurrent chromatography. The lower layer of the solvent systems (3) and (4) was used as the stationary phase for the normal-phase development, and it was pumped into the columns prior to the loading of the sample solution. The upper layer was used as the stationary phase for the reversed-phase development. The solvent systems for the separations in Sections III.A.1,2 were selected based on the partition coefficients of the constituent polyphenols between the upper phase and lower phase of the solvent system [15] (Table 1). The solvent systems for the samples in Sections III.B.1–4 were selected based on the results of thin-layer chromatography (TLC) on cellulose plates.

3. Combination of Normal-Phase and Reversed-Phase Developments

Normal-phase development of a fraction obtained from reversed-phase development (or vice versa) sometimes gave good separation of each component in a complex mixture [15]. An example of the application of this combined technique is described in Section III.B.4 (fractionation of polyphenols of *Liquidambar formosana* leaves).

4. Combination with Gel Column Chromatography

When the fractionation with CPC did not give complete separation of each component, the mixture in a fraction from CPC was usually subjected to gel column chromatography on Toyopearl HW-40 (coarse, fine, superfine grades;

Table 1 Partition Coefficients of Polyphenols from Licorice, Tea, and *Woodfordia fruticosa*

Polyphenols	Solvent system[a]	Partition coefficient (upper phase/lower phase)
Licorice polyphenols	1	
Licophranocoumarin		0.37
Glycycoumarin		0.22
Glycyrrhisoflavone		0.15
Tea polyphenols	2	
(−)-Epigallocatechin gallate		7.3
(−)-Epigallocatechin gallate		6.9
(−)-Epicatechin gallate		3.9
(−)-Epicatechin		2.9
Caffeine		0.27
Tannins from *W. fruticosa*	3	
Woodfordin C		0.36
Oenothein B		0.19

[a]See Section II.A.2.

Tosoh), Sephadex LH-20 (Pharmacia), or MCI-gel CHP 20P (Mitsubishi Chemical Industries). Though final purification was often effected by gel column chromatography, the fractionation with CPC prior to the purification on the gel column markedly improved the efficiency of the latter [13–15].

5. *Analysis of Fractions*

The residues, obtained upon evaporation of solvent from the fractions of CPC, were analyzed by HPLC with UV detection at 280 nm, and also by diode-array UV detection. Apparatus: LC-6A system (Shimadzu), L-6000 (Hitachi), or M-45J (Waters), equipped with a YMC A 312 (ODS) (150 × 6 mm, Yamamura-kagaku), LiChrospher 100RP-18 (5 μm, 250 × 4 mm), or Superspher Si-60 (4 μm, 125 × 4 mm, Merck) column, a UV monitor, SPD-6A (Shimadzu) or SF-1205A (ATTO), and a diode-array detector, MCPD-350PC, System II (Otsuka Electronics). The column was kept at 40°C in an oven and was eluted with 0.05 M H_3PO_4–0.05 M KH_2PO_4–methanol (2:2:1, v/v/v). The purity of isolated components was examined also by ^1H and ^{13}C nuclear magnetic resonance (NMR) spectroscopy using a Varian VXR 500 instrument (500 MHz for ^1H and 126 MHz for ^{13}C).

B. Preparative Fractionation in Larger Scale

1. Apparatus

CPC Apparatus (3) Model L-90 containing 12 partition cell cartridges (Type 1000E, 900 ml in total) (Sanki Engineering) was used. The solvents were pumped into the columns rotating 700 rpm, with a pump, Model CPC-LBP-II (Sanki). The flow rate was set at 3 ml/min. Fractions (10 g) were collected with a fraction collector, Model SF-160K (Advantec, Tokyo). Eluates were monitored with UV absorbance at 280 nm, and every 10 fractions were analyzed by HPLC [12].

2. Solvents

The solvent systems were as follows: solvent (1) (in Section II.A.2) for licorice polyphenols, solvent (2) for tea polyphenols, and solvent (3) for hydrolyzable tannins.

Analysis of the fractions obtained from large-scale CPC and their further purification by column chromatography were essentially the same as those of small-scale CPC described in Sections II.A.3–5.

III. APPLICATION OF CPC TO FRACTIONATION OF PLANT POLYPHENOLS

Mixtures of polyphenols of various chemical structures with wide diversity in molecular weights, contained in the extracts from many species of plants used as medicines or beverages, were subjected to fractionation with CPC [13,14]. Some examples of the fractionation with CPC are described in this chapter. Improvements of the separating efficiency by combining several techniques, and also fractionation in larger scale, are also described [15].

A. Fractionation of Plant Polyphenols of Small Molecular Weight with Higher Efficiency

Fractionation of polyphenols of small molecular weight by CPC is generally rather easier than that of polyphenols of large molecular weight. The development in the fractionation of the former can be carried out with less polar solvents, which can be evaporated quickly after collection, resulting in shortening of the total time required for CPC.

1. Bioactive Polyphenols of Licorice

Medicinal Effects of Licorice and Its Chemical Constituents. Licorice, the root or rhizome of *Glycyrrhiza* species of plants, is a crude drug widely used in the world, particularly in oriental medicine [16]. Glycyrrhizin, a triterpenoid glycoside, is known as the main component of licorice. This compound, manufactured by extraction from licorice, is also used as a medicine for treating stomachic ulcers [17]. There are several different *Glycyrrhiza* species of plants from which the crude drug, licorice, is prepared. Traditionally in oriental medicine, the quality of each kind of licorice from different origin has been regarded as markedly different, although most of them are rich in glycyrrhizin. Recent chemical analyses of several kinds of licorice of different origin have revealed notable differences in phenolic constituents among the licorice specimens [18]. Some examples of these phenolic compounds are shown by the chemical structures in Fig. 1. The recent research on licorice constituents has also revealed potent antioxidant activities of various extents of these phenolic constituents, in several experimental systems such as the inhibition of activities of xanthine oxidase (XOD) [19] and of monoamine oxidase (MAO) [20]. Noticeable inhibition of leucotrienes production in arachidonic acid metabolism in human polymorphonuclear neutrophils was exhibited by some of these phenolic compounds (e.g., licochalcones A and B, constituents of the licorice specimen produced in Xinjiang region of China, assignable to *Glycyrrhiza inflata*) [21,22].

These new findings strengthened the requirement of higher efficiency in the fractionation and purification of the phenolic constituents in licorice that are supplied for further detailed investigation of biological activities.

Fractionation of Phenolics in Licorice from Northwest China. The fractionation of licorice phenolics has generally been performed on the solid supports of column chromatography because of their comparatively low affinity to the solid supports. When the phenolics in the licorice from several origins were subjected to column chromatography with silica gel, however, irreversible adsorp-

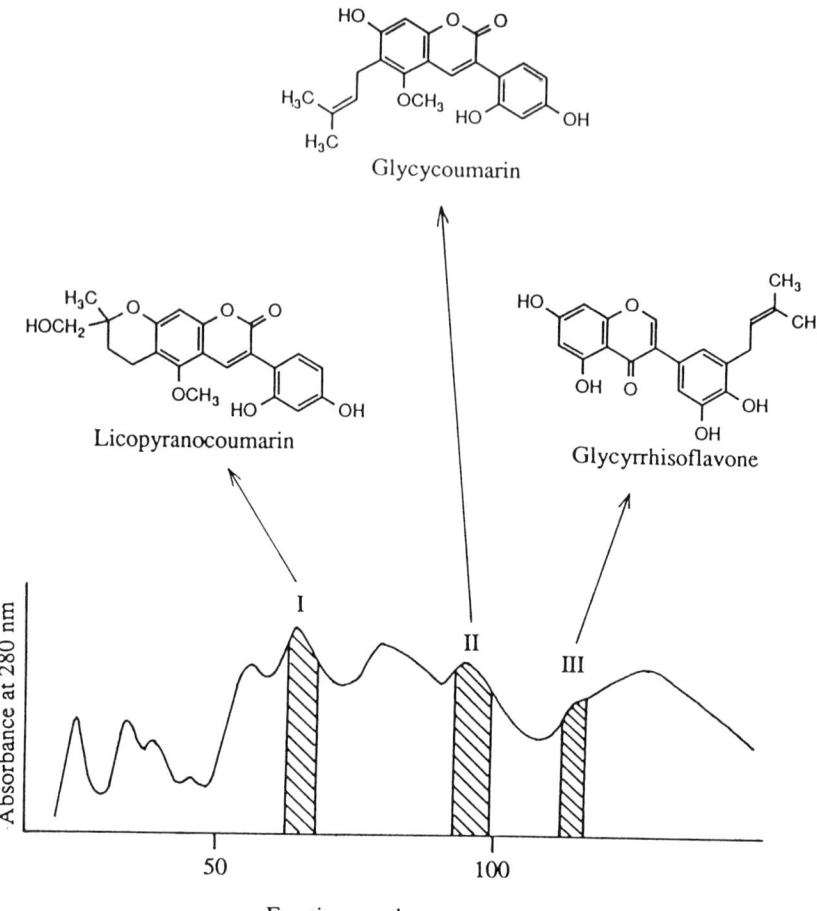

Figure 1 CPC of NW-licorice extract.

tion of some of these phenolics on silica gel inevitably occurred to various extents. Their fractionation by CPC was then attempted, and the licorice imported from the northwest region of China (abbreviated here as NW-licorice, from the plant species assignable to *Glycyrrhiza uralensis* [18]) was subjected to CPC.

Preparation of Extract. The pulverized NW-licorice (500 g) was first defatted with hexane, and then extracted with ethyl acetate at room temperature. The bioactive phenolics in the ethyl acetate extract (23.9 g) obtained after filtration and evaporation were separated by CPC [15].

Fractionation by Small-Scale CPC. The upper aqueous phase of solvent system (1) was used as the mobile phase (reversed-phase development), based on the partition coefficients in Table 1. The result obtained by the fractionation of the ethyl acetate extract (1 g) of NW-licorice is shown in Fig. 1. The flavonoidal constituents in each of the fractions I–III were finally purified by column chromatography on MCI gel CHP-20P, and they were respectively characterized as licopyranocoumarin, glycycoumarin, and glycyrrhisoflavone [20].

Fractionation by CPC of Larger Scale. Fractionation of a larger amount (30 g) phenolics in the same licorice extract was performed using the 1000E cartridges (apparatus 3), resulting in slight lowering of efficiency in the separation [15].

Fractionation of Phenolics in Licorice from the Xinjiang Region of China

Preparation of Extract. The aqueous acetone extract of Xinjiang licorice was partitioned between water and ether. The aqueous layer was further extracted with ethyl acetate and n-butanol. Among these extracts and water-soluble portion, the ether extract showed the highest tanninlike binding activity, which was evaluated based on the binding to hemoglobin [23].

Fractionation by CPC. The phenolic compounds in the ether extract (3 g) were fractionated by CPC, using solvent system (1) in normal-phase development, into four fractions (I–IV) (Fig. 2). The tanninlike activity was shown only by fraction IV. This fraction was further purified by column chromatography over silica gel to give licochalcone B of a significant tanninlike activity and 7,4′dihydroxyflavone. Similar rechromatography of fraction II afforded licochalcone A [22].

Biological Activities of Isolated Phenolics. Among the phenolics thus isolated from licorice of different origins, licochalcones A and B at concentrations of 10^{-3}–10^{-7} M were found to exhibit strong inhibitory effects on the formation of 5-HETE, 5,12-diHETE, leucotrienes B4 and C4 in the A 23187–induced arachidonate metabolism in human polymorphonuclear neutrophils [21]. In addition to licochalcone A, glycycoumarin, glycyrrhisoflavone, and licopyranocoumarin also inhibited the cytopathic activity of a human immunodeficiency virus [24]. The licorice flavonoids, including licopyranocoumarin, glycyrrhisoflavone, and licochalcones A and B, showed inhibitory effects on xanthine oxidase (XOD) and monoamine oxidase (MAO), which are regarded as participating in the oxidative damage of living tissue [19,20]. Radical scavenging activity of these phenolics on 1,1-diphenyl-2-picrylhydrazyl radical, and superoxide anion radical generated from xanthine–XOD system were also demonstrated [23,25]. It is probable that these antioxidative activities of licorice flavonoids underlie the medicinal effects of licorice.

Comparison of CPC with Preparative HPLC. These results of the fractionations with CPC show that a single operation of CPC does not always lead to final purification, and sometimes preparative HPLC appears to effect better

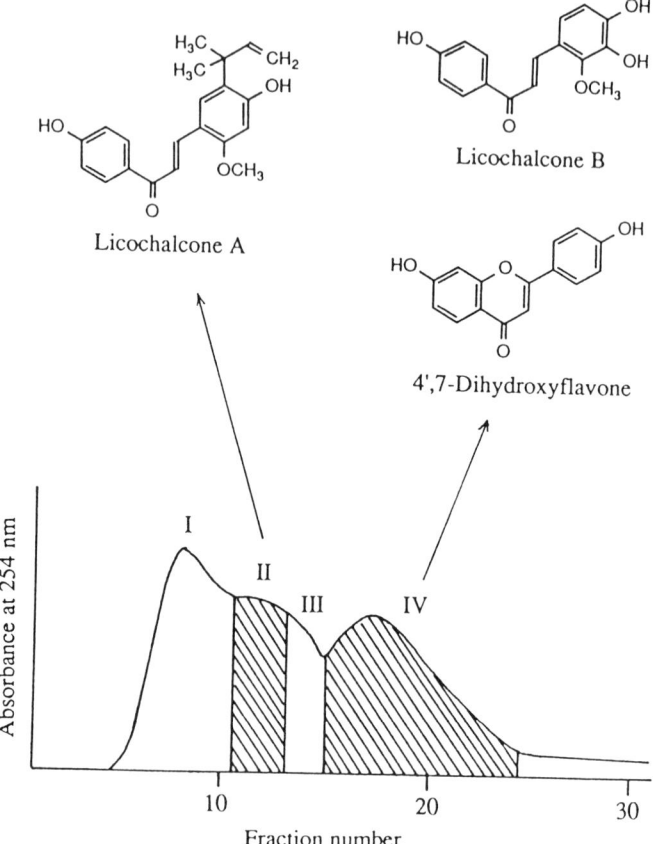

Figure 2 CPC of Xinjiang licorice extract.

separation when extensive sample loss does not occur on the HPLC column. This sample loss, accompanied by the contamination of solid support, which prevents repeated application of the column, is one of the serious problems in preparative HPLC of polyphenols. Although this problem is not too severe for the chromatography of licorice phenolics, it is more serious for the polyphenols of larger molecular weight described later. Besides the loss due to adsorption, structural transformation of large-molecular weight polyphenols occurs in the column sometimes. The CPC technique can be applied to the separation of fairly large samples (10–30 g or more with 1000E cartridges) of large-molecular-weight tannins, such as tetrameric hydrolyzable tannins, exemplified by trapanin B [26], nobotanin K [22], and sanguiin H-11 [27], without suffering their loss. This is a great advantage of CPC over the other methods employing solid supports, such as column chromatography and preparative HPLC.

2. *Tea Polyphenols*

Tea has been the most widely used beverage in the world for much of human history. It was regarded as a medicine in ancient times, and recent advances in the study of biological and pharmacological activities of tea constituents, particularly those of polyphenolic constituents, have again generated strong interest in the effects of tea on health [15,28].

Chemical Differences in Tea Polyphenols Due to Differences in Preparation. Genuine tea, prepared from the leaves of *Camellia sinensis* (some local drinks, such as those called "bush tea," prepared from entirely different plants containing different chemical constituents, are excluded here) can be classified into three types: nonfermented tea (green tea), partially fermented tea (oolong tea and several others), and fermented tea (black tea). The main difference in chemical constituents among these three types of tea is in the chemical structures of polyphenolic constituents (caffeine is unaffected by fermentation). The main polyphenol in fresh tea leaf (and also in green tea prepared for beverage), (−)-epigallocatechin gallate, suffers structural transformation into oligomers and polymers upon fermentation. The polyphenols in fermented teas are complex mixtures that cannot be represented by a single compound. Therefore, the investigation of pharmacological activities of tea polyphenols in recent years has focused on (−)-epigallocatechin gallate, which occupies the main part of green-tea polyphenols [5,6].

Pharmacological Activities of Tea Polyphenols. (−)-Epigallocatechin gallate (EGCG), the main polyphenolic constituent of green tea, also represents the properties and activities of green-tea tannin. In spite of its small molecular size (m.w. 458), it shows strong binding activity to proteins and basic organic compounds that is comparable with that of many tannins of larger molecular size [29]. Among the pharmacological activities of EGCG, the inhibition of tumor promotion has been investigated most extensively.

EGCG potently inhibited tumor promotion in the two-stage carcinogenesis on mouse skin induced with 7,12-dimethylbenzo[*a*]anthracene (DMBA) as initiator and teleocidin as promoter [5]. In the experiments using 12-*O*-tetradecanoyl-phorbol-13-acetate (TPA) as promoter, EGCG and some hydrolyzable tannins inhibited the binding of TPA to the receptor prepared from the particulate of mouse skin [5]. Oral administration of EGCG remarkably inhibited the induction of the duodenal cancer of rats by *N*-ethyl-*N'*-nitrosoguanidine. Potent antimutagenic effects of EGCG and many other polyphenolic compounds on several carcinogens were also found [6].

Fractionation of Green-Tea Polyphenols. As shown in Table 1 (p. 103), the partition coefficients of four tea polyphenols between upper phase and lower phase of solvent system (2) which were all higher than 1, were significantly different from each other, while the coefficient of caffeine was much lower. Good separation of these polyphenols from each other and from caffeine, by the normal-phase development, was therefore expected.

In an initial experiment of fractionating tea polyphenols with CPC, a relatively small amount of the aqueous extract (1 g) from green-tea leaves was subjected to CPC, using smaller cartridges (apparatus 1), and developing with solvent system (2) in "descending mode." The result is shown in Fig. 3 [15]. The major three peaks in the elution curve are those due to caffeine, (−)-epicatechin gallate (ECG), and EGCG. The purity of the polyphenolic compounds and caffeine isolated in this way is shown by the HPLC profiles in Fig. 3 [15].

A larger amount (10 g) of the aqueous extract of green tea was then subjected to CPC using apparatus 3, developing with the same solvent system. The result was similar to that of the separation experiment with 1 g extract, as shown in Fig. 3. Lowering of separation efficiency due to the increase of the sample was not observed. This single operation yielded 1 g of (−)-epigallocatechin gallate of more than 80% purity, which gave pure EGCG upon crystallization from aqueous solution.

B. Fractionation of Polyphenols of Larger Molecular Size

1. Comparison of Separating Efficiency with CPC and DCCC of Polyphenols in Lythrum anceps

The tannin-rich herb of *Lythrum anceps* Makino (Lythraceae), grown in Japan and China, is a medicinal plant used for treatment of diarrhea, edema, and other ailments. At the beginning of the investigation of the constituents in this plant, fractionation was attempted with DCCC, and then CPC was applied to improve the efficiency in the fractionation. The extract of this plant was prepared in the following way.

Preparation of Extract. Fresh leaves (1.4 kg) were homogenized in acetone–water (1:1, v/v), and the filtrate was concentrated in vacuo at <40°C. After extracting with diethyl ether, the mother liquor was extracted with ethyl acetate. The mother liquor was evaporated in vacuo, and the residue was extracted with methanol.

Fractionation by CPC. A portion (4.3 g) of the residue (77 g) obtained upon evaporation of methanol was dissolved in the lower layer of solvent system (3) (10 ml) and then pumped into the columns initially at a rate of 0.2 ml/min. This rate was slowly increased to 1.0 ml/min in 1 hour, and 10-g portions of the eluent were collected.

Fractionation by DCCC. The tannins in another portion of the residue (3 g) from evaporation of the methanol solution were fractionated by droplet counter-current chromatography, with solvent system (3), developing initially at a rate

FRACTIONATION OF PLANT POLYPHENOLS

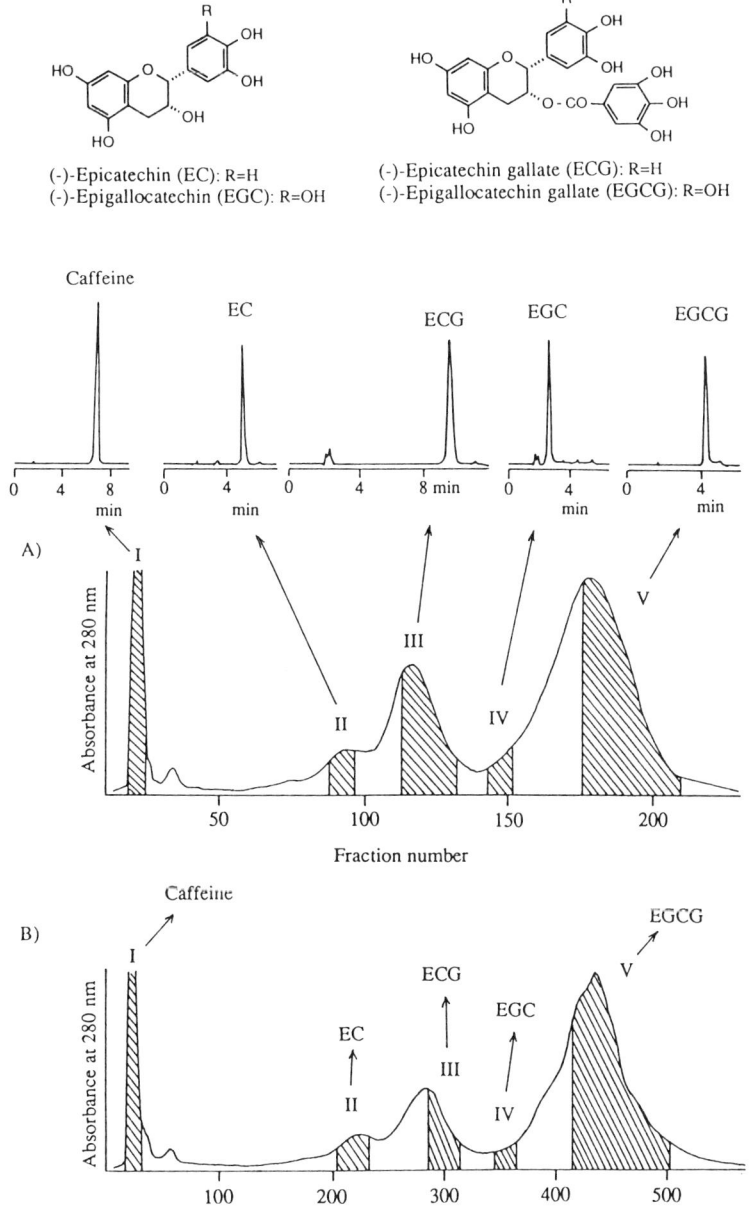

Figure 3 (A) CPC of tea extract in a small scale, and HPLC profiles of combined fractions; (B) CPC of tea extract in a larger scale.

of 0.2 ml/min, which was increased up to 0.8 ml/min in 1 hour, 10-g fractions being collected.

Comparison of Results. As shown in Fig. 4, the separation of polyphenols by CPC was achieved within a quarter of the time required for DCCC. The extent of separation of each component with these two techniques were similar to each other. Fractions in the groups III and IV (Fig. 4) in CPC of the ethyl acetate extract from *L. anceps* contained a dimeric hydrolyzable tannin (oenothein B [30], 38 mg) and a *C*-glycosidic tannin (castalagin [31], 85 mg), respectively. These two compounds are poorly transferred to the mobile phase, and collection of the components in the groups III and IV by DCCC was therefore essentially impracticable. Even collection of the components in the group II, which contains granatin A and 3-galloylgranatin A [32], required more than 60 hours of the development in DCCC. Another *C*-glucosidic hydrolyzable tannin (vescalagin [33], 51 mg) was retained in the stationary phase.

2. *Polyphenols of* Stachyurus praecox

Plant Extract. Fruits, leaves, and twigs of *Stachyurus praecox* (Stachyuraceae), grown in Japan and China, have been used as a dye and also as an antidiarrheic and a diuretic. It has been regarded as rich in tannin without knowing its content. The extraction of fresh leaves (2.5 kg) was carried out analogously to that of *L. anceps* [34].

Fractionation by CPC. A portion (3 g) of the ethyl acetate extract (46 g) was subjected to CPC using solvent system (4) for the normal-phase development. The flow rate was slowly increased in the same way as for the separation of the extract of *L. anceps*.

The collected fractions were gathered in three groups: group I (fractions 1–17, 0.6 g, containing praecoxin C [34]); group II (fractions 18–41, 1 g, containing praecoxins A, B, D, casuarictin, tellimagrandin I, 1,2,6-tri-*O*-galloyl-β-D-glucose, rugosin C [35], guavin A [36], and rugosin F, which is a dimeric hydrolyzable tannin [37]); group III (fractions 42–70, 1.4 g, containing strictinin, pedunculagin, and casuarinin, which is a *C*-glucosidic tannin [38]). It is notable that praecoxin C was separated from rugosin C (Fig. 5) in this way, because praecoxin C has a labile depside group and is apt to be transformed into rugosin C under mild conditions [34].

3. *Highly Water-Soluble Condensate of a Polyphenol with Ascorbic Acid in* Geranium thunbergii

Tannins in *Geranium thunbergii*. The herb of *Geranium thunbergii* is one of the most popular folk medicines and is also an official drug in Japan, mainly used for treating diarrhea and other intestinal disorders. It is a tannin-rich plant, and the main component is a yellow nonirritant crystalline tannin called geraniin [39]. A large portion of the polyphenolic constituents in this plant, and also in most of *Geranium* species of plants, is occupied by geraniin (its content in dry

FRACTIONATION OF PLANT POLYPHENOLS

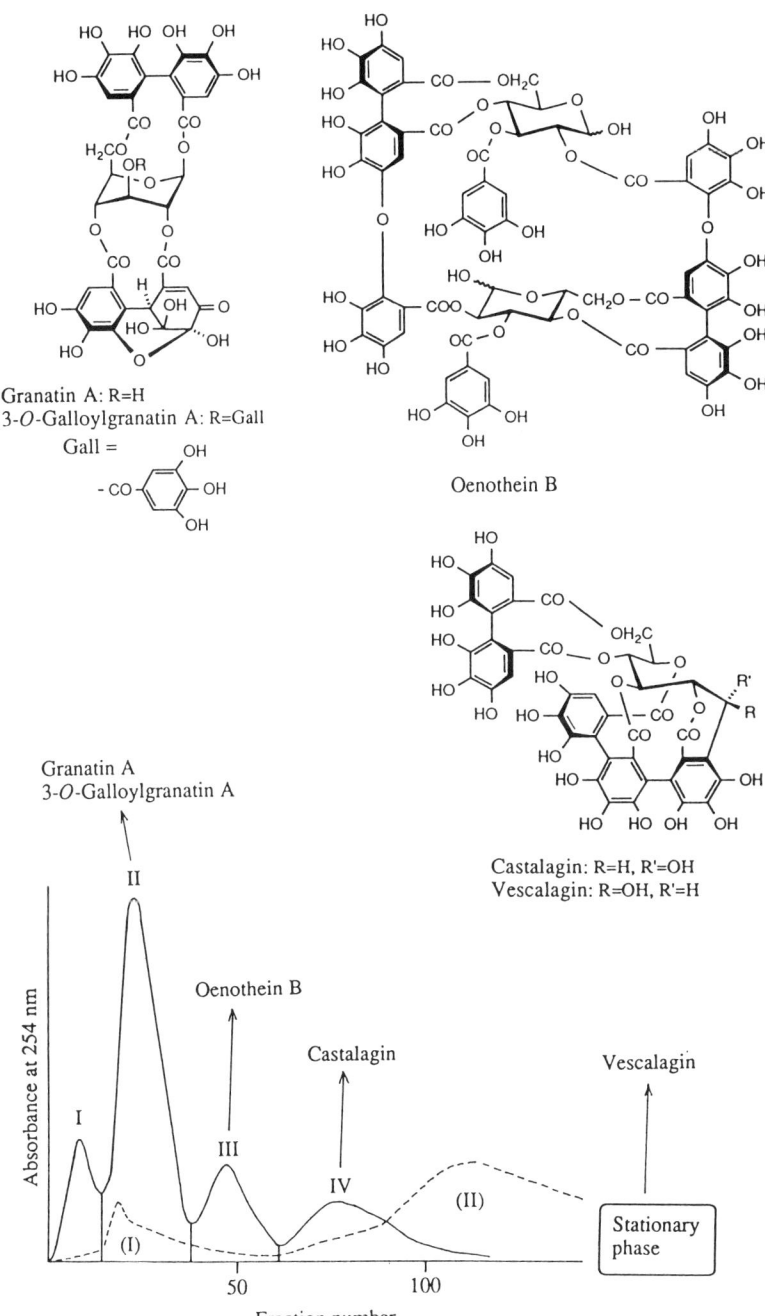

Figure 4 CPC of *Lythrum anceps* extract.

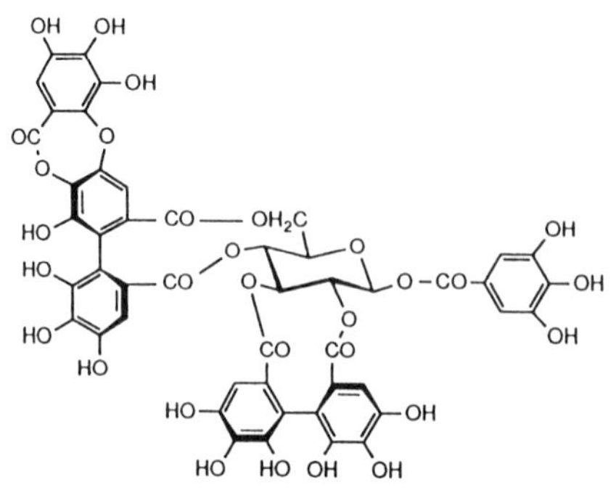

Praecoxin C

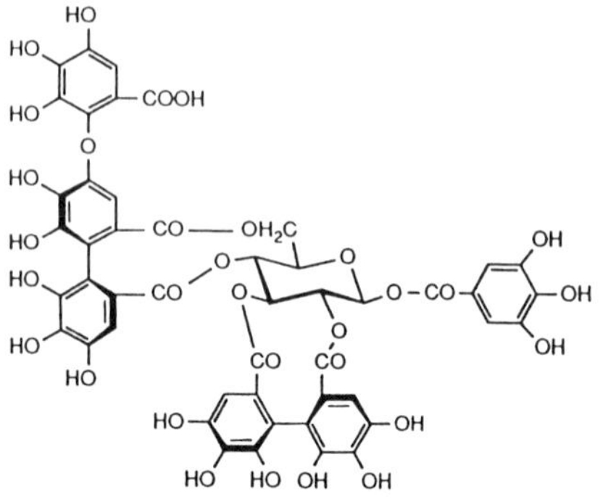

Rugosin C

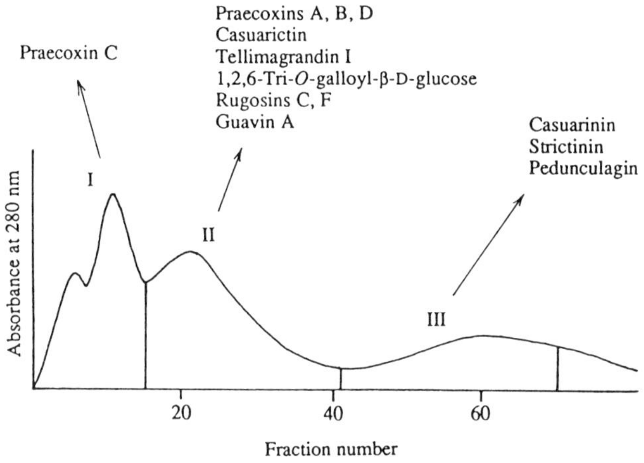

Figure 5 CPC of *Stachyurus praecox* extract.

leaf is over 10%). It is unusual that a compound occupies such a large part of the tannin content in species of plant, since the tannins in most of the tannin-rich plants are mixtures of many polyphenolic compounds.

Geraniin in *G. thunbergii*, however, is accompanied by smaller amounts of other polyphenolic compounds, among which are some extremely water-soluble polyphenolic compounds, such as the condensate of geraniin with ascorbic acid [40]. The isolation of these compounds required removal of geraniin beforehand by crystallizing out from the extract. Quick development in the chromatography, which cannot be done by DCCC, was also required for the separation of this kind of highly water-soluble compounds.

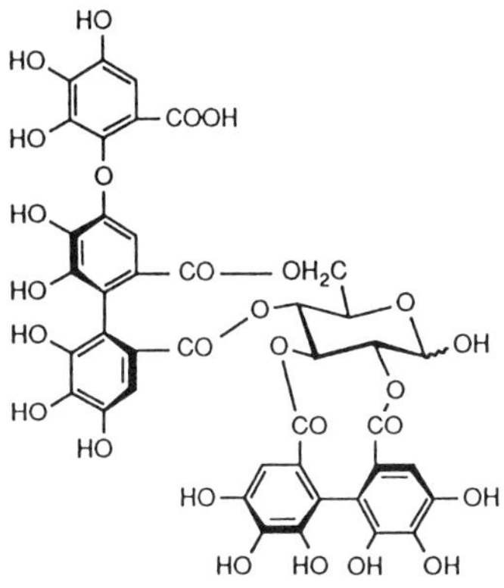

FRACTIONATION OF PLANT POLYPHENOLS

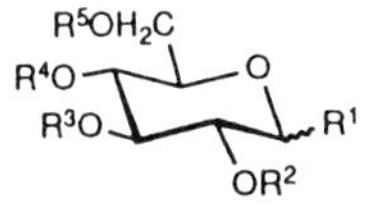

Praecoxin B: R^1=OH, R^2R^3=HHDP, R^4=R^5=Gall
Casuarictin: R^1=(β)-O-Gall, R^2R^3=R^4R^5=HHDP
Tellimagrandin I: R^1=OH, R^2=R^3=Gall, R^4R^5=HHDP
1,2,6-Tri-O-galloyl-β-D-glucose: R^1=(β)-O-Gall, R^2=R^5=Gall, R^3=R^4=H
Strictinin: R^1=(β)-O-Gall, R^2=R^3=H, R^4R^5=HHDP
Pedunculagin: R^1=OH, R^2R^3=R^4R^5=HHDP

Praecoxin A

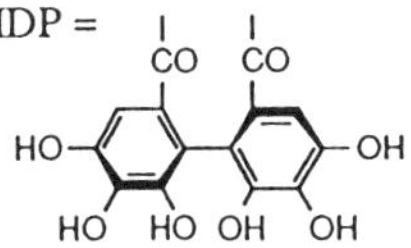

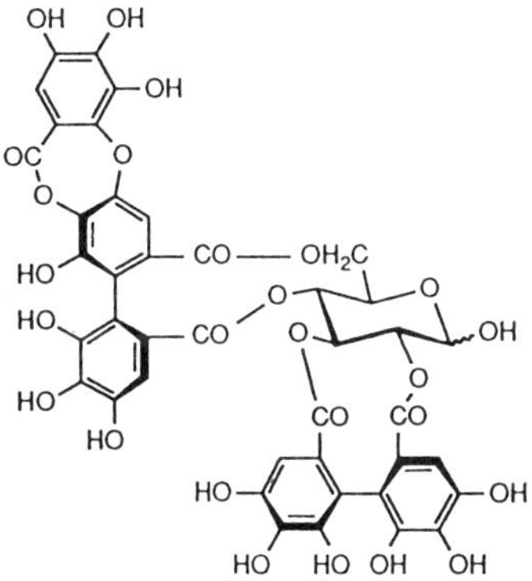

Praecoxin D

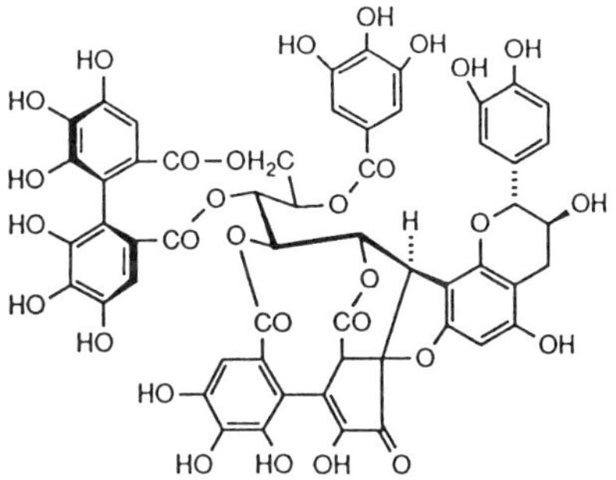

Guavin A

Preparation of Extract. The ethyl acetate extract (8.5 g) was obtained from the dried leaves (70 g) of *G. thunbergii* in a way analogous to that employed for the extraction from *L. anceps*. Geraniin was crystallized out of aqueous methanol solution of this extract by concentration followed by seeding with geraniin crystals. The mother liquor was evaporated, and 2 g of the residue (4.4 g) was subjected to CPC.

Fractionation by CPC. The residue (2 g) was fractionated in apparatus 1 by normal-phase development with solvent system (3), collecting 5-g portions of the eluent. As shown in Fig. 6, the group II fractions contained geraniin, which still remained in the residue after the main portion of geraniin was crystallized out. The group III fractions gave elaeocarpusin, a condensate of geraniin with ascorbic

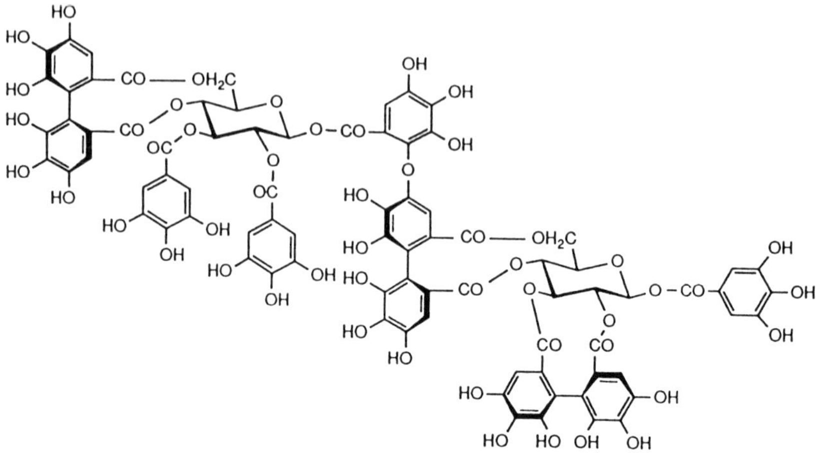

Rugosin F

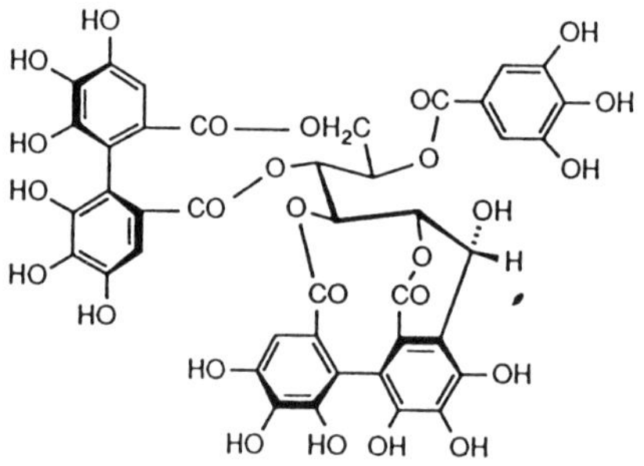

Casuarinin

Figure 5 Continued

acid, which is very soluble in water [41]. When kept in an aqueous solution for several hours, it decomposed to yield geraniin and corilagin. This labile water-soluble compound was easily synthesized by just leaving an acidic aqueous solution of 1:5 mixture (molar ratio) of geraniin and ascorbic acid [41].

4. *Combined Application of Normal-Phase and Reversed-Phase Developments to Polyphenols of* Liquidambar formosana

Preparation of Extract. Leaves of *Liquidambar formosana* (Hamamelidaceae), grown in the southeast region of China, have been used as a medicine for the treatment of disorders of the digestive organs [42]. Isolation of its unknown polyphenolic constituents with CPC was attempted. Preparation of the ethyl

FRACTIONATION OF PLANT POLYPHENOLS

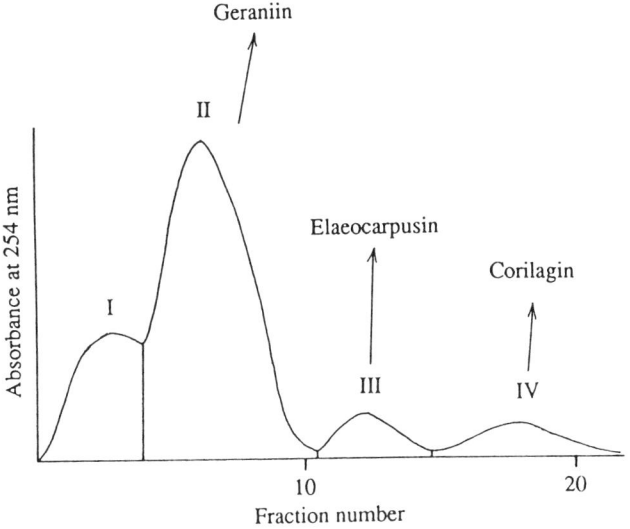

Figure 6 CPC of water-soluble portion of *Geranium thunbergii*.

acetate extract (8.7 g) from fresh leaves (2.9 kg) was carried out analogously to that from *L. anceps*.

Reversed-Phase Development in CPC. A portion (3 g) of the ethyl acetate extract was fractionated by CPC, using solvent system (3), by the reversed-phase development, at the initial flow rate 0.2 ml/min, which was increased to 1.0 ml/min in 1 hour. As shown in Fig. 7, the fractions from this development for about 25 hours were collected into three groups, I, II, and III.

As expected from the results of the fractionation of *L. anceps* extract, a *C*-glucosidic tannin (casuarinin) was in the group I, which was eluted first, and a dimeric hydrolyzable tannin (isorugosin D [43]) was in the second group. Although each group of fractions contains two or three polyphenolic components, this result shows that two structurally similar components, that is, pedunculagin and tellimagrandin I, and also casuarictin and tellimagrandin II [37], can be separated from each other by a single reversed-phase development.

Normal-Phase Development in CPC. The residue (0.7 g) obtained upon evaporation of the combined fractions of group I (Fig. 7) obtained twice by reversed-phase developments in CPC of the ethyl acetate extract (from 5 g of the ethyl acetate extract in total) was subjected to normal-phase development using the same solvent system at the same flow rate as above, yielding two groups of fractions, I-1 and I-2.

This normal-phase CPC of the polyphenols in the group I fractions from the reversed-phase development induced successful separation of two components, casuarinin (77 mg) and pedunculagin (0.4 g), as illustrated in Fig. 7 [13].

C. Fractionation of Oligomeric Hydrolyzable Tannins

Each oligomeric hydrolyzable tannin has a molecule composed of two or more monomeric hydrolyzable tannin molecules [1,2]. The first oligomer found in nature was agrimoniin, from *Agrimonia pilosa* [44], and more than 150 oligomers, up to tetramers, have been isolated [8,45]. These oligomeric hydrolyzable tannins are biogenetically regarded mostly as the products from intermolecular C–O oxidative coupling among two or more monomeric hydrolyzable tannins, although some oligomers having *C*-glucosidic structures are regarded as the products from C–C coupling [8]. They are classified into several types according to the structures of the linking unit between monomers, and some of them have noticeable chemical structures such as those of macrocyclic oligomers [8].

1. Macrocyclic Antitumor Oligomers from Woodfordia fruticosa

The first macrocyclic oligomer was oenothein B, a dimer isolated from *Oenothera erythrosepala* (Onagraceae) [30], which showed potent host-mediated antitumor activity [11]. *Woodfordia fruticosa* Kurz. (Lythraceae) is widely grown in southeast Asia, and its dried flowers are popularly used as a traditional medicine

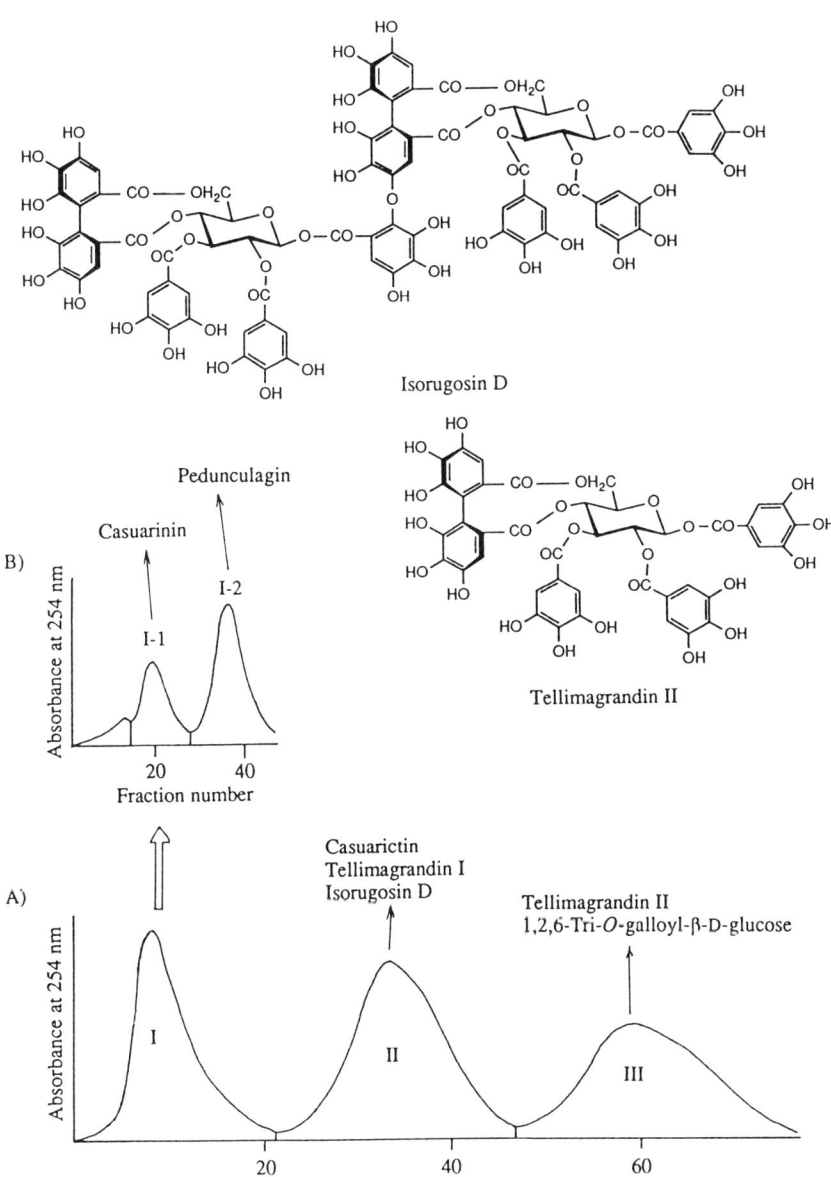

Figure 7 (A) CPC of extract of *Liquidambar formosana* in reversed-phase development; (B) CPC of extract of *L. formosana* in normal-phase development.

(Jamu medicine), called Sidowayah, for treatment of dysentery, sprue, rheumatism, dysuria, and hematuria in Indonesia and Malaysia. This plant has also been found to be rich in the oligomeric hydrolyzable tannins, including oenothein B, and other macrocyclic hydrolyzable tannin oligomers (woodfordins C–F), which are accompanied by several other related monomers [46–48].

Preparation of Extract. The extraction of dried flower (1 kg) of *W. fruticosa* was carried out in a way analogous to that of *L. anceps* described above [46]. The hydrolyzable tannin oligomers with macrocyclic structure in this flower are highly polar, and they are hardly extractable with ethyl acetate. After the extraction with the organic solvents, they were mostly found in the *n*-butanol and aqueous extracts. The main tannins in the aqueous extract were oenothein B and woodfordin C.

Fractionation by CPC. Oenothein B and woodfordin C are the dimers structurally differing from each other only in the presence or absence of a galloyl group on one of the anomeric hydroxyl groups [46]. In spite of such a small structural difference, these dimers showed a considerable difference of the partition coefficients in solvent system (3) (Table 1). Based on their coefficients (0.36 for woodfordin C and 0.19 for oenothein B), the separation of these tannins on CPC (apparatus 1) was achieved by normal-phase development using the upper phase as the mobile phase, as shown in Fig. 8. Each tannin thus obtained was spectroscopically pure. Nontannin materials such as amino acids, sugars, proteins, and inorganic materials were left in the stationary phase. This facile removal of nontannin materials, which are abundantly present in the aqueous extract, is a noticeable advantage of CPC to column chromatography with solid support [15].

Host-Mediated Antitumor Activity of Oligomers. Oenothein B, woodfordin C, and other macrocyclic oligomers exhibited a remarkable host-mediated antitumor activity upon intraperitoneal administration into mice, on fourth day before intraperitoneal inoculation of sarcoma 180 cells [11,46]. Among the active tannins, oenothein B showed the most potent activity [49]. This antitumor effect of oligomers is attributable to the host–immune defense mechanism, based on activation of macrophages releasing interleukin-1-like activity [50]. Oenothein B also exhibited antitumor effects against MM2 ascites tumors and Meth-A solid type tumors in mice [51].

2. Labile Oligomers of Heterocentron roseum

Hydrolyzable tannin oligomers are strongly adsorbed on various solid supports such as polyvinyl and polystyrene resin and Sephadex LH-20, which are often used for column chromatographic separation of monomeric and oligomeric hydrolyzable tannins. They are usually retained in a column until the elution at the final stage of gradient development, which was performed a week or more later. Some

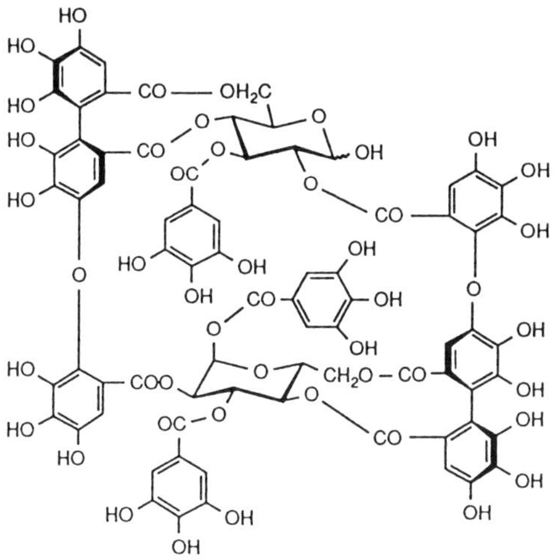

FRACTIONATION OF PLANT POLYPHENOLS

Woodfordin C

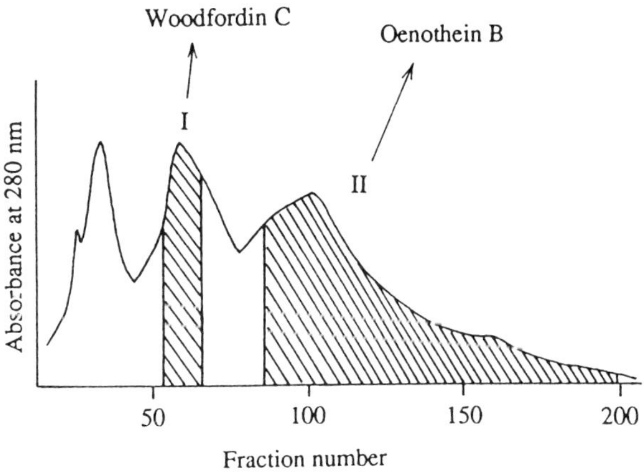

Figure 8 CPC of water-soluble portion of *Woodfordia fruticosa*.

oligomers are susceptible to hydrolysis during the development and therefore cannot be subjected to repeated column chromatography. The CPC technique was successfully applied to purification of such labile trimeric and tetrameric hydrolyzable tannins (nobotanins J and K) of *Heterocentron roseum* (Melastomataceae) [22].

Preparation of Extract. The *n*-butanol extract (8.1 g) was obtained from the fresh leaves (2.1 kg) of *H. roseum* in a way analogous to that for extraction of tannins from *L. anceps*. It was fractionated by column chromatography over DIAION HP-20 developing with water and aqueous methanol (20% → 40% → 60% → 80% MeOH) in stepwise gradient mode. The 40% MeOH eluate, rich in hydrolyzable tannins including oligomers, yielded crude nobotanin J (1.1 g) upon separation from coexisting several monomers and dimers by further chromatography on Toyopearl HW-20 (coarse grade). The last fraction eluted with 70% acetone from the Toyopearl HW-40 column gave a tetramer, nobotanin K, which showed a single peak in normal-phase HPLC. Reversed-phase HPLC demonstrated, however, that it was still contaminated by other tetrameric analogs.

Attempts to Separate a Trimer on Gel-Column. Attempted purification of crude nobotanin J by further column chromatography over Sephadex LH-20, using aqueous ethanol as eluant, gave a small amount of pure nobotanin J. Most of the fractions, however, were inevitably contaminated by a substantial amount of monomers and dimers, presumably produced at least in part produced by partial hydrolysis of nobotanin J during the development. Actually, complete decomposition of nobotanin J into pedunculagin (monomer) and nobotanin H (dimer) occurred when an aqueous solution of nobotanin J was kept at 37°C for three weeks (Fig. 9) [22].

Fractionation by CPC. In order to minimize this undesirable hydrolysis during the final purification of nobotanin J, the CPC technique, which enables quick development in the absence of solid support, was employed. The crude nobotanin J (1.1 g), which was obtained from column chromatography over Toyopearl HW-40 as described above, was subjected to CPC in the normal-phase mode using solvent system (3), giving the chromatogram shown in Fig. 10. Although the residue (510 mg; nobotanin J of 88% purity) from fraction II still contained a small amount of monomer and dimer, as revealed by HPLC, fraction III afforded pure nobotanin J (154 mg), which showed a single peak in both normal- and reversed-phase HPLC. The ^{1}H NMR spectrum of nobotanin J thus obtained showed that this product was pure enough for structural elucidation.

Based on the successful purification of nobotanin J by CPC, this technique was then applied to purification of crude nobotanin K, which was presumed to have the labile nobotanin J moiety as a structural unit in its tetrameric molecule. Nobotanin K is even more polar than trimeric nobotanin J, and more strongly adsorbed on solid supports as described earlier. The reversed-phase development

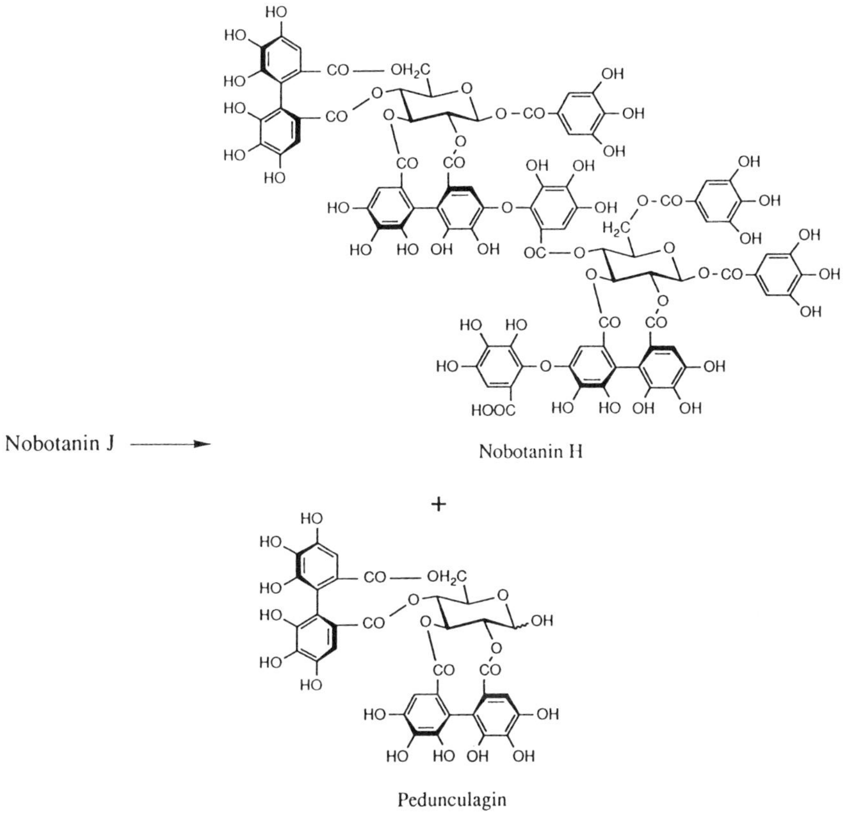

FRACTIONATION OF PLANT POLYPHENOLS

Nobotanin J ⟶ Nobotanin H

+

Pedunculagin

Figure 9 Decomposition of nobotanin J.

of CPC thus performed using solvent system (3) effectively yielded purified nobotanin K (Fig. 11). Nobotanin K obtained from fraction II in this way was free from concomitant, as revealed by a single peak in the reversed-phase HPLC.

3. Hydrolyzable Tannins from Coriaria japonica *and* Barringtonia asiatica

The CPC technique combined with column chromatography was applied to hydrolyzable tannins from *Coriaria japonica* A. Gray (Coriariaceae). The *n*-butanol soluble portion of the aqueous acetone extract of the leaf of *C. japonica* was first fractionated by column chromatography over Sephadex LH-20. The tannin-rich fraction eluted with 70% EtOH was subjected to CPC using solvent system (3) in normal-phase development, and fractions were collected into five groups based on profiles in HPLC. Each fraction group was finally purified by the column chromatography over Sephadex LH-20 to afford nine dimeric hydrolyzable tan-

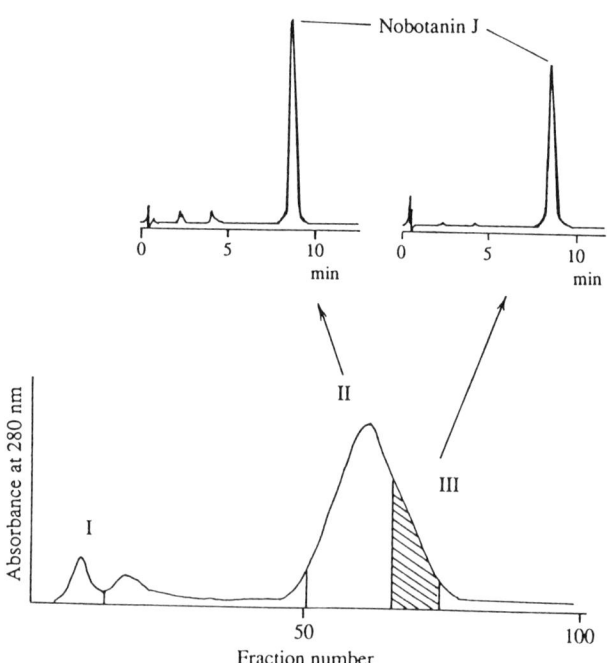

Figure 10 CPC of crude nobotanin J, and HPLC profiles of combined fractions.

FRACTIONATION OF PLANT POLYPHENOLS

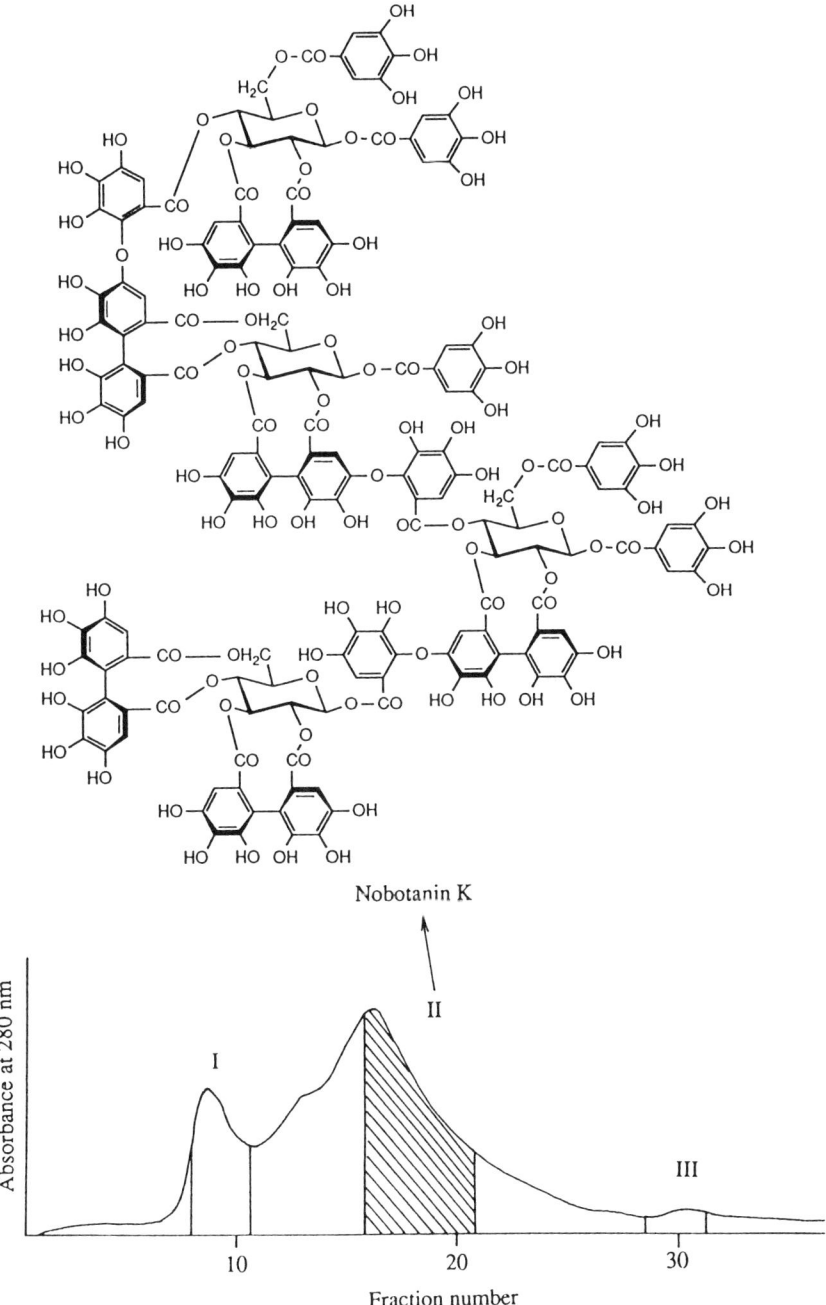

Figure 11 CPC of crude nobotanin K.

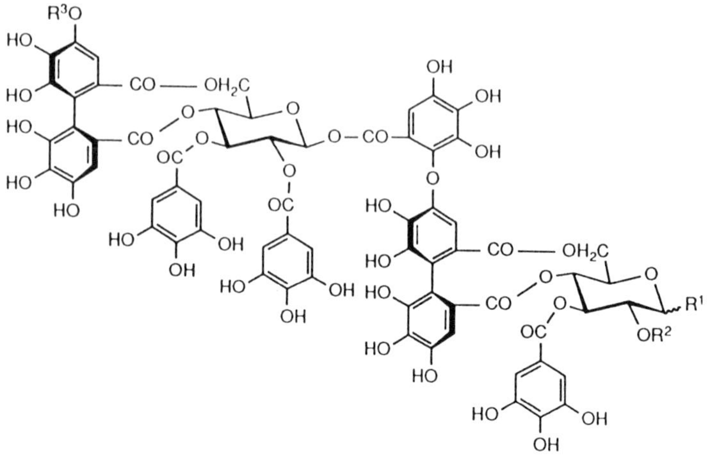

Rugosin D: R^1=(β)-O-Gall, R^2=Gall, R^3=H
Rugosin E: R^1=OH, R^2=Gall, R^3=H
Coriariin D: R^1=OH, R^2=Gall, R^3=GA
Coriariin E: R^1=OH, R^2=R^3=H

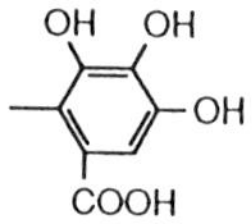

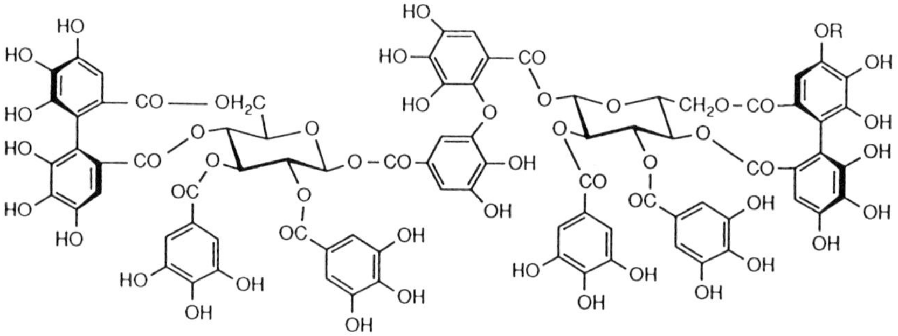

Coriariin A: R=H
Coriariin C: R=GA

Figure 12 Structures of hydrolyzable tannins from *Coriaria japonica*.

FRACTIONATION OF PLANT POLYPHENOLS

Coriariin G: R=(β)-O-Gall
Coriariin H: R=OH

Coriariin I

nins (rugosins D and E, and coriariins A, C, D, E, G, H, and I), among which coriariins G, H, and I were found to have unique structures possessing a sedoheptulose residue in each molecule [52,53].

A new dimeric ellagitannin, barringtin A, along with schimawalin A, was similarly isolated from a tropical plant, *Barringtonia asiatica* Kurz. (Lecythidaceae), and its structure was elucidated, as shown by Figure 13 [54].

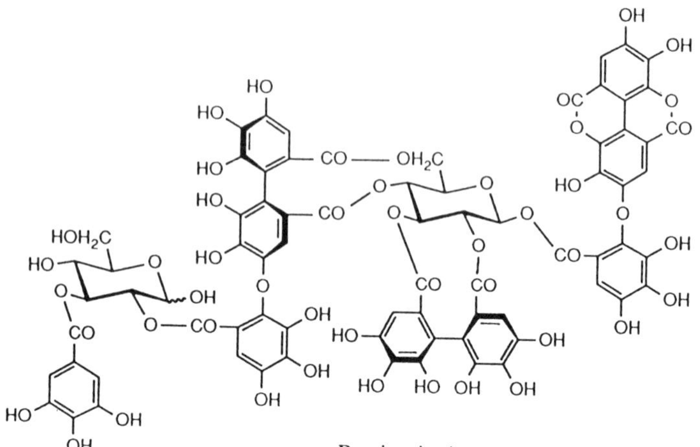

Barringtin A

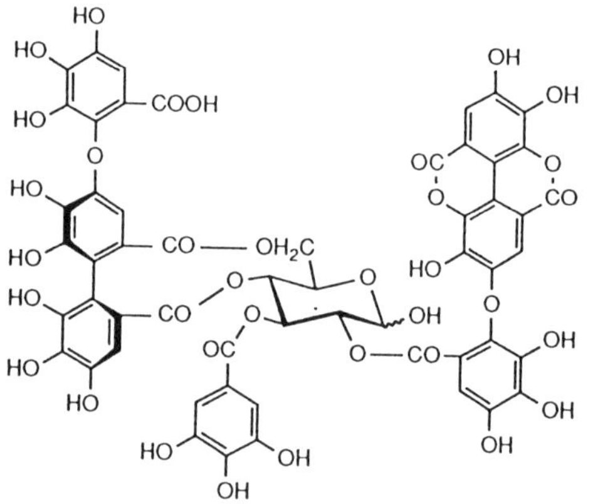

Schimawalin A

Figure 13 Structures of hydrolyzable tannins from *Barringtonia asiatica*.

IV. CONCLUSIONS

Centrifugal partition chromatography is an effective tool for the separation and purification of biologically active polyphenols of various types (flavonoids and hydrolyzable tannins of large molecular weight, etc.), extracted from medicinal plants, foods, and beverages. This modern liquid–liquid chromatography of short developing time is particularly valuable for separation and purification of polyphenols of large molecular weight and high polarity, which cannot be achieved by the column chromatography with solid support because of strong adsorption and decomposition of labile polyphenols. It can be efficiently applied to polyphenols of small molecular weight. The CPC technique also has advantages in its wide range of selecting solvent systems, and in facile alteration of the development modes (normal- and reversed-phase), performable in accordance with the polarities of coexisting polyphenols to be separated. Complex mixtures of polyphenols in a plant extract cannot always be separated by CPC alone, however. Combination of this technique with other chromatographic techniques, such as preparative HPLC and column chromatography with solid support with appropriate properties, is usually required for final purification of complex polyphenols.

REFERENCES

1. E. Haslam, *Plant Polyphenols*, Cambridge University Press, Cambridge (1989).
2. T. Okuda, T. Yoshida, and T. Hatano, *Economic and Medicinal Plant Research* (H. Wagner and N. R. Farnsworth, eds.), Academic Press, London, pp. 129–165 (1991).
3. T. Okuda, T. Yoshida, and T. Hatano, *Planta Medica*, 55:117 (1989).
4. R. E. Kreibich, *Chemistry and Significance of Condensed Tannins* (R. W. Hemingway and J. J. Karchesy, eds.), Plenum Press, New York, pp. 457–478 (1989).
5. S. Yoshizawa, T. Horiuchi, H. Fujiki, T. Yoshida, T. Okuda, and T. Sugimura, *Phytotherapy Res.*, 1:44 (1987).
6. Y. Fujita, T. Yamane, M. Tanaka, K. Kuwata, J. Okuzumi, T. Takahashi, H. Fujiki, and T. Okuda, *Jpn. J. Cancer Res.*, 80:503 (1989).
7. *The Merck Index*, Tenth Edition, The Merck & Co., Inc., Rahway, New Jersey, p. 4216 (1983).
8. T. Okuda, T. Yoshida, and T. Hatano, *Phytochemistry*, 32:507 (1993).
9. T. Okuda, T. Yoshida, and T. Hatano, American Chemical Society Symposium Series 507, *Phenolic Compounds in Food and Their Effects on Health*, II (M.-T. Huang, C.-T. Ho, and C. Y. Lee, eds.), American Chemical Society, Washington, D.C., pp. 87–97 (1992).
10. T. Okuda, T. Yoshida, and T. Hatano, *Antioxidant Polyphenols in Oriental Medicine*, (K. Yagi, ed.) CRC Press, Boca Raton, pp. 333–346 (1993).
11. K. Miyamoto, N. Kishi, R. Koshiura, T. Yoshida, T. Hatano, and T. Okuda, *Chem. Pharm. Bull.*, 35:814 (1987).
12. T. Okuda, T. Yoshida, and T. Hatano, *J. Nat. Prod.*, 52:1 (1989).
13. T. Okuda, T. Yoshida, T. Hatano, K. Yazaki, R. Kira, and Y. Ikeda, *J. Chromatogr.*, 362:375 (1986).
14. T. Okuda, T. Yoshida, and T. Hatano, *J. Liq. Chromatogr.*, 11:2447 (1988).
15. T. Okuda, T. Yoshida, T. Hatano, K. Mori, and T. Fukuda, *J. Liq. Chromatogr.*, 13: 3637 (1990).
16. S. Shibata and T. Saitoh, *J. Indian. Chem. Soc.*, 55:1184 (1978).
17. K. Takagi, S. Okabe, and R. Saziki, *Jpn. J. Pharmacol.*, 19:418 (1969).
18. T. Hatano, T. Fukuda, T.-Z. Liu, T. Noro, and T. Okuda, *Yakugaku Zasshi*, 111:311 (1991).

19. T. Hatano, T. Yasuhara, T. Fukuda, T. Noro, and T. Okuda, *Chem. Pharm. Bull.*, *37*:3005 (1989).
20. T. Hatano, T. Fukuda, T. Miyase, T. Noro, and T. Okuda, *Chem. Pharm. Bull.*, *39*:1238 (1991).
21. Y. Kimura, H. Okuda, T. Okuda, and S. Arichi, *Phytotherapy Res.*, *2*:140 (1988).
22. T. Yoshida, T. Hatano, and T. Okuda, *J. Chromatogr.*, *467*:139 (1989).
23. T. Hatano, H. Kagawa, T. Yasuhara, and T. Okuda, *Chem. Pharm. Bull.*, *36*:2090 (1988).
24. T. Hatano, T. Yasuhara, K. Miyamoto, and T. Okuda, *Chem. Pharm. Bull.*, *36*:2286 (1988).
25. Y. Aga, K. Tsutsumi, T. Hatano, and T. Okuda, Abstract Papers, 39th Annual Meeting of the Japanese Society of Pharmacognosy, September, 1992, Tokyo, p. 141.
26. T. Hatano, A. Okonogi, K. Yazaki, and T. Okuda, *Chem. Pharm. Bull.*, *38*:2707 (1990).
27. T. Okuda, T. Yoshida, T. Hatano, M. Iwasaki, M. Kubo, M. Yoshizaki, and N. Naruhashi, *Phytochemistry*, *31*:3091 (1992).
28. C.-T. Ho, C. Y. Lee, and M.-T. Huang, eds., *Phenolic Compounds in Food and Their Effects on Health*, I, II, American Chemical Society, Washington, D.C. (1992).
29. T. Okuda, K. Mori, and T. Hatano, *Chem. Pharm. Bull.*, *33*:1424 (1985).
30. T. Hatano, T. Yasuhara, M. Matsuda, K. Yazaki, T. Yoshida, and T. Okuda, *J. Chem. Soc., Perkin Trans.*, *1*:2735 (1990).
31. W. Mayer, H. Seitz, and J. C. Jochims, *Liebigs Ann. Chem.*, *721*:186 (1969).
32. The tannin reported as granatin B in Ref. 13 has been revealed to be 3-galloylgranatin A based on detailed spectroscopical analysis.
33. W. Mayer, H. Seitz, J. C. Jochims, K. Schauerte, and G. Schilling, *Liebigs Ann. Chem.*, *751*:60 (1971).
34. T. Hatano, K. Yazaki, A. Okonogi, and T. Okuda, *Chem. Pharm. Bull.*, *39*:1689 (1991).
35. T. Hatano, N. Ogawa, T. Yasuhara, and T. Okuda, *Chem. Pharm. Bull.*, *38*:3308 (1990).
36. The structure of guavin A has been assigned as shown in Fig. 5, based on the spectral analogy to malabathrin E [T. Yoshida, F. Nakata, K. Hosotani, A. Nitta, and T. Okuda, *Chem. Pharm. Bull.*, *40*:1727 (1992)].
37. T. Hatano, N. Ogawa, T. Shingu, and T. Okuda, *Chem. Pharm. Bull.*, *38*:3341 (1990).
38. T. Okuda, T. Yoshida, M. Ashida, and K. Yazaki, *J. Chem. Soc., Perkin Trans.*, *1*:1765 (1983).
39. T. Okuda, T. Yoshida, and T. Hatano, *J. Chem. Soc., Perkin Trans.*, *1*:9 (1982).
40. T. Okuda, T. Yoshida, T. Hatano, and Y. Ikeda, *Heterocycles*, *24*:1841 (1986).
41. T. Okuda, T. Yoshida, T. Hatano, Y. Ikeda, T. Shingu, and T. Inoue, *Chem. Pharm. Bull.*, *34*:4075 (1986).
42. S.-C. Cheng, *Chinese Medicinal Herbs of Hong Kong*, Vol. 2, Commercial Press, Hong Kong, 32.
43. T. Hatano, R. Kira, T. Yasuhara, and T. Okuda, *Chem. Pharm. Bull.*, *36*:3920 (1988).
44. T. Okuda, T. Yoshida, M. Kuwahara, M. U. Memon, and T. Shingu, *J. Chem. Soc., Chem. Commun.*, 163 (1983).

45. T. Okuda, T. Yoshida, and T. Hatano, *Heterocycles*, *30*:1195 (1990).
46. T. Yoshida, T. Chou, A. Nitta, K. Miyamoto, R. Koshiura, and T. Okuda, *Chem. Pharm. Bull.*, *38*:1211 (1990).
47. T. Yoshida, T. Chou, M. Matsuda, T. Yasuhara, K. Yazaki, T. Hatano, and T. Okuda, *Chem. Pharm. Bull.*, *39*:1157 (1991).
48. T. Yoshida, T. Chou, A. Nitta, and T. Okuda, *Chem. Pharm. Bull.*, *40*:2023 (1992).
49. K. Miyamoto, M. Nomura, T. Murayama, T. Furukawa, T. Hatano, T. Yoshida, R. Koshiura, and T. Okuda, *Biol. Pharm. Bull.*, *16*:379 (1993).
50. K. Miyamoto, T. Murayama, M. Nomura, T. Hatano, T. Yoshida, T. Furukawa, R. Koshiura, and T. Okuda, *Anticancer Res.*, *31*:37 (1993).
51. K. Miyamoto, M. Sasakura, E. Matsui, R. Koshiura, T. Murayama, T. Hatano, T. Yoshida, and T. Okuda, *Jpn. J. Cancer Res.*, *84*:99 (1993)
52. T. Hatano, S. Hattori, and T. Okuda, *Chem. Pharm. Bull.*, *34*:4533 (1986).
53. T. Hatano, R. Yoshihara, S. Hattori, N. Yoshizaki, T. Shingu, and T. Okuda, *Chem. Pharm. Bull.*, *40*:1703 (1992).
54. S. Kobayashi, T. Hatano, T. Yoshida, T. Okuda, C.-F. Lu, L.-L. Yang, and K.-Y. Yen, Abstract Paper (II) 112th Annual Meeting of the Pharmaceutical Society of Japan, March 1992, Fukuoka, p. 208.

6
Centrifugal Partition Chromatography in Assay-Guided Isolation of Natural Products
A Case Study of Immunosuppressive Components of *Tripterygium wilfordii*

Jan A. Glinski and Gary O. Caviness
Boehringer Ingelheim Pharmaceuticals, Inc., Ridgefield, Connecticut

I. INTRODUCTION

The development of new drugs continues to rely extensively on the discovery of new chemical structures from natural sources such as plants, fungi, bacteria, and algae. These organisms produce secondary metabolites, which are diversified nonprotein compounds involved in chemical signaling and defense. Some of these metabolites possess biological properties qualifying them for therapeutic use. Historically, plants have been the first, and until recently, the dominant source of medical remedies. Advances in biological and separation sciences have facilitated the transition between the use of crude herbal medicines and the exploitation of the purified active constituents of the herbs. Pure substances are easily standardized, delivered, and monitored, facilitating the study of the molecular interactions between a drug and its biological target. The role of natural products as structural templates for medicinal chemistry is steadily gaining importance. The exploration of semisynthetic derivatives and synthetic analogs leads to new drugs of enhanced potency and selectivity, as well as lower toxicity. Plant-derived drugs and their synthetic modifications have a sizable representation in all drug categories. Well-known examples include drugs for various heart problems such as digitalis, reserpine, quinidine; anesthetics based on cocaine and morphine; antileukemic *Vinca* alkaloids and the clinically promising anticancer diterpene taxol;

antimalarial quinine and artemisinine; antiasthmatic and ophthalmic drugs based on atropine. These, however, are only a few well-known plant-derived drugs selected from a very long list. Fungal and bacterial fermentations have revolutionized medicine since the mid-1940s through the contribution of thousands of antibiotics. The search for new antibiotics eventually led to development of high-throughput screening programs, which gradually evolved into integral part of discovery systems in large pharmaceutical companies and became instrumental to discovery of other than antibiotic drugs. In the area of immunosuppression alone, several potent agents have been found, including cyclosporin A—a drug that made heart transplantation possible—and two structurally related macrolides, FK506 and rapamycin.

The recent advances in molecular biology, biochemistry, and biotechnology have led to the discovery of novel enzymes, receptors, and biochemical pathways. Small molecules can effectively control disease by binding to these enzymes or receptors and interfering with the normal function of these proteins. Natural products, with their time-proven potential for generating structurally novel leads, are increasingly exploited in screening programs for the discovery of diverse mechanism-based activities. Crude extracts of natural products that are found to be active must be subjected to assay-guided fractionation until an active component is isolated and its structure can be determined. This exploratory fractionation is performed on complex mixtures containing substances of undetermined properties and typically involves suboptimal experimental chromatographic conditions. Throughout the entire process of isolation, utmost care should be exercised to avoid destruction of potentially labile chemical entities. Currently, separation techniques rely extensively on solid-phase chromatographic adsorbents such as silica gel, alumina, or reverse-phase adsorbents, in spite of their known lack of chemical inertness. Routine purifications on silica gel usually afford yields of 70–90%, but much more severe losses of valuable materials due to irreversible adsorption are not uncommon. In addition, the aforementioned reactivity of the solid-phase adsorbents occasionally causes chemical transformations resulting in the isolation of artifacts.

Countercurrent chromatography is based on liquid–liquid partitioning and presents the best alternative to circumvent the problems associated with solid-phase adsorbents and preserves the chemical integrity of mixtures subjected to fractionation. Technological refinements of countercurrent instrument designs, especially the use of centrifugal force to enhance phase separation, have led to a new generation of versatile instruments. The centrifugal partition chromatograph (CPC) models from Sanki Laboratories (Kyoto, Japan) are representative of this group of instruments. They are designed entirely with inert materials compatible with virtually all solvents. In the model LLN, used by the authors, the actual liquid–liquid partitioning takes place in interchangeable cartridges. The rotating chamber can accommodate up to 12, analytical or preparative, cartridges. In the most recently evolved line of the CPC instruments, the cartridges are superseded by rotors, which can be viewed as a one-piece integrated cartridge. Depending on the model, they are suitable for work with any quantity—up to 1500 g for the largest (LLI-100) model. With these advantages, the CPC is gaining popularity as a method of choice in the area of natural products, and especially in supporting the bioassay-guided fractionation of extracts.

The single most important element for successful use of CPC is an appropriate two-phase solvent system. For this reason, designing solvent systems is the focal point of the isolation operation, and it consumes the most time and attention. Assay-guided fractionation requires that active portions of an extract go through a sequence of consecutive fractionations. These steps are traditionally performed using solid-phase chromatographic adsorbents, the variety of which is very limited. The reliance on solid-phase adsorbents can, in fact, be replaced entirely with liquid–liquid countercurrent fractionations. In order to achieve efficient separations, the sequence should employ solvent systems that are diverse in individual solvent content and proportion. This strategy parallels the use of different solid adsorbents for consecutive steps in conventional chromatography. The active fraction composition is simplified through a "cascade" of CPC fractionation runs. Since each different solvent system contributes its unique separation pattern, sequential application greatly improves the chance of isolating an active component in a pure state.

Section II of this chapter is intended to be a practical guide for researchers beginning to use centrifugal partition chromatography for fractionations of complex mixtures of organic compounds. Natural product extracts, because of their complexity, present the most difficult challenge. The discussion focuses on approaches to designing isocratic solvent systems for CPC fractionation and deals with their basic physical and chemical characteristics.

Although the main theme of this chapter is that isolation of an active component may rely exclusively, or at least predominantly, on countercurrent fractionation, another important theme is presented. The determination of the partition coefficient values used for isolating compounds of interest at the analytical level may be crucial for designing efficient scaled up isolation processes. Such processes would employ the biphasic solvent systems in a batch mode. Using this process would significantly reduce the predicted cost of purifying components from complex mixtures.

Section III describes the application of countercurrent methodology to an assay-guided isolation of bioactive components from a complex extract of the Chinese medicinal plant *Tripterygium wilfordii* Hook (TWH). Each fractionation step is illustrated with a discussion of practical aspects of the solvent system selection, run parameters, and the outcome of the fractionation. As a result of a multistep cascade of countercurrent partitioning, two most potent immunosuppressive components of the extract, triptolide and tripdiolide, were isolated.

II. DESIGNING SOLVENT SYSTEMS FOR CPC

A. The Basics

An appropriate solvent system is the critical element determining successful fractionation in centrifugal partition chromatography. Initially, the design of a solvent system for countercurrent chromatography may seem to be difficult and time consuming. Dealing with multisolvent equilibria is not a natural skill for a chemist; however, the manipulations with multisolvent biphasic systems soon become an inventive and rewarding experience. An important problem for the beginning is, however, how to start designing of a solvent system for a given mixture. Fortunately, all biphasic solvent systems that have been developed for other, earlier types of countercurrent instruments are also applicable to CPC. The most comprehensive source of information on various aspects of countercurrent chromatography with a rich collection of literature citations is a book by Conway [1]. A series of educational analytical articles on the fundamental aspects of CPC was published by Berthod [2–8] and by Armstrong [9]. An interesting approach to designing novel solvent systems was presented by Gluck [10]. Ito, the inventor of countercurrent designs utilizing centrifugal force, discussed the correlation of the hydrodynamic properties of biphasic solvent systems in countercurrent chromatography with interfacial tension, viscosity, and the difference between the phases' densities [11]. He has also introduced a classification of biphasic solvent systems based on their settling time.

Natural products chemists have been interested in countercurrent chromatography for two reasons. First, it provides another mode for the separation of extracts in addition to the well-established use of solid-phase adsorbents. Second, it provides extremely mild conditions for fractionation. As a result, a substantial volume of literature devoted to applications of countercurrent chromatography to natural products has accumulated. There are many examples of solvent systems useful for separating compounds belonging to various groups of natural products. Applications of countercurrent chromatography for separation of natural products are discussed in a special volume of *Chromatographic Sciences* [12–15] and in the extensive reviews by Hostettmann [16,17] and by Marston [18]. Examples of fractionating natural product extracts using cascades of sequential CPC separations have been presented by Glinski et al. [19].

This chapter is purposely limited to discussing isocratic solvent systems, and it bypasses recent developments in gradient methods [20]. The authors believe the time-consuming optimization of fractionation parameters, including solvent gradient, is fully warranted for repetitive processes. In reality, however, each assay-guided fractionation is a unique event, unlikely to be repeated. Each extract has different characteristics, including the solubility and polarity range of its components, that require development of somewhat "personalized" fractionation conditions. There is not a single universal gradient solvent system capable of fractionating all kinds of extracts. Thus, for assay-guided fractionations of extracts it is simpler and more practical to use an isocratic, instead of a gradient, solvent system.

Important factors for designing solvent systems can be combined into three groups: (1) partition coefficient P; (2) physical factors—dissolving capacity and phase stability; and (3) chemical considerations. These aspects are discussed in Sections II.B–D.

1. Composition and Manipulation of Solvent Systems

The composition of biphasic solvent systems for countercurrent chromatography is based on a simple principle: an organic hydrophobic solvent (often EtOAc, $CHCl_3$, or CH_2Cl_2) is combined with water and a modifying solvent (often a polar organic solvent that is miscible with water, e.g., lower alcohols). A modifying solvent of intermediate polarity partitions between the organic solvent and water, thus changing the properties of both. The modified phases are now more similar (this is also called the "bridging" effect [21]): the organic phase acquires some polar character, while the aqueous phase acquires some capacity for dissolving organic compounds. For practical purposes, a greatly simplified picture of the interactions between a biphasic solvent system and chemical compounds can be used in which the dissolved material is the subject of intermolecular hydrophobic and polar attraction–repulsion interactions.

A sample mixture that partitions excessively into one of the phases requires the addition of a modifying solvent to adjust the partitioning to a desired level. Corrective actions can be classified as either direct or indirect. In a *direct* corrective action, the addition of a modifying solvent will *increase* the overall dissolving capacity of a solvent system as a result of enhanced polar interactions within the organic phase and hydrophobic interactions in the aqueous phase. Among the solvents that enhance solubility, some will have a stronger effect on the organic phase, while others will primarily affect the aqueous phase. To shift the partition ratio in favor of the organic phase, solvents such as EtOH, isopropanol (iPrOH), acetone, acetonitrile, acetic acid, and butyl alcohol can be used. To shift partitioning toward an aqueous phase, methanol, formic acid, dimethyl formamide (DMF) dimethyl sulfoxide (DMSO), or formamide can be used.

In a different approach, the partitioning of a mixture may be influenced using an *indirect* corrective action—by deliberately *decreasing* concentration of the sample in one of the phases. Among solvents that work by decreasing solubility in organic phase are nonpolar solvents—such as hexane, heptane, or carbon tetrachloride. Decreasing concentration in the aqueous phase can be accomplished by increasing the water content or by substituting buffers or inorganic salts solutions for water. Each change of composition should be followed by a measurement of the new partition coefficient, which will suggest a direction for the next modification. Several consecutive additions of modifying solvents are usually needed before the desired partition ratio is obtained.

Increments of modifying solvents often combine preferentially with one phase. Figure 1 illustrates the changes of the volume ratio of the upper and lower phases for EtOAc–H_2O and CH_2Cl_2–H_2O resulting from addition of several commonly used modifying solvents. By monitoring how this ratio is affected, the impact of the change on the partition coefficient P can be roughly predicted. The phase that combines with alcohol will acquire better solubilizing properties. In effect, partitioning of a sample will shift to this phase. In this context, it is interesting to notice the differences in hydrophobicity between lower alcohols such as MeOH, EtOH, and iPrOH. Methanol combines mostly with an aqueous phase, ethanol partitions about equally between aqueous and organic, while isopropanol preferentially joins the organic phase. Once a solvent system has been found, it can be prepared in a volume required for a run and the phases separated using a separatory funnel.

2. Conducting Fractionations

There are only two primary controls on the CPC instrument: one for adjustment of the flow rate, another for selection of the ascending or descending mode. In ascending mode, the upper phase is used as the mobile phase; and in the descending mode, the lower phase is used as the mobile phase. The use of the organic phase as the mobile phase makes recovery of the eluted material by evaporation easier. Furthermore, the organic phase can be directly applied to thin-layer chromatography (TLC) plates for analysis.

In a typical CPC run, the cartridges are first filled with stationary phase in ascending mode, at a rotational speed of 300 rpm. Then, the direction of slow is either reset to descending mode, for use of a mobile lower phase, or left unchanged, for use of a mobile upper phase. Mobile phase is pumped into the system after a rotational speed of 600–1500 rpm (depending on the solvent system and the number of cartridges used) is achieved, and this continues until phase volumes within the cartridges reach a state of equilibrium. At this point, the pump speed is adjusted to obtain the maximal flow rate without causing excessive elution of the stationary phase. The highest resolution of sample components has been experimentally traced to the fastest flow rate [4]. Fast flow rates, however, may require higher rotational speed to prevent the elution of emulsified stationary phase. A low level of continuous elution of the stationary phase is normal and acceptable for some solvent systems. The diminishing volume of the stationary phase creates an effect similar to that of gradient elution, producing more compressed elution peaks.

Components elute in order of increasing partition coefficient $P_{stat/mob}$, and collection of mobile phase is typically carried out until components with P of about 10 are eluted. Beyond this point, elution is uneconomical, resulting in broad peaks of low concentration. The recovery of components with $P > 10$ that remained uneluted can be most conveniently achieved by switching the solvent

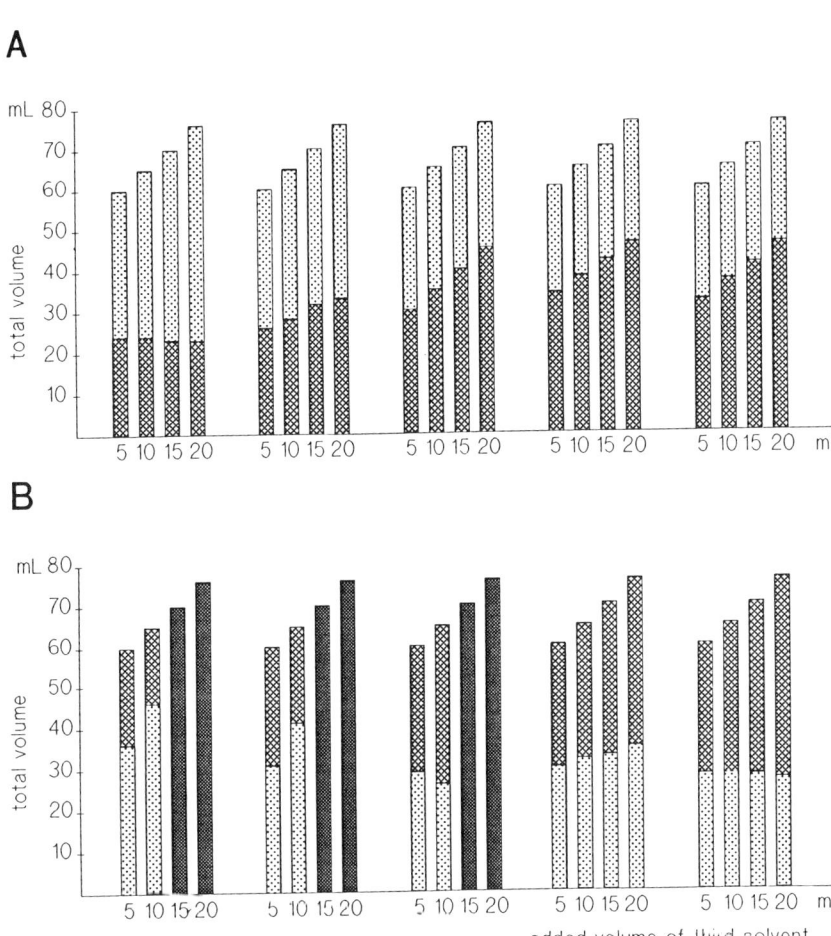

Figure 1 The effect of a third solvent on the phase equilibrium of a two-phase solvent system (▭ aqueous, ▨ organic, ■ homogeneous). For (A) the initial mixture consisted of 25 ml of methylene chloride and 25 ml of water. For (B) the initial mixture consisted of 25 ml of ethyl acetate and 25 ml of water.

intake tube from mobile to stationary phase and pumping methanol (at 300 rpm) through the cartridges until the stationary phase is completely replaced. This method of replacing the stationary phase is superior to reversing the direction of flow, which produces some cross-contaminated fractions.

The total volume of stationary phase required for a CPC fractionation is approximately the same for every run if the same type and number of cartridges are used. The volume of mobile phase required for elution of components up to a predetermined value of P can be easily estimated from a basic chromatographic equation:

$$v_{ret} = v_0 + P v_{stat} \qquad (1)$$

where v_{ret} is retention volume, v_0 dead volume, and v_{stat} stationary phase volume. Substituting for v_0 with $v_{total} - v_{stat}$ results in another form of the same equation:

$$v_{ret} = v_{total} + (P - 1) v_{stat} \qquad (2)$$

B. The Partition Coefficient P and Its Determination

The partition coefficient P represents the ratio of the amount of a compound, or a given mixture, that partitions into each phase of a biphasic solvent system. Throughout this chapter, P is expressed as a ratio of material that partitioned into stationary versus mobile phase ($P_{stat/mob}$). The value of P for an individual chemical entity allows a precise determination of elution volume. On the other hand, for mixtures, P represents only a weighed average of individual components and does not allow prediction of their distribution between the phases. Crude extracts of natural products often contain a large spectrum of compounds: from the extremely hydrophobic (hydrocarbons), to the extremely polar (phenolics, glycosides, saccharides), including ionic species (organic acids, inorganic salts, etc.) as well. A P determined for an extract containing mostly hydrophobic and polar portions with little of intermediate-polarity components will be heavily influenced by the ratio of the hydrophobic content to polar content of the extract. These types of mixtures are not well suited for countercurrent fractionation, since they will produce two major and not well-resolved fractions, one hydrophobic, one hydrophilic, with little material in between. In order to avoid such situations and to take full advantage of CPC fractionation, only portions of the crude extract with a limited range of polarity, such as those resulting from partitioning of an extract between an organic solvent and water, should be used for fractionation. Experience demonstrates that more than 90% of all runs with solvent systems selected on the basis of acceptable P have afforded satisfactory separation of components. Thus, P appears to be a factor that characterizes well the relationship between a sample and a particular solvent system, and therefore determination of P values should precede every experiment.

Countercurrent fractionation is most successful for those components of a sample which have a P within the range of $0.2 < P < 10$. Components with $P < 0.2$ will be grouped near the front of eluent in a small volume without being well separated. If $P > 10$, the eluted peaks will be broad and fractionation will require an excessive volume of mobile phase, not to mention the long hours needed to complete a run. Based on extensive experience, crude complex mixtures are best fractionated when the value of P measured for the mixture is between 0.5 and 2. For mixtures containing only a few components, P between 2 and 3 is often sufficient to provide satisfactory separation.

Various laboratories have developed different approaches for determining the suitability of solvent systems for countercurrent fractionation. A method favored by the authors for determining P is based on a comparison of the weight of the material that has partitioned into each phase (see Section III.H.2). This method is inherently advantageous over other methods in that it provides good projection of mass distribution during a CPC fractionation run. Furthermore, the method is unbiased by certain properties of some components, such as the effects of chromophors or the ability to visualize components on TLC plates. A single researcher can determine the P for about 20 different solvent systems in one day.

Another method for assessing P is based on applying equal volumes of the upper and lower phases to a TLC plate and comparing the intensities of the TLC spots. The comparison may be performed either visually or densitometrically. This method is most useful for compounds that can be separated from other components on a TLC plate.

Measurement of light absorption by UV/VIS spectrophotometry at preselected characteristic wavelengths may quickly provide information for selected components. Several variations of this method are possible. The method, of course, will not apply to many chromophore-lacking components.

A method proposed by Hostettmann and coworkers [12] represents a different approach to finding a solvent system for countercurrent fractionation. The method relies upon the measurement of Rf values for compounds separated on silica gel plates developed in the organic phase of a biphasic solvent system. The solvent system is determined to be usable for Rf values between 0.4 and 0.6, and separation can be achieved by using either the more polar aqueous or the less polar organic layer as the mobile phase. If the Rf values are either too low ($<0.2-0.3$) or too high ($>0.7-0.8$), no elution will take place in a reasonable amount of time. Other interesting methods for determination of partition coefficients are given by Conway [1].

C. Dissolving Capacity and Stability of Biphasic Solvent Systems

1. Dissolving Capacity

Measurements of the partition coefficient P by weight comparison (see Section III.H.2) are more easily attained at relatively low sample concentration, close to

1 or 2%. Actual fractionation runs may, however, require much larger sample loads. Attempts to prepare more concentrated sample solutions for injection may be difficult, despite seemingly ideal P, as each solvent system retains a limited dissolving capacity. Division of the sample into a number of smaller portions is a simple solution, but it increases the time required to fractionate the sample. Alternatively, a search for a new solvent system with a better dissolving capacity should be considered.

Solvent systems that do not effectively solubilize a sample often contain high concentrations of solvents from opposite extremes of polarity, for example, a hydrophobic solvent such as hexane or CCl_4 on one end, and water on the other. Solubility can be improved by reducing the concentration of these solvents or by incorporating solvents of intermediate polarity. For example, simultaneous addition of equal volumes of isopropanol (modification of organic phase) and methanol (modification of aqueous phase) (see Section II.A and Fig. 1) will often increase the dissolving capacity of each phase, while changing P negligibly. Alternatively, the addition of EtOH may have an equivalent impact, since EtOH partitions into both phases more equally. This conservative approach is unlikely to help in a case if a substantially higher sample solubility is desired.

Natural products extracts often contain compounds with very low solubilities belonging to diverse classes of metabolites, such as triterpenes, sterols, flavonoids, anthraquinones, and the glycosides of these compounds. The best way to cope with a problem of low solubility is to design a solvent system that incorporates a higher content of solvents known for their good solubilizing properties. Thus, solvents such as alcohols, chloroform, methylene chloride, ethyl acetate, dimethyl formamide (DMF), dimethyl sulfoxide (DMSO), and formamide should be considered. The use of the last three solvents is rarely reported in the literature [10,19]. These solvents partition primarily into the aqueous phase; however, they can dramatically increase the solubility of many otherwise difficult to dissolve compounds. In our laboratory, DMF has been used on numerous occasions, replacing MeOH in hexane–EtOAc–MeOH–water. The advantage of increased dissolving capacity heavily outweighs the disadvantage caused by the rather high boiling points of these solvents (for DMF: bp_{39} 76°C, $bp_{3.7}$ 25°C; for DMSO: bp_{17} 83°C, $bp_{2.82}$ 47.4°C; $bp_{0.79}$ 30°C; for formamide: bp_{20} 122.5°C; bp_1 70.5°C [22]). The boiling points suggest that evaporation of DMF and DMSO, using a water aspirator, will require temperatures exceeding 80°C, but regular oil vacuum pumps require ambient or only slightly elevated temperatures. In addition to evaporation, alternative methods of compound recovery from the eluent could be considered. For example, evaporation of volatile solvents will leave a sample in a small volume of DMF. Dilution with water may precipitate the sample and enable its removal by filtration or by reextraction with a more convenient organic solvent.

2. Stability of the Phases

In extracts of natural products, active components are often present at very low concentrations. For final confirmation of activity and for structural elucidation, sufficient quantities, usually a few milligrams, should be isolated. To obtain these few milligrams, it is often necessary to start with large, multigram samples, which necessitates exploration of the upper limit of a sample load. In the absence of solubility problems, disturbance of the phase equilibrium may be anticipated. A large sample behaves similarly to a modifying "bridging" solvent and is likely to produce a drastic change in the mobile/stationary phase volume ratio. In extreme cases, homogenization of the phases may occur. If this homogenization takes place during a run, a "plug" originating from local homogenization and extending for several mixing chambers within a cartridge is all that is needed to ruin a fractionation. Several general precautions should be considered during preparation of a sample for injection to prevent the possibility of this not-so-rare event:

1. Prepare the sample solution for loading using both phases, best in a 1:1 (v/v) ratio, and at a concentration generally not exceeding 10%. Using two phases aids in detecting the possible effects a sample may have on the phase equilibrium. If the composition of the phases is easily disturbed by a sample, homogenization is likely to occur while dissolving a sample.

2. Preequilibrating the cartridges helps to diminish the possibility of internal homogenization. Furthermore, it helps in early detection of numerous problems, such as leakage or a tendency for replacement of stationary with mobile phase.

3. Maintain a constant temperature in the rotating chamber. Prolonged runs may elevate the chamber temperature, and unless internal refrigeration is not used, this may compromise the phase equilibrium up to the point of homogenization. A simple test based on warming to about 45°C a sample solution prepared in a mixture of both phases at injectable concentration addresses the likelihood of temperature-induced homogenization during fractionation. On rare occasions, a sample that appears not to disturb the phase equilibrium may induce homogenization during a fractionation. An example given in Section III.C demonstrates one of these unusual phenomena.

D. Chemical Considerations

Most common solvents are fairly nonreactive and thus applicable for CPC use, but there are some exceptions. An example of the solvents that present a recognized possibility for reacting with sample components are carbonyl-containing solvents, such as ethyl acetate or acetone, which are incompatible with amines. The fractionation of alkaloid-containing mixtures may catalyze the condensation reaction of these solvents. Furthermore, ethyl acetate and other esters, in basic pH, are prone to hydrolysis, which produces acetic acid. Under these conditions, the same mixture of alkaloids may produce different elution patterns depending on the degree of partial protonation of these alkaloids by the released acetic acid.

Simple ethers such as Et_2O or $(i\text{-}Pr)_2O$ are rather undesirable components of solvent systems for CPC. They have high vapor pressure, which causes pumping disturbances and also tends to significantly increase surface tension, leading in turn to a higher back pressure. A better substitute for these solvents is t-butyl methyl ether, which additionally does not form dangerous peroxides, unlike Et_2O and $(i\text{-}Pr)_2O$.

Long-term storage of chloroform and its exposure to light may generate HCl. The use of chloroform contaminated with HCl has a rich history of generating numerous proven, as well as suspected, artifacts from natural sources. Isolation of triptochloride from an extract of *Tripterygium wilfordii* is a typical illustration of this problem [23,24].

Atmospheric oxygen is a contaminant of solvents of underestimated destructiveness. The high redox potential of oxygen in solution makes it very reactive toward organic compounds. The extent of this damage is the greatest for dilute samples, in which the ratio of dissolved oxygen to organic compounds is the greatest. The oxidative damage can be further augmented by the presence in extracts of photodynamic agents (e.g., common plant metabolites such as chlorins from degradation of chlorophyll, *bis*-anthraquinones, furocumarins, quinolizidine alkaloids) that can convert O_2 into singlet oxygen upon exposure to visible or UV light. Degassing solvents prior to fractionation, avoidance of exposure to daylight, and prompt evaporation of eluted fractions can minimize the chances for oxidation.

Small samples (less than 50 mg) require the most care in handling. They require the same volume of mobile phase for elution as loads many times larger and thus are most susceptible to contamination by the solvent contaminants. As a basic precaution, only solvents of the highest purity should be used for their fractionation.

E. Systems for Sequential Fractionation

A mixture of compounds subjected to CPC fractionation will produce different elution patterns in different solvent systems. Thus, each sequentially applied solvent system may be viewed as another dimension in the separation and used to further fractionate the coeluted components. This can be repeated as many times as is required for isolation of an active component, without resorting to solid–phase adsorbents.

A distribution pattern of components of mixtures is difficult to foresee, especially for those of unknown composition such as extracts of natural products. What cannot be done for a mixture of unknowns can be approached for compounds of known structure with the help of basic analysis of intermolecular interactions between solvents and solutes. In solution, solvent molecules assume positions oriented around the hydrophobic and polar parts of the dissolved compound molecules. A change in the composition of a solvent system causes a compound's surrounding to change, which, in turn, will be reflected by a new value for the partition coefficient. For example, the P of compounds containing easily ionizable groups such as carboxylic acids, amines, or phenols will be strongly effected by a change of pH occuring in the vicinity of their pK_a. Many other functional groups can be affected by pH; however, the direction of change of P is more difficult to predict. It can also be anticipated that solvents containing either aromatic rings or unsaturated bonds will display stronger affinity for compounds having similar structural elements. Similarly, chlorinated solvents, because of their high electron polarization, should affect the partition coefficients of many compounds.

A large selection of biphasic solvent systems to choose from would certainly help potential users of CPC. Expansion of the current, rather limited, collection with innovative designs of solvent systems would be a valuable investment. There are two ways to expand the collection of isocratic solvent systems:

1. Incorporate more "exotic" solvents.
2. Aim for more complex compositions, utilizing traditional time-tested solvents.

The main reservation about rarely used solvents stems from a fear of possible unforeseen reactivity toward components of extracts. In contrast, the known biphasic solvent systems typically used for countercurrent partitioning employ common laboratory solvents of well-known reactivity. As is often true for complex mixtures, solvent systems composed of a larger number of individual solvents may offer a dissolving capacity superior to those based on a smaller number of individual solvents.

New solvent systems can be designed by using a larger number of individual solvents, with some of them combined into blocks of constant composition. For example, compositions containing six individual solvents, A, B, C, D, E, and F, for the sake of simplicity of manipulation may be considered as a four-solvent system: $A:(nB + C):D:(E + mF)$, with its four blocks combined at a ratio $w:x:y:z$. In order to achieve a desired P for a given mixture, the ratio of parts $w:x:y:z$ is adjusted, but the parameters n and m remain constant. An example of a new solvent system developed according to these principles, given in Sections III.D and III.E, was demonstrated to be easy to manipulate and was successfully used for the separation of several samples.

F. Implications for Designing Scaled Up Isolation

Assay-guided isolation of natural product extracts usually leads to fairly small quantities of active materials, but often sufficient to elucidate a structure and to conduct a series of in vitro assays. If the enzymatic or biological tests look encouraging, additional quantities of the active component are needed for more extensive evaluation, which possibly will include some in vivo assays. At such an early stage of discovery, initiation of a synthetic program is not yet warranted. Instead, a larger-scale isolation from the same source appears to be more advisable. Processing several hundred grams of an extract can present significant problems if the technical infrastructure for handling these quantities is not already in existence. The limited capacity of chromatographic equipment is a particularly serious obstacle. Under these common circumstances, simply scaling up a procedure that was successfully used during the exploratory stage is unlikely to be economical. An efficient, practical scale-up scheme should provide initial enrichment through a series of simple nonchromatographic steps and, after reducing substantially the mass of an active fraction, may be followed with the fewest possible number of chromatographic steps.

Fractionating a given mixture by CPC produces a unique and different elution pattern for each solvent system. Analysis of the fractionation provides values of the partition coefficients P, which may serve as a highly reliable prediction for the behavior of single components in the mixtures subjected to a batch-type partitioning between the phases.

It is suggested here that biphasic solvent systems developed for the CPC can be easily adapted for a large-scale nonchromatographic enrichment of the active component. In this procedure, a mixture is partitioned between the phases of the biphasic solvent systems in a batch mode. The idea is not entirely new, since some solvent systems such as hexane−(MeOH + 5−10% H_2O) and hexane−acetonitrile have been commonly employed for this purpose. Both are helpful in separating crude extracts into nonpolar and polar portions. The full potential of this approach, however, has not been exploited. The most desirable solvent systems are those which while partitioning concentrate an active component into one phase and a significant portion of the other components into another phase. For example, if an active constituent has a partition coefficient $P_{upper/lower}$ equal to 7, then by partitioning a mixture between equal volumes of both phases a sevenfold enrichment of the active component may be attained. By collecting and replacing the enriched upper phase with an equal volume of fresh phase, the lower phase will be depleted again seven times. As a result, after two partitionings 98% of the active component will be separated from the material retained by the lower phase. Repeating the batch partitioning steps in two-phase solvent systems is justified as long as the gains achieved by reducing the weight of an active fraction outweigh any alternative, such as a chromatographic step.

The final price of natural products is largely determined by the cost of isolation operations. By reducing this cost, many therapeutically useful natural products could become significantly less expensive and, thus, more accessible. The expansion advocated here of the batch-type enrichment of active components has been extensively studied in our laboratory in conjunction with several scale-up isolation projects. In particular, it has been successfully employed for the initial steps of the large-scale isolation of triptolide and tripdiolide, both present in dry plant material at less than 3 ppm, from 100 kg of *Tripterygium wilfordii* root.

III. IMMUNOSUPPRESSIVE COMPONENTS OF *TRIPTERYGIUM WILFORDII* (HOOK)

A. Background

Tripterygium wilfordii Hook (TWH), of the *Celastraceae* family, is a perennial vine native to southern China, where it is also known under the common names of Lei Gong Teng (Thundergod vine), Threewing nut, or Mang Cao. Use of this plant in Chinese folk medicine was first recorded about 2000 years ago in the *Saint Peasant's Scripture of Materia Medica*. The potent poison of the powdered roots has been exploited by Chinese farmers as an insecticide to protect their crops against insects. In 1972 Kupchan and coworkers [25] isolated three cytotoxic diterpene triepoxides from TWH, including (1) triptolide and (2) tripdiolide (Fig. 2), using bioassay (in vivo P388 leucemia)-guided fractionation. The examination of TWH has since concentrated on the exploration of the immunosuppressive (rheumatoid arthritis and other autoimmune diseases), anticancer, antiinflammatory, and antispermatogenic (male contraceptive) properties. Subsequent research has attributed the biological activity to several components isolated from the extract. Both triptolide and tripdiolide are considered the most components of the ethyl acetate–soluble part of the extract, known as "Leigonteng" preparation and used for treatment of autoimmune diseases. Triptolide was recently also evaluated in vivo as being potentially useful agent in the prevention of organ transplant rejection [26]. Among other bioactive components are an alkaloid, wilforine,

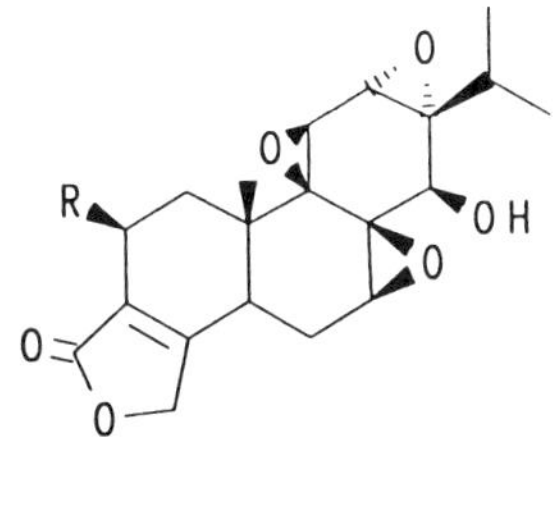

1 TRIPTOLIDE R=H
2 TRIPDIOLIDE R=OH

Figure 2 The structures of two potent immunosuppressants isolated from *Tripterygium wilfordii*.

shown to be effective against rheumatoid arthritis in clinical testing [27]; a methine–quinone, celastrol, described as an antiproliferative agent [28]; a lipid peroxidation inhibitor [29], tripterine, which inhibits numerous immune responses in vitro as well as in vivo [30,31]; and a pentacyclic, triptotriterpenic acid A, with antiinflammatory in vivo properties [32].

The polar portion of the TWH root extract, which is soluble in aqueous methanol, is known as "T2" preparation. Because it has pronounced immunosuppressive and antiinflammatory properties [33], it has also been successfully used in the clinical treatment of rheumatoid arthritis [34]. Some triterpene glycosides of ill-defined structure are considered to be the active principles of this preparation [35–37].

To gain a better understanding of the components of the TWH extract that contribute significantly to the immunosuppressive properties, a thorough evaluation of the extract was attempted. The strategy devised called for conducting an extensive fractionation of the extract to produce a large number of fractions for evaluation in a bioassay. The relative concentration of minor components in each fraction would be increased, enhancing the probability of detecting biologically active components. The method chosen for fractionation was based on a nondestructive countercurrent chromatographic technique and performed using a centrifugal partition chromatograph. Bioassay-guided fractionation of the active fractions should lead to the isolation and identification of active components. A mixed lymphocyte reaction (MLR) assay using human blood leukocytes was selected to monitor the immunosuppressive activity and guide the direction of fractionation. The MLR assay measures the proliferation of T- and B-lymphocytes. In the body, these cells are responsible for the immune response, and the assay may be viewed as a simple in vitro model for tissue transplant rejection.

B. Extraction of Plant Material

Finely powdered root of *T. wilfordii* [38] (2 kg) was percolated with 12 L of 95% EtOH. Evaporation of the ethanol yielded 80 g of an initial extract, designated E1, which was powdered and added in portions to a flask containing ethyl acetate and water. The contents of the flask were stirred vigorously for several minutes before the organic layer was decanted and replaced with a fresh portion of EtOAc. The extraction of the aqueous layer with EtOAc was repeated several times until the aqueous layer no longer contained significant amounts of EtOAc-soluble material. This procedure afforded three components: ethyl acetate–soluble E2, water-soluble E3, and an insoluble gum, E4. Results from the MLR assay indicated immunosuppressive activity in E2 and, to a lesser degree, in E4. A thin-layer chromatography on silica gel revealed a complex mixture of components in E2, while E4 primarily contained components belonging to the polar end of the TLC spectrum. On thin-layer chromatograms, the components of a mixture show up as spots, and the density of the spots is an indication of the relative concentration of the component. Minor components may be present in concentrations below the detection cutoff. Thus, using TLC, it was not possible to determine if the activity of E2 and E4 was caused by the same components or by different ones. In view of the impressive complexity of the extract and an initial absence of any reference samples, both E2 and E4 were investigated separately.

The following describes the assay-guided fractionation of the E2 extract carried out through a cascade of CPC fractionations: Triptolide (1) was isolated after five sequential runs, and tripdiolide (2) was isolated after seven sequential runs (Figs. 3 and 4). The potency of both triptolide and tripdiolide in the MLR assay were of similar magnitude, each with EC_{50} determined to be in the range of 1–10 pg/ml. The range reflects differences in the sensitivity of human lymphocytes derived from various donors.

C. Initial Fractionation of the Ethyl Acetate Extract E2

1. CPC Run #1: Fractionation of E2

Solvent systems containing chlorinated solvents were examined for the initial countercurrent fractionation. They often provide superior solubility that helps prevent precipitation of less soluble components during fractionation. A frequently used solvent system incorporating chlorinated solvents is based on CCl_4–CH_2Cl_2–MeOH–H_2O and can be used to separate compounds of a fairly wide spectrum of polarity, from hydrophobic to intermediate. The adjustments of P can be accom-

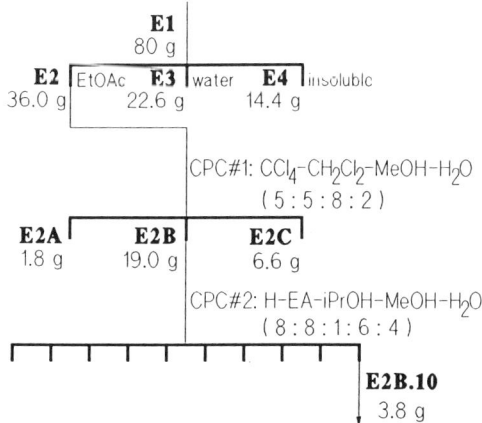

Figure 3 Extraction of the *Tripterygium wilfordii* root and initial fractionation of the extract (2 kg of ground root extracted with 95% EtOH).

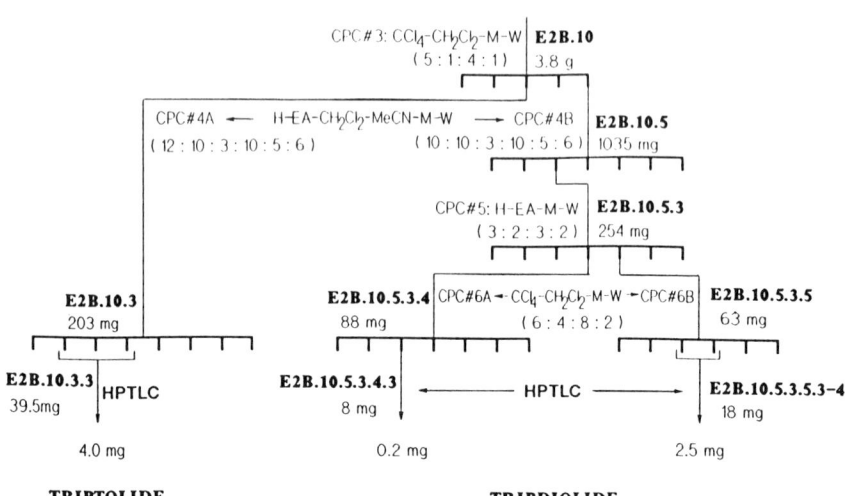

Figure 4 Fractionation scheme leading to the isolation of triptolide and tripdiolide.

plished by changing the ratio of CCl_4 to CH_2Cl_2 or of MeOH to H_2O. Thus, a higher CCl_4 to CH_2Cl_2 ratio will push compounds into the aqueous phase due to enhanced hydrophobic repulsion. Changing the ratio of MeOH to H_2O from the standard ratio of 8:2 may work only within very narrow limits. Less than 2 parts of water may cause homogenization, while above 3 parts of water a difference in hydrophobicity between the phases often becomes too large for practical fractionation. In addition, solubility of organic compounds in the upper phase may become insufficient.

Of the three solvent systems tested, solvent system A gave an optimal P value for initial fractionation. It had superior dissolving capacity due to the highest MeOH/H_2O ratio and was selected for CPC run #1 (Table 1). Experimental results determined that a mixture consisting of 1 ml of the upper phase and 1 ml of the

Table 1 CPC Run #1: Designing a Solvent System for E2

Solvent system	CCl_4	CH_2Cl_2	MeOH	H_2O	P
A	5	5	8	2	1.0
B	5	5	6	2	2.1
C	5	5	8	3	2.1

lower phase can dissolve at least 480 mg of E2 (equivalent to 24 g in 100 ml of injectable solution) without disturbing the phase volume ratio. With this assurance of the phases' stability, the first run was conducted in descending mode (lower phase as mobile) using preparative cartridges loaded through a 100-ml sample chamber with 17 g of E2 dissolved in a mixture of equal volumes of both phases. The run proceeded normally in the beginning, but it soon became apparent that the partitioning was flawed. Eluted fractions consisted of homogenized upper and lower phases, indicating that the two-phase system had been disrupted. A TLC analysis confirmed that only a very crude separation had been achieved. From the results of CPC run #1, it appears that the presence of the most hydrophobic, as well as the most polar, components of E2 were responsible for maintaining the phases. When the most hydrophobic components were carried away with the mobile phase and polar components were left behind in the stationary phase, the mixture of components of intermediate polarity induced homogenization of the phase. A "plug" formed from the homogenized phases, preventing further fractionation. The same problem reoccurred during a second attempt, when the loaded sample was reduced to 11 g. Although the phenomenon described is unique in the authors' experience, CPC users should be aware of the potential problem when dealing with concentrated crude extracts (see Section II.C).

This experiment, while largely unsuccessful, did succeed in separating E2 into three subfractions:

E2A—1.8 g; an inactive, oily mixture of the least polar components collected from the first five test tubes (only from the first run).

E2B—19 g; highly active in the MLR ($EC_{50} < 0.1$ μg/ml) and collected from the homogenized fractions. Most of the components seen on TLC plates were located here.

E2C—6.6 g; inactive and recovered from the first cartridges containing unhomogenized stationary phase. These components were much more polar than E2B, requiring a significantly more polar TLC solvent system for analysis.

2. CPC Run #2: Fractionation of E2B

The E2B fraction was still very complex, and its components were spread over a wide range of polarity. In this situation, it is necessary to compromise resolution with a reasonable time of a run. A solvent system with P slightly below 1 would provide the widest distribution of components. Variations of a versatile solvent system based on mixtures of hexane–EtOAc–MeOH–H_2O were explored (see Section II.A.2). To obtain a higher P, the ratio of hexane/EtOAc needs to be increased. Hexane acts *indirectly* (see Section II.A), increasing the hydrophobicity of the upper phase and expelling more polar components into the lower phase. Alternatively, increasing the water content makes the lower phase repulse the more hydrophobic compounds into the upper phase. Modification of this solvent system with the addition of small volumes of isopropanol has a pronounced *direct* effect on the organic phase, producing a lower value of P (Table 2). Solvent systems F and G were considered for the run; however, G had better dissolving properties, and its phases separated faster without forming an emulsion.

Fraction E2B, 17.4 g, was divided in two equal portions, which were fractionated in ascending mode. The run was conducted using solvent system G pumped at a flow rate of 21% and maintaining a rotational speed of 900 rpm. The sample was injected into preparative cartridges through a 100-ml sample loading chamber filled with a solution of 8.7 g of E2B in 100 ml of a mixture of both phases. Eluent was collected into test tubes in increments of 25 ml at 4 minute intervals. After a TLC analysis of the test tube contents, those with similar composition were combined into 10 fractions (Fig. 5). Fraction E2B.10, 3.8 g, recovered from test tubes 70–85, containing stationary phase, was the most potent in MLR assay.

3. CPC Run #3: Fractionation of E2B.10

For fractionation of E2B.10, variations of the system containing chlorinated solvents used for CPC run #1 were examined (Table 3). The addition of hexane strongly increases the hydrophobic character of the organic (lower) phase. Hexane is less miscible with MeOH–H_2O mixtures and thus is more effective than carbon tetrachloride in expelling these polar solvents from the organic phase. When mixing low-density solvents such as hexane with high density solvents such as carbon tetrachloride, however, caution should be exercised since certain ratios may cause a reversal of the upper and lower phases.

For fractionation of E2B.10, solvent system A was chosen to be used in descending mode (denser phase as mobile). The sample, 3.5 g, was injected through a 50-ml sample chamber. The run was performed using preparative cartridges at a flow rate of 25%, producing 58 atm of back pressure. The eluent

Table 2 CPC Run #2: Designing a Solvent System for E2B

Solvent system	C_6H_{14}	EtOAc	MeOH	H_2O	iPrOH	P
A	4	4	3	1	0	1.40
B	3	4	3	1	0	1.33
C	3	5	3	1	0	1.45
D	3	5	2	1.5	1	0.68
E	5	5	5	3	0	0.63
F	6	4	5	3	0	0.91
G	4	4	3	2	0.5	0.91

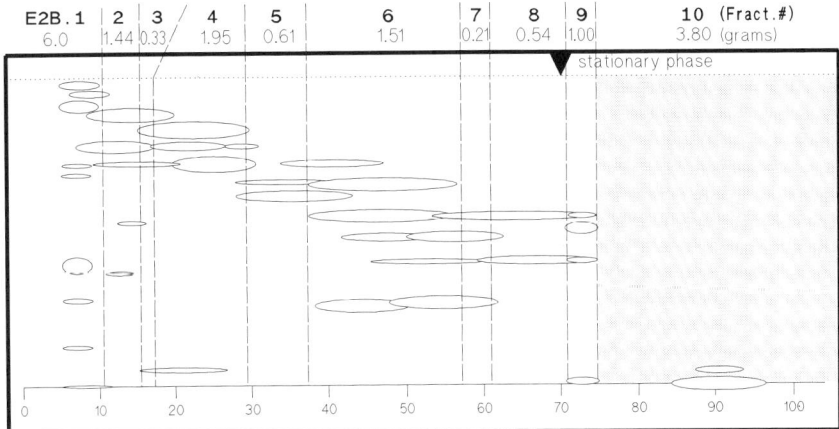

Figure 5 Thin-layer chromatogram of the CPC fractionation of E2B; 12 1000E cartridges; 25 ml/test tube; solvent system: hexane–EtOAc–iPrOH–MeOH–H$_2$O (4:4:0.5:3:2); mode: ascending; TLC: silica gel; hexane–EtOAc–HCOOH (10:10:1); active fraction: shaded area.

was collected every 4.2 minutes in 19 ml increments. After a TLC analysis (Fig. 6), the test tube contents were combined into five fractions. Immunosuppressive activity was detected in two fractions, E2B.10.3, 203 mg, test tubes 48–78, and E2B.10.5, test tubes 95–113 (stationary phase), 1035 mg. The two active fractions contained at least two different active components separated and were further investigated using independent pathways.

D. Isolation of Triptolide

1. CPC Run #4A: Fractionation of E2B.10.3

To further fractionate E2B.10.3, a novel solvent system designed for the ascending mode was explored and evaluated (Table 4). One of its components is acetonitrile, a solvent whose molecules contain a triple bond. An interaction between molecules of acetonitrile and the dissolved compounds, especially those that also

Table 3 CPC Run #3: Designing a Solvent System for E2B.10

Solvent system	C$_6$H$_{14}$	CCl$_4$	CH$_2$Cl$_2$	MeOH	H$_2$O	P
A	0	10	2	8	2	1.67
B	0	8	4	8	2	1.11
C	3	5	5	8	2	0.77

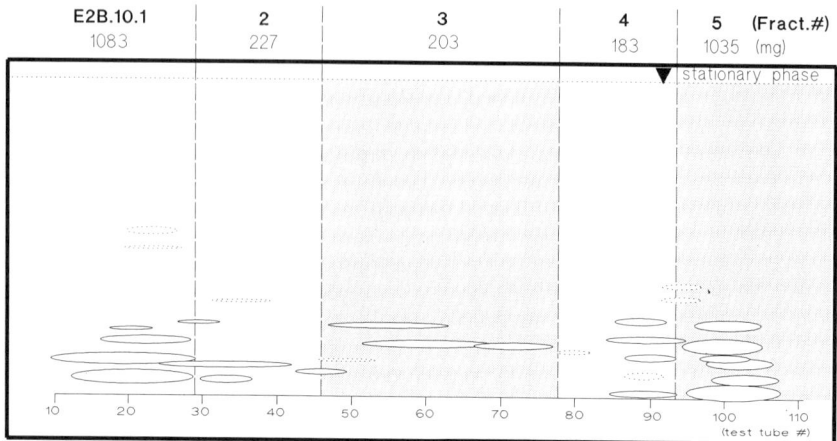

Figure 6 Thin-layer chromatogram of the CPC fractionation of E2B.10; 12 1000E cartridges; 25 ml/test tube; solvent system: CCl_4–CH_2Cl_2–MeOH–H_2O (5:1:4:1); mode: ascending; TLC: silica gel; hexane–EtOAc–HCOOH (5:5:0.1); active fraction: shaded area.

contain multiple bonds, is likely to affect significantly their partition coefficients. A series of initial experiments necessary to get acquainted with this novel solvent system suggested that its design should include a fixed proportion of EtOAc–MeCN—MeOH (10:10:5). Three other solvents, hexane, methylene chloride, and water, were used as variables to adjust the composition of the system to obtain the

Table 4 CPC Run #4A: Designing a Solvent System for E2B.10.3

Solvent system	C_6H_{14}	CH_2Cl_2	EtOAc	MeCN	MeOH	H_2O	P	V_{up}/V_{lw}
A	10	5	10	10	5	5	ND[a]	2.3
B	10	5	5	10	5	5	6.25	1
C	10	3	10	10	5	7	1.0	1.7
D	15	0	10	10	5	7	ND[a]	0.5
E	15	2	10	10	5	7	6.67	1.2
F	12	3	10	10	5	6	2.50	1.4
G	15	3	10	10	5	6	4.17	1.5
H	15	3	10	10	5	7	1.25	2
I	17	3	10	10	5	7	3.57	2

[a]ND = no data.

desired P (3.3 in this case). An increase in the number of parts of methylene chloride has a very powerful effect decreasing P (compare E versus H); however, it should not exceed 5 parts. Otherwise, the densities of both phases are becoming very similar, which could lead to reversal of densities during actual CPC fractionation. Adding or subtracting the number of parts of hexane affects hydrophobicity of the upper phase, conveniently allowing fine adjustment of P (compare F versus G). Addition of water causes an opposite effect (compare G versus H). Below the lowest limit of 5 parts of water, the solvent system comes dangerously close to the homogenization point. A partition coefficient for solvent system A was not determined because of too high, and for D because of too low, ratio of upper to lower phases. This versatile solvent system was used frequently throughout the project. For fractionation of the fraction E2B.10.3, version I was chosen on the basis of a suitable P value.

Fractionation of E2B.10.3, 162 mg, was carried out using solvent system I in ascending mode with analytical cartridges. The sample was injected through a 5-ml sample loop. The run was conducted at 6% flow rate and 1000 rpm, producing 50 atm of back pressure. Eluent was collected into test tubes in 8 ml increments. After a TLC analysis, the eluent was combined into eight fractions (Fig. 7). Immunosuppressive activity was detected in two adjacent fractions: E2B10.3.3, 8.8 mg, test tubes 41–42; and in E2B.10.3.4, 19.5 mg, test tubes 43–46.

2. Final Purification of Triptolide

A TLC of the active fractions E2B.10.3.3–4 produced a greatly simplified picture, indicating the presence of six components. Since some bands stretched between the fractions 3 and 4, they were the likely candidates for the active compounds. Considering the small quantities of material left for further fractionation, it appeared justified to carry out the last purification step with TLC plates. This would involve smaller volumes of solvents, therefore lowering the probability of introducing the solvents' contaminants, including dissolved O_2. The fractions were applied to high-performance TLC (HPTLC) plates and developed in hexane–EtOAc–CH_2Cl_2 (1:8:1). Detection was performed by cutting off the edges of the plate and visualizing the bands with a spray reagent. Fraction E2B.10.3.3, 8 mg, was separated using four 10 × 20 cm silica HPTLC plates horizontally loaded, to obtain triptolide (1) Rf 0.47, 2.0 mg. Some of the material, however, failed to migrate. Extraction of the band of origin allowed recovery of an additional 1.0 mg of material, indicating a strong adsorption of the mixture to silica gel at the spot of origin. Despite apparent sharp TLC resolution, the MLR and mass spectrometry data revealed that some of triptolide adsorbed (smeared) to silica on the entire length between the origin spot and Rf 0.47. The tailing on silica gel could be theoretically avoided, or at least alleviated, through selection of a better eluting TLC solvent system.

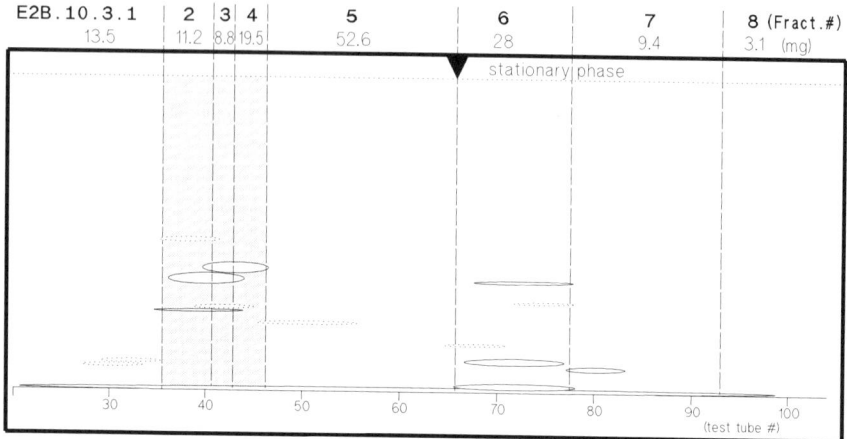

Figure 7 Thin-layer chromatogram of the CPC fractionation of E2B.10.3; 12 250W cartridges; 8 ml/test tube; solvent system: hexane–EtOAc–CH_2Cl_2–MeCN–MeOH–H_2O (12:10:3:10:5:6); mode: ascending; TLC: silica gel; hexane–EtOAc–HCOOH (5:5:0.1); active fraction: shaded area.

Unfortunately, a series of experiments with other eluting solvents have not been more successful. Separation of E2B.10.3.4, 19 mg, was performed similarly using five HPTLC plates and yielded an additional 3.5 mg of triptolide. All batches of triptolide were pooled and recrystallized from MeOH to give 4.0 mg of crystals melting at 235–237°C [39], the identity of which was confirmed by the spectral and x-ray analyses to be of triptolide.

E. Isolation of Tripdiolide from E2B

1. CPC Run #4B: Fractionation of E2B.10.5

Two active fractions were derived from E2B.10: E2B.10.3, which afforded triptolide, and E2B.10.5, which contained component(s) of higher polarity. For fractionation of the latter, two compositions of a previously developed solvent system were explored (compare with CPC run #3) (Table 5).

Solvent system *B* gave a satisfactory value of *P*. When, however, 100 mg of the sample was dissolved in a mixture of 1 ml of lower and 1 ml of upper phase, a serious change of the v_{up}/v_{lw} ratio, from 1 to 0.25, occurred. To minimize the phase disturbance during fractionation, the E2B.10.5 sample, 950 mg, was dissolved in a mixture of 15 ml of the phases and injected into analytical cartridges through a glass sample chamber. The run was conducted in ascending mode starting with a flow rate of 6.6% at a rotational speed of 900 rpm. To avoid excessive spill of the stationary phase, the run parameters were changed to a low

Table 5 CPC Run #4B: Designing a Solvent System for E2B.10.5

Solvent system	C_6H_{14}	CH_2Cl_2	EtOAc	MeCN	MeOH	H_2O	P
A	12	3	10	10	5	6	5.0
B	10	3	10	10	5	6	2.86

3.7% and 1100 rpm, producing 40 atm of back pressure. The eluent was collected in 8 ml increments every 8.5 minutes. After a TLC analysis, the eluent was combined into seven fractions. Immunosuppressive activity was confined to the third fraction, E2B.10.5.3, 254 mg, test tubes 54–72 (Fig. 8).

2. *CPC Run #5: Fractionation of E2B.10.5.3*
Three solvent systems similar to those used for CPC run #2 (see Table 2), but better suited for partitioning of more polar components (lower hexane/EtOAc ratio), were explored (Table 6).

Solvent system *B*, with a desired *P* value and satisfactory solubilizing properties, was selected for the run. The sample, E2B.10.5.3, 250 mg, was injected through a sample loop in a volume of 4 ml. The run was conducted in ascending mode using analytical cartridges at 6% flow rate and 900 rpm,

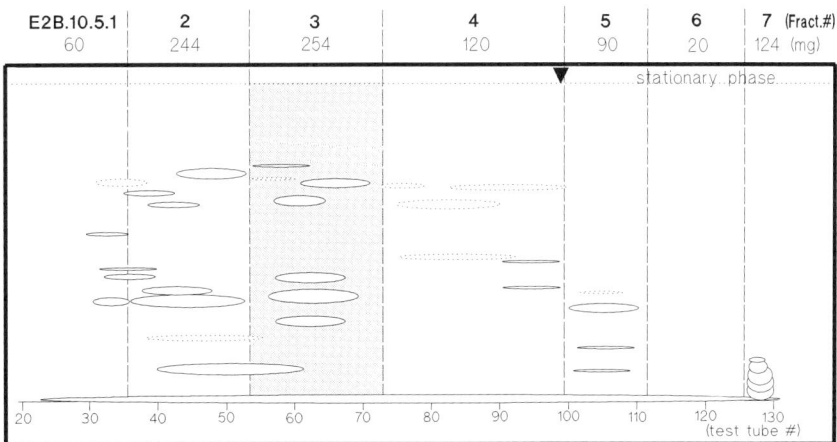

Figure 8 Thin-layer chromatogram of the CPC fractionation of E2B.10.5; 12 250W cartridges; 8 ml/test tube; solvent system: hexane–EtOAc–CH_2Cl_2–MeCN–MeOH–H_2O (10:10:3:10:5:6); mode: ascending; TLC: silica gel; hexane–EtOAc–HCOOH (5:5:0.1); active fraction: shaded area.

Table 6 CPC Run #5: Designing a Solvent System for E2B.10.5.3

Solvent system	C_6H_{14}	EtOAc	MeOH	H_2O	P
A	2	7	3	2	1.52
B	3	7	3	2	2.50
C	4	7	3	2	5.56

producing back a pressure of 50 atm. The eluent was collected every 4.5 minutes in 8 ml increments. The run proceeded smoothly, but any attempts to increase this low flow rate caused increased elution of the stationary phase. The first 20 test tubes contained a total of 40 ml of the eluted stationary phase in addition to the expected 120 ml of the stationary phase eluted as a void volume. After a TLC analysis, the eluent was combined into seven fractions, with activity located in E2B.10.5.3.4, 87.6 mg, test tubes 51–62, and E2B.10.5.3.5, 62.6 mg, test tubes 63–80 (Fig. 9). The activity in adjacent fractions could be caused by a single component or by more than one active component. In this case, because of the significant complexity of these fractions, it was decided to investigate the two fractions independently.

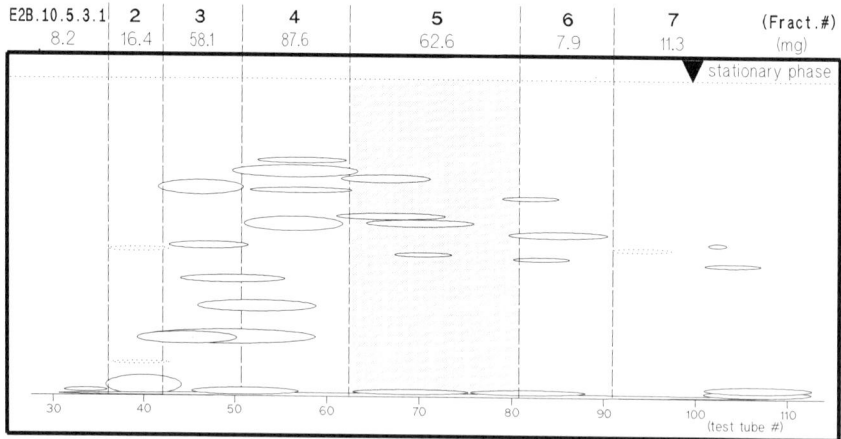

Figure 9 Thin-layer chromatogram of the CPC fractionation of E2B.10.5.3; 12 250W cartridges; 8 ml/test tube; solvent system: hexane–EtOAc–MeOH–H_r2O (3:7:3:2); mode: ascending; TLC: silica gel; EtOAc–CH_2Cl_2–iPrOH (8:2:1); active fraction: shaded area.

CPC IN ISOLATION OF NATURAL PRODUCTS

Table 7 CPC Run #6A: Designing a Solvent System for E2B.10.5.3.4

Solvent system	CCl_4	CH_2Cl_2	MeOH	H_2O	P
A	0	10	8	3	>20
B	3	7	8	3	1.18
C	5	5	8	2	0.67
D	6	4	8	2	0.48

3. CPC Run #6A: Fractionation of E2B.10.5.3.4

A series of solvent systems similar to the ones used for CPC run #1 and #3 were examined for this fraction (Table 7).

The run was conducted using solvent system D in descending mode, using analytical cartridges at a flow rate 6.5% and 400 rpm, which produced 40 atm of back pressure. The sample E2B.10.5.3.4, 82 mg, was injected through a 5-ml sample loop. The eluent was collected in 8 ml volume increments. After a TLC analysis, the test tubes were combined into six fractions, with activity located in the fraction E2B.10.5.3.4.3, 8.4 mg, test tubes 62–77 (Fig. 10).

4. CPC Run #6B: Fractionation of E2B.10.5.3.5

This fraction was separated with the same solvent system as used for E2B.10.5.3.4. Run conditions were similar; however, the flow rate was reduced to 6.0% and the rotational speed was raised to 500 rpm, which helped to decrease the volume of eluted stationary phase. Fractions E2B.10.5.3.5.3–4, from test tubes 53–57 and 58–80, were active in the MLR assay and yielded a total of 13.9 mg after both fractions were pooled (Fig. 11).

5. Final Purification of Tripdiolide

The active fractions obtained from CPC separations 6A and 6B had greatly simplified compositions and were eluted with similar volumes of the mobile phase. CPC fractionation is highly reproducible, and finding activity in fractions eluted with similar volumes suggests the presence of the same active constituent. A TLC comparison made in several solvent systems against a reference sample of tripdiolide revealed matching spots present in E2B.10.5.3.4.3 and E2B.10.5.3.5.3–4. The components were separated using silica gel HPTLC plates loaded horizontally with 3–4 mg of the mixture and developed in hexane–CH_2Cl_2–EtOAc–Me_2CO–MeCN (1:1:1:2:0.8). After visualizing the edges of the plates, the bands were scraped off and eluted with EtOAc–MeOH (1:1). In all three fractions, tripdiolide formed the uppermost band with an Rf 0.59. The remaining bands, including the band of origin, were scraped off, eluted, and reapplied to an HPTLC plate, which

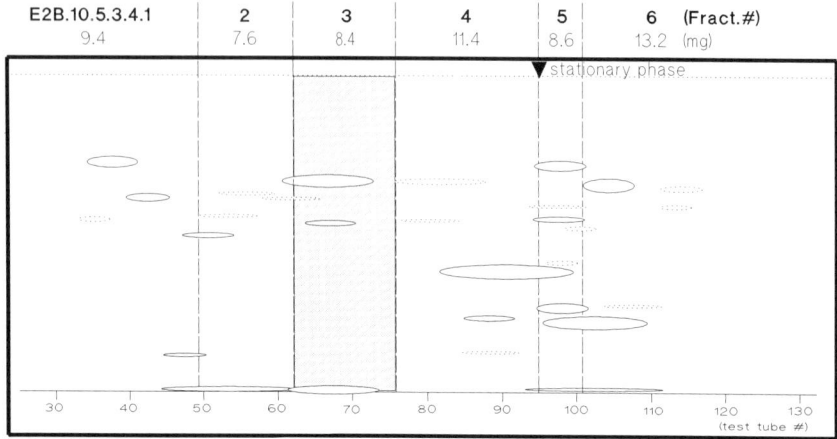

Figure 10 Thin-layer chromatogram of the CPC fractionation of E2B.10.5.3.4; 12 250W cartridges; 8 ml/test tube; solvent system: $CCl_4-CH_2Cl_2-MeOH-H_2O$ (6:4:8:2); mode: descending; TLC: silica gel; hexane–CH_2Cl_2–EtOAc–acetone–MeCN (1:1:1:2:1); active fraction: shaded area.

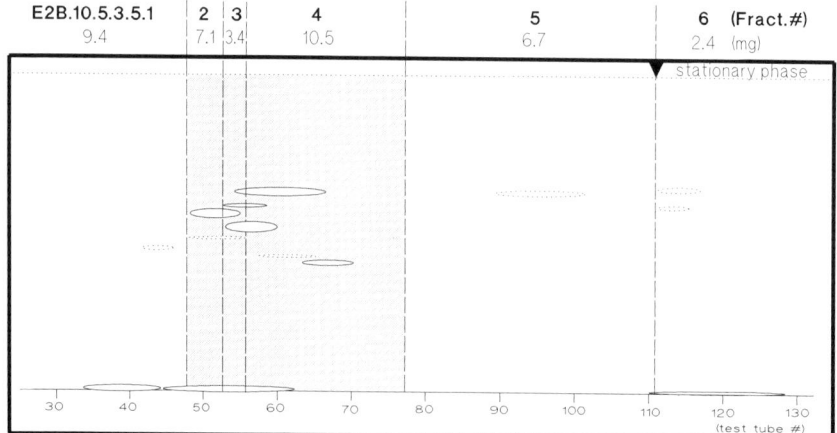

Figure 11 Thin-layer chromatogram of the CPC fractionation of E2B.10.5.3.5; 12 250W cartridges; 8 ml/test tube; solvent system: $CCl_4-CH_2Cl_2-MeOH-H_2O$ (6:4:8:2); mode: descending; TLC: silica gel; hexane–CH_2Cl_2–EtOAc–acetone–MeCN (1:1:1:2:1); active fraction: shaded area.

yielded an additional 0.5 mg of tripdiolide. This "tailing" phenomenon was observed before during purification of triptolide (Section III.D) and was not eliminated after extensive experimentation with several different TLC solvent systems. It was responsible for the lower yield of the valuable active material and for generating confusing data suggesting the presence of activity in the bands of otherwise inactive compounds. In conclusion, the silica gel TLC is an unsuitable medium for purification of the triepoxides triptolide and tripdiolide.

Fraction E2B.10.5.3.4 yielded 0.2 mg of tripdiolide, and E2B.10.5.3.5.3 and E2B.10.5.3.5.4 yielded 2.5 mg. These batches were combined and recrystallized from acetone–ethyl ether to produce crystals melting at 227–229°C. The acquired mass spectral and nuclear magnetic resonance (NMR) data were consistent with those published for tripdiolide [39,40].

F. Triptolide and Tripdiolide from E4A

The original partitioning of the evaporated ethanol extract E1 between EtOAc and H_2O afforded an insoluble gum, named E4. The results of MLR assay indicated that E4 also had immunosuppressive properties. Since triptolide and tripdiolide are easily soluble in EtOAc, it was assumed that the MLR activity of E4 was caused by components different from either of them. Subsequent investigations indicated that E4 consisted of approximately 10% w/w of monomeric compounds and the remaining 90% consisted of an undefined polymeric gum (MLR-inactive). The monomeric compounds turned out to be EtOAc-soluble, but during the original partitioning between EtOAc and water they obviously became occluded by the resinous gum. They were separated from the polymeric matter by dissolving E4 into the smallest possible volume of methanol and precipitating the polymers by adding nine volumes of CH_2Cl_2. Evaporation of the clear solution gave 1.6 g of the mixture, designated E4A. The composition of E4A was far too complex to determine without further fractionation if the activity resulted from the presence of triptolide or tripdiolide or, perhaps, from other active compounds [41,42].

The experience gained from fractionation of E2B was useful for designing a fractionation scheme for the investigation of E4A. An initial fractionation of E4A was performed using preparative cartridges, and the same solvent system previously used for E2B.10.5 was selected (see Section III.E): Hex–CH_2Cl_2–EtOAc–MeCN–MeOH–H_2O (10:3:10:10:5:6), P 3.57. Eluent, on the basis of TLC analysis, was combined into 11 fractions. Because of the complexity of E4A, a comparison of the active fractions with the reference samples was inconclusive during initial CPC steps, and so fractions were defined along the edges of main constituents. MLR assay indicated that two distinct fractions were active: first, E4A.5, 60 mg, test tubes 33–40, and E4A.6, 44 mg, test tubes 41–48. A second active band eluted in E4A.8, 73 mg, test tubes 79–85, and in E4A.9, 88 mg, test tubes 86–89. A TLC comparison with the samples of triptolide and tripdiolide isolated from the E2B part of the extract gave a positive match for triptolide in E4A.5–6 and for tripdiolide in E4A.8–9. This helped to precisely define the borders of active fractions during subsequent fractionations, which finally afforded 1.4 mg of triptolide from E4A.5–6 and 4.0 mg of tripdiolide from E4A.8–9. No other active compounds were found in the E4A part of the extract.

G. Note on the TLC Illustrations

The authors intended to give to the readers an accurate chronologic record as well as the story of an assay-guided isolation. The search was unbiased, because at the time the reference samples of triptolide and tripdiolide were not available. For the same reason, the TLC positions of their bands within the active fractions was not known till the very last fractionations. The illustrations of the TLC analysis of the CPC fractionation supplement the text by providing a visual record of the results of fractionation, instead of photocopies of the actual TLC plates, which can be very difficult to reproduce. Drawings emulating the TLC plates were made with help of computer software (Drawing Gallery, Hewlett Packard). In these drawings, strings of spots are depicted by ellipses that simulate eluted bands to the closest approximation.

The sequence of TLC plots illustrating fractionations may appear confusing, because the number of displayed components does not always decrease from step to step. This dramatic effect of "multiplication" of components results from an increased concentration of the components, which before the enrichment did not pass the detection threshold. Furthermore, each CPC fractionation was resolving the previously unresolved bands. This effect lasted through the enrichment sequence until an active component became predominant in an subsequent fraction.

H. General Methods and Conventions

1. CPC Instrument

All countercurrent fractionations were conducted with a centrifugal partition chromatograph, Model LLN, from Sanki Laboratories, Inc. (Kyoto, Japan). CPC runs were routinely performed using a full set of 12 either analytical (250W) or preparative (1000E) cartridges.

2. Partition Coefficient Measurements

The partition coefficients were measured using the following procedure: Several milliliters of an experimental solvent system was prepared and thoroughly equilibrated. An aliquot of a sample mixture, usually close to 10 mg was partitioned in a vial between 1 ml of the upper phase and 1 ml of the lower phase. The lower phase was then transferred to another vial using a pipette. The separated phases were evaporated to dryness under a stream of nitrogen on a heating block. More accurate results are often obtained if the vials with the residue are additionally dried in vacuo for a few minutes before the final weight is determined.

3. Prerun Preparation

The cartridges were equilibrated with the mobile phase prior to injecting a sample. The optimization of the two run parameters, flow rate and rotational speed, were experimentally determined during that stage. The numerical value of a given flow rate reflects a percentage of the maximum flow indicated by a calibrated dial. Solutions of samples were prepared for injection in a mixture of both phases. Volumes above 5 ml required the use of a glass sample chamber for injection. By inverting the position of the sample chamber, the stationary phase was always pumped in first, followed by the mobile phase, for both ascending and descending modes of separation.

4. Work Up of the Fractions and TLC Analysis

Automated collection of the eluent into test tubes began at the moment dissolved sample left the sample loop (for Teflon tubing loops) or immediately after the stationary phase was pushed out of the sample glass chamber. The test tube contents were combined into fractions based on the analysis of the TLC patterns. Eluent was either taken directly from test tubes and spotted on the TLC plates, or sometimes, prior to spotting, a small aliquot was first dried down and then redissolved in a smaller volume of an organic solvent to obtain suitable concentration. A centrifugal vacuum evaporator, Speedvac A290 (Savant Instruments, Inc., Farmingdale, NY), equipped with suitable rotors or a Reacti-Therm III Heating Module (Pierce, Rockford, IL) with a nitrogen blowing port, was used for the evaporation of multiple samples. The eluent was analyzed by TLC after being spotted horizontally on 10 × 20 cm silica gel HPTLC plates (Merck, Kieselgel 60 $F_{254}S$; Cat no. 15696) and developed in one or more appropriate solvent systems, each time optimized for the best separation. Developed plates were dried and sprayed with a sensitive general application reagent prepared by dissolving 25 g of ammonium molybdate $(NH_4)_6Mo_7O_{24} \times 4H_2O$, 10 g ceric sulfate, and 100 ml of concentrated sulfuric acid in water and brought to a volume of 1 L. Plates were visualized by heating at 125°C for 0.5–2 minutes.

After each fractionation run, a portion of the active fraction was set apart for future reference. Thus, the final isolated amounts of triptolide and tripdiolide are smaller than their actual content in the plant sample.

Human peripheral blood lymphocytes were used in the mixed lymphocyte reaction (MLR) assay. Responder cells were cultured with irradiated stimulator cells for 5 days, and proliferation was assessed by [^{3}H]thymidine incorporation.

ACKNOWLEDGMENT

The authors express their gratitude to the following colleagues: Dr. Randy Barton and Mr. John Ksiazek, both from Boehringer Ingelheim Pharm. (BIPI), for supporting the fractionation with MLR assay; Drs. Peter Lipsky and Xuelian Tao from the University of Texas, Dallas, for supplying the plant material and performing biological evaluation of the fractions; Ms. Grace Migaki for critical review of the manuscript and helpful suggestions; Dr. Jack Cazes from Sanki Laboratories, Inc. and Dr. John Proudfoot (BIPI) for their comments on the content of this chapter. Finally, our thanks to the colleagues from the mass spectrometry and NMR laboratories: Mr. Roger Dinallo, Mr. Scott Leonard, Ms. Tracy Saboe, Mr. Gordon Hansen, Mr. Walter Davidson, and Mr. Scot Campbell for spectral determinations of the isolated compounds.

REFERENCES

1. W. D. Conway, *Countercurrent Chromatography: Apparatus, Theory, and Application*, VCH Publishers, Inc. (1990).
2. A. Berthod, *J. Chromatogr.*, 550:677–693 (1991).
3. A. Berthod and D. W. Armstrong, *J. Liq. Chromatogr.*, 11:547–566 (1988).
4. A. Berthod and D. W. Armstrong, *J. Liq. Chromatogr.*, 11:567–584 (1988).
5. A. Berthod, J. D. Duncan, and D. W. Armstrong, *J. Liq. Chromatogr.*, 11:1171–1186 (1988).
6. A. Berthod and D. W. Armstrong, *J. Liq. Chromatogr.*, 11:1187–1204 (1988).
7. A. Berthod, Y. I. Han, and D. W. Armstrong, *J. Liq. Chromatogr.*, 11:1441–1456 (1988).
8. A. Berthod and D. W. Armstrong, *J. Liq. Chromatogr.*, 11:1457–1474 (1988).
9. D. W. Armstrong, *J. Liq. Chromatogr.*, 11:2433–2446 (1988).
10. S. J. Gluck and M. P. Wingeier, *J. Chromatogr.*, 547:69–78 (1991).
11. Y. Ito and W. D. Conway, *J. Chromatogr.*, 301:405–414 (1984).
12. K. Hostettmann and A. Marston, Natural products isolation by droplet countercurrent chromatography, *Countercurrent Chromatography: Theory and Practice* (N. B. Mandava and Y. Ito, eds.), Marcel Dekker, *Chromatographic Science*, 44, pp. 465–492 (1988).
13. I. Kubo, G. T. Marshall, and F. J. Hanke, Rotational locular countercurrent chromatography for natural products isolation, *Countercurrent Chromatography: Theory and Practice* (N. B. Mandava and Y. Ito, eds.), Marcel Dekker, *Chromatographic Science*, 44, pp. 493–507 (1988).
14. R. C. Bruening, F. Derguini, and K. Nakanishi, Centrifugal droplet countercurrent chromatography in natural product chemistry, *Countercurrent Chromatography: Theory and Practice* (N. B. Mandava and Y. Ito, eds.), Marcel Dekker, *Chromatographic Science*, 44, pp. 509–524 (1988).
15. D. G. Martin, Countercurrent chromatography for drug discovery and development, *Countercurrent Chromatography: Theory and Practice* (N. B. Mandava and Y. Ito, eds.), Marcel Dekker, *Chromatographic Science*, 44, pp. 565–581 (1988).

16. K. Hostettmann, *Planta Medica*, *39*:1 (1980).
17. N. El Tayar, A. Marston, A. Bechalany, K. Hostettmann, and B. Testa, *J. Chromatogr.*, *469*:91–99 (1989).
18. A. Marston, I. Slacanin, and K. Hostettmann, *Phytochemical Anal.*, *1*:3–17 (1990).
19. J. A. Glinski, J. R. Mikell, and G. O. Caviness, *J. Liq. Chromatogr.*, *13*:3625–3635 (1990).
20. A. Foucault and K. Nakanishi, *J. Liq. Chromatogr.*, *12*:2587–2600 (1989).
21. J. Cazes, *American Laboratory*, September (1990).
22. *Merck Index*, 11th ed.
23. X. Lu, P. Ma, Y. Chen, Ch. Zhang, Y. Zhang, Z, Zhang, L. Sheng, S. Li, D. An, et al., *Zhongguo Yixue Kexueyuan Xuebao*, *12*:157–161 (1990).
24. D. Yu, D. Zhang, H. Wang, and X. Liang, *Chin. Chem. Lett.*, *2*:399–402 (1991).
25. S. M. Kupchan, W. A. Court, R. G. Dailey, C. J. Gilmore, and R. F. Bryan, *J. Am. Chem. Soc.*, *94*:7194–7195 (1972).
26. S. Yang, H. Gao, S. Xie, W. R. Zhang, and Z. Long, *Int. J. Immunopharmacol.*, *14*:963–969 (1992).
27. Z. Xia and J. Chen, *Zhongguo Yaoxue Zazhi*, *25*:266–267 (1990).
28. W. Zhang, D. Pan, L. Zhang, and Y. Shao, *Yaoxue Xuebao*, *21*:592–598 (1986).
29. H. Sassa, Y. Takaishi, and H. Terada, *Biochem. Biophys. Res. Commun.*, *172*:890–897 (1990).
30. L. X. Zhang, F. K. Yu, Q. Y. Zherng, Z. Fang, and D. J. Pan, *Acta Pharmaceut. Sin.*, *25*:573–577 (1990).
31. W. M. Xu, L. X. Zhang, Z. H. Cheng, W. Z. Cai, H. H. Miao, and D. J. Pan, *Acta Pharmaceut. Sin.*, *26*:641–645 (1991).
32. C. Zhang, Y. Zhang, X. Lu, Y. Chen, P. Ma, Y. Yin, and L. Xu, *Zhongguo Yixue Kexueyuan Xuebao*, *8*:204–206 (1986).
33. X. Tao, L. S. David, and P. E. Lipsky, *Arthritis Rheum.*, *34*:1274–1281 (1991).
34. P. Y. Yan, G. J. Zhi, and M. Z. Jiang, *Zhongxii Jiehe Zazhi*, *5*:280–283 (1985).
35. J. Zheng, J. Fang, K. Gu, L. Xu, J. Gao, H. Guo, Y. Yu, and H. Sun, *Zhongguo Yixue Kexueyuan Xuebao*, *9*:317–322 (1987).
36. W.-Z. Gu, S. R. Brandwein, and S. Banerjee, *J. Rheumatol.*, *19*:682–688 (1992).
37. X. Li and M. R. Weir, *Transplantation*, *50*:82–86 (1990).
38. *Tripterygium wilfordii* root was obtained from Fujian province, People's Republic of China.
39. J. P. Kutney, G. M. Hewitt, G. Lee, K. Piotrowska, M. Roberts, and S. J. Rettig, *Can. J. Chem.*, *70*:1455–1480 (1992).
40. J. P. Kutney, G. M. Hewitt, T. Kurihara, P. J. Salisbury, R. D. Sindelar, K. L. Stuart, P. M. Townsley, W. T. Chalmers, and G. G. Jacoli, *Can. J. Chem.*, *59*:2677–2683 (1981).
41. P. C. Ma, J. R. Zheng, and X. Y. Lu; patent: *Faming Zhuanli Shenquing Gon* CN 1052859 A (1991); CA:*116*(14)136234a.
42. S. Z. Qian, J. R. Zheng, and X. Y. Lu; patent: *Faming Zhuanli Shenquing Gon* CN 1052861 A (1991); CA:*116*(14)136233z.

7
Liquid–Liquid Partition Coefficients
The Particular Case of Octanol–Water Coefficients

Alain Berthod
Laboratoire des Sciences Analytiques, Centre National de la Recherche Scientifique, Université de Lyon 1, Villeurbanne, France

I. INTRODUCTION

In countercurrent chromatography (CCC), the solute retention mechanism is very simple. It depends on only one physicochemical parameter: the liquid–liquid partition coefficient P. The retention volume V_R is

$$V_R = V_M + PV_S = V_T + (P - 1)V_S \qquad (1)$$

with V_M, V_S, V_T the mobile phase, stationary phase, and total internal apparatus volumes, respectively. V_M and V_S are not independent; if a volume V_S of stationary phase liquid is retained inside the apparatus, the remaining room is occupied by enough mobile phase so that $V_M + V_S = V_T$. The partition coefficient depends on the biphasic liquid system used. When the liquid system composition is adjusted to change the selectivity and/or the retention volumes of different solutes, it is the partition coefficients of the solutes that are changed.

This chapter focuses on the definition, properties, uses, and determination of liquid–liquid partition coefficients. This subject is not particular to centrifugal partition chromatography (CPC). Most of the scientific information presented herein is valid in CCC whatever apparatus is used. The first part recalls briefly the thermodynamic theory of partition coefficients. The second part stresses the importance of the octanol–water partition coefficient. The last part deals with the determination of P with special emphasis on octanol–water partition coefficient determination using CCC.

II. SOLUTE PARTITION BETWEEN TWO LIQUID PHASES

A. Did You Say Partition Coefficients?

CCC uses two-phase liquid systems. When a solute is introduced in such a system, it *partitions* between the two phases; that is, part of the solute goes in one phase and the rest goes in the other phase. The *International Union of Pure and Applied Chemistry* defined in the note IUPAC 87b the *distribution ratio* as "the total analytical concentration of the substance in the organic phase to its total analytical concentration in the aqueous phase usually measured at equilibrium" [1]. The distribution ratio is a common parameter in solvent extraction; it uses total analytical concentrations of the solute in both phases. The distribution ratio is not a constant. If any reaction, such as dimerization or ionic dissociation, occurs in a liquid phase, the distribution ratio becomes concentration-dependent.

The *distribution constant*, or *partition coefficient*, refers to a particular molecular form: it is a constant for diluted solutions. Although IUPAC recommends the use of the term *distribution constant*, abbreviated K, the term *partition coefficient*, abbreviated P, is preferred in CCC and will be used throughout this chapter.

B. Liquid Polarity and Solubility

The polarity of a solvent is related to several physicochemical properties of the molecules. Differences in the electronegativity of the atoms constituting the solvent molecule create a permanent *dipole moment*, denoted μ, given in Debye units. An external electric field can induce a molecular electric dipole. Then, the solvent molecules can reorient due to this induced dipole, which changes the relative electrical permittivity in the solvent. The *dielectric constant* quantifies this effect. Table 1 lists the dipole moment, dielectric constant, and other selected physicochemical properties of solvents commonly used in CCC.

Another facet of solvent polarity is related to solubility. Solvents with similar polarities have a high mutual solubility. Water is the common solvent with the highest polarity. Table 1 lists the water solubility in solvents and the solvent solubility in water for polarity comparison. A high water solubility is observed with polar solvents, and vice versa. Hildebrand [2] introduced the solubility parameter δ defined as the work necessary to separate two solvent molecules:

$$\delta = \sqrt{\frac{\Delta H_V - RT}{V_m}} \quad (2)$$

where ΔH_V is the solvent molar heat of vaporization and V_m is its molar volume. The δ parameter is included in Table 1. It provides excellent solubility predictions for nonpolar solutes in nonpolar solvents.

The dipole moment, the dielectric constant, the δ parameter, and water solubility all are parameters related to the solvent polarity. There is, however, no direct relationship among these parameters. The term *polarity* itself has, until now, not been precisely defined. Beside the δ parameter of Hildebrand, Snyder [3] established a ϵ_0 polarity scale based on the eluting power of the solvent used as a mobile phase in thin-layer chromatography. Reichardt [4] proposed a polarity parameter E_T based on the transition energy for the longest-wavelength solvatochromic absorption band of a pyridinium-N-phenoxyde betaine dye. The Snyder ϵ_0 scale ranges from 0, a completely apolar solvent, to 10.2 for water, the most polar solvent. The Reichardt E_T scale is normalized using tetramethylsilane as the completely apolar solvent with $E_T = 0$ and water with $E_T = 100$. Table 1 lists the δ, ϵ_0, and E_T polarity parameters. The δ parameter considers the London intramolecular forces; the two other scales try to take in account dipole–dipole, dipole-induced dipole, charge transfer, and hydrogen-bonding interactions.

Figure 1 compares the three polarity indexes. It shows that there is only approximate correspondence between the three polarity scales. Solvents such as diethylacetate, dimethyl formamide, and dimethyl sulfoxide are given as polar by the Snyder index with $\epsilon_0 > 5.5$, and as intermediate or low polarity solvents by the Reichardt index and the delta parameter (Table 1). Similarly, the δ parameter of benzene (18.8) is higher than the one of acetone (18.6), while the two other indexes give acetone as much more polar than benzene. This clearly shows that it is not possible to define the polarity of a solvent with only one parameter.

C. A Little Thermodynamics

As just reviewed, when two solvents have similar polarities, they have a high mutual solubility or they are completely miscible. This also is not an absolute rule. Table 1 shows that the polarity of dioxane is lower than that of butanol. Dioxane is miscible with water in all proportions; butanol is not. Dioxane is a hydrogen bond acceptor; butanol is a hydrogen bond donor. Free energy changes in molecule interactions can explain such solubility differences. The complete study of the thermodynamics of solutions is not the topic of this chapter and can be found in Refs. 2 and 4–6.

If we consider two practically immiscible solvents, 1 and 2, they form two phases of one solvent saturated in the other. When a solute A is introduced in such a biphasic liquid system, it partitions between the two phases. Assuming ideal mixtures, in the solvent 1 phase, the Gibbs free energy of A, or *chemical potential*, μ_{1A}, is expressed by

$$\mu_{1A} = \mu_{1A}{}^0 + RT \ln x_{1A} \tag{3}$$

in which $\mu_{1A}{}^0$ is the standard chemical potential of A at infinite dilution in phase 1. In the solvent 2 phase, the chemical potential, μ_{2A}, is

$$\mu_{2A} = \mu_{2A}{}^0 + RT \ln x_{2A} \tag{4}$$

Table 1 Physicochemical Properties, Polarity, and Water Solubility for Selected Solvents

Solvent	M.W. (dalton)	Dipole moment (Debye)	Dielectric constant	Hildebrand	Polarity Reichardt	Snyder	Solubility (%w/w) solv. in water	wat. in solvent
Acetic acid	60	1.74	6.15	20.6	64.8	5.1	∞	∞
Acetone	58	2.69	20.7	18.6	35.5	5.1	∞	∞
Acetonitrile	41	3.44	37.5	24.1	46	5.8	∞	∞
Benzene	78	0	2.28	18.8	11.1	2.7	0.18	0.063
1-Butanol	74	1.75	17.5	27.2	60.2	3.9	7.8	20.1
2-Butanol	74	1.64	15.8	25.2	50.6	4	12.5	44.1
Chloroform	119.4	1.15	4.9	18.9	25.9	4.1	0.815	0.056
Cyclohexane	84	0	2.02	15.8	0.6	0.2	0.006	0.01
Decahydronaphthalene	138	0	2.15	15.6	1.5	0.2	<0.02	0.0063
Dibutylphtalate	278	2.82	6.44	17.4	20	4.6	0.01	0.46
1,2-Dichloroethane	99	1.86	10.4	20.4	32.7	—	0.81	0.187
Dichloromethane	85	1.14	8.9	20	30.9	3.1	1.6	0.24
Diethyl ether	74	1.15	4.34	15.4	11.7	2.8	6.9	1.3
Diethylcarbonate	118	0.9	2.82	18	19.4	5.5	hydrolyze	1.4
Dimethoxyethane	90	1.71	7.2	22.8	23.1	—	∞	∞
N,N-Dimethyl acetamide	87	3.72	37.8	21.6	40.1	6.5	∞	∞
Dimethyl formamide	73	3.86	36.7	24.2	40.4	6.4	∞	∞
Dimenthyl sulfoxyde	78	4.3	48.7	24	44.4	7.2	∞	∞

LIQUID–LIQUID PARTITION COEFFICIENTS

Dioxane	88	0.45	2.25	20	16.4	4.8	∞	∞
Di-isopropyl ether	102	1.32	3.88	14.6	9.8	—	1.2	0.57
Ethanol	46	1.66	26.6	26	65.4	4.5	∞	∞
Ethyl acetate	88	1.88	6	18.2	22.8	4.4	8.7	3.3
Furfural	92	3.5	38	23.6	50	2.8	8.2	6.3
Isooctane	114	0	1.94	15.4	0.9	0.1	0.0002	0.006
Methanol	32	2.87	32.7	29.3	76.2	5.1	∞	∞
Methyl ethyl ketone	72	2.76	15.2	19.2	32.7	4.7	24	10
Methyl-tert-butyl ether	88	1.32	4.5	15.1	14.8	2.5	4.8	1.5
N-Methylpyrrolidone	99	4.1	32	22.4	35.5	6.7	∞	∞
n-Heptane	100	0	1.92	15.2	1.2	0.1	0.0003	0.01
n-Hexane	36	0.08	1.88	15	0.9	0.1	0.001	0.01
Octanol	130	1.76	10.3	20.9	54.3	3.4	0.054	4.1
n-Pentane	72	0	1.84	14.9	0.9	0	0.004	0.009
1-Pentanol	88	1.82	14.7	22.1	56.8	3.7	2.2	7.5
1-Propanol	60	3.1	20.3	24.4	61.7	4	∞	∞
2-Propanol	60	1.66	19.9	23.7	54.6	3.9	∞	∞
Propylene carbonate	102	4.9	65	27.3	49.1	6.1	17.5	8.3
Tetrachloromethane	154	0	2.24	17.6	5.2	1.6	0.08	0.008
Tetrahydrofuran	72	1.75	7.6	18.2	20.7	4	∞	∞
Toluene	92	0.31	2.38	18.3	9.9	2.4	0.074	0.03
Tributylphosphate	266	3.07	8.1	15.3	27.5	4.6	0.039	4.7
111-Trichloroethane	133	1.7	7.3	17.9	26.9	3.2	0.132	0.034
112-Trichloro, 122-trifluoroethane	187	—	2.4	14.5	0.1	0.1	0.017	0.011
Water	18	1.87	80.1	48.6	100	10.2	—	—

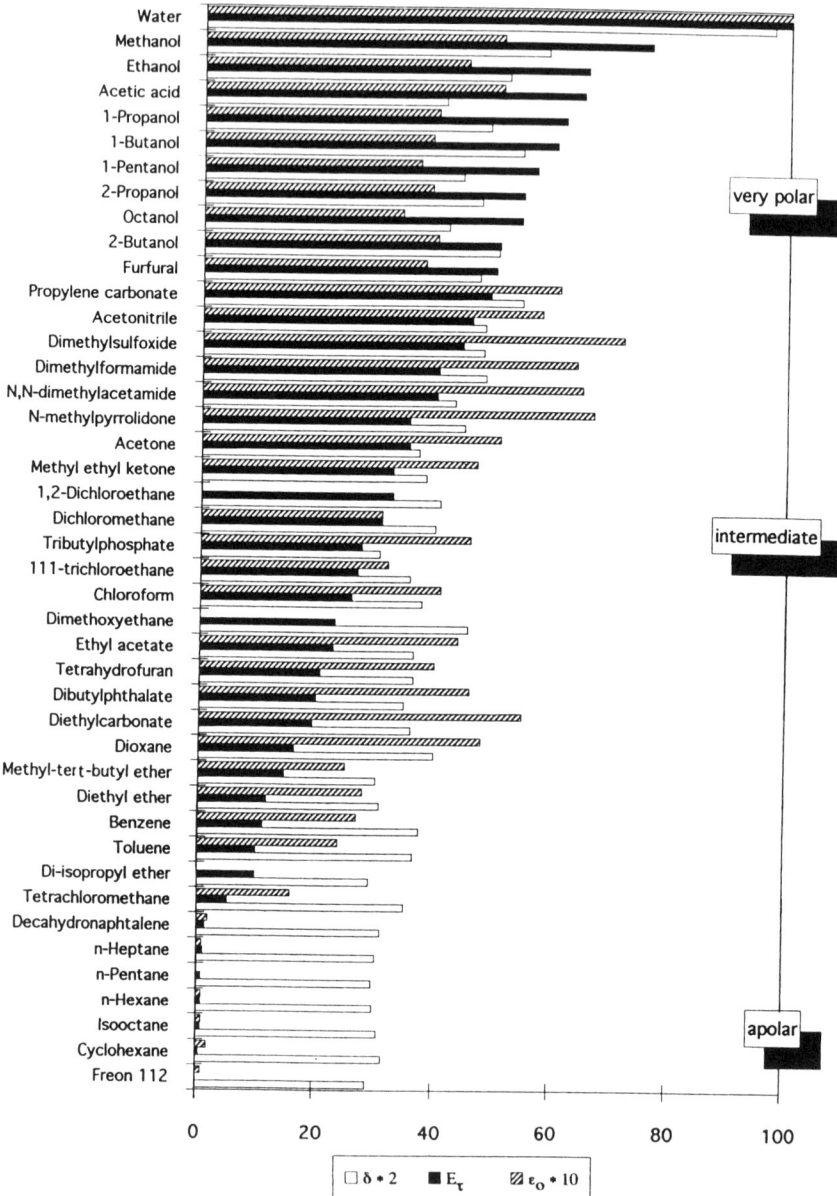

Figure 1 Comparison of selected solvent polarity. The Reichardt index was used to sort the solvents from water, the most polar solvent, to freon 112, the least polar.

LIQUID–LIQUID PARTITION COEFFICIENTS

If the chemical potential is not identical in the two phases, mass transfer of A occurs, and the mole fractions x change so that the chemical potential of A becomes equal in both phases; that is, the equilibrium is reached. Then

$$\mu_{1A}^0 - \mu_{2A}^0 = RT \ln \frac{x_{2A}}{x_{1A}} \tag{5}$$

in which x_{2A}/x_{1A} is the distribution ratio, expressed by

$$\frac{x_{2A}}{x_{1A}} = P_{2/1} = \exp \frac{\mu_{1A}^0 - \mu_{2A}^0}{RT} \tag{6}$$

In the case of nonideal mixtures, the mole fractions x should be replaced by activities, $a = xf$, in which f is the activity coefficient. The distribution ratio is constant only if the activity coefficient is constant, which is not true in concentrated solutions.

Partition coefficients are usually expressed as molarity ratios. Molar solubilities c_A and mole fractions x_A are related as follows:

$$c_A = \frac{\rho x_A}{M_S + (M_A - M_S) x_A} \tag{7}$$

where ρ is the solution density in g/cm³ and M_S and M_A are the molecular weight of the solvent and the solute, respectively, in g/mole to form c_A in mole/dm³. For octanol–water partition coefficients and diluted solutions,

$$(M_A - M_S) x_A \ll M_S \tag{8}$$

and

$$\frac{\rho}{M_S} \sim \frac{1}{V_S} \tag{9}$$

where V_S is the solvent molar volume. This gives

$$P_{o/w}^x = P_{o/w}^c \frac{V_o}{V_w} \tag{10}$$

in which the superscripts x and c refer to the mole fraction and concentration partition coefficients, respectively. The subscripts o and w refer to the octanol and water phases, respectively. At 20°C, the molar volumes of water and octanol are respectively 18 and 158.4 cm³/mole. Equation 10 can be rewritten as

$$\log P_{o/w}^x = \log P_{o/w}^c + 0.944 \tag{11}$$

In the following, the octanol–water partition coefficient $P_{o/w}^c$ will be written P_{oct}.

D. Effect of Chemical Reactions

If a chemical reaction occurs in one liquid phase, the concentration in unmodified solute decreases and the distribution ratio changes. A chemical reaction can be a self-association of the solute, such as the dimerization of organic acids in low-polarity solvents or the dissociation of ionizable compounds in very polar solvents. In the case of octanol–water partition coefficients, the dissociation of ionizable solutes in the aqueous phase is the most important chemical reaction encountered and will be developed herein.

1. Acid Solutes

If a solute, AH, can dissociate in the aqueous phase, part of it becomes an anion A^- according to

$$AH \Leftrightarrow A^- + H^+ \quad \text{and} \quad K_a = \frac{|A^-||H^+|}{|AH|} \quad (12)$$

Only the molecular form AH can go into the organic (octanol) phase with a concentration $|AH|_o$. The distribution ratio is noted P_{app} for "apparent" partition coefficient:

$$P_{app} = \frac{|AH|_o}{|AH|_w + |A^-|} \quad (13)$$

Expressing $|A^-|$ as $K_a |AH|_w/|H^+|$ (Eq. 12), we get

$$P_{app} = \frac{|AH|_o}{|AH|_w} \times \frac{1}{1 + K_A/|H^+|} = \frac{P_{oct}}{1 + K_A/|H^+|} \quad (14)$$

Figure 2 shows the plot of the $\log P_{app}$ versus pH for a hypothetical solute with $P_{oct} = 100$ and three possible pK_a values: 2, 5, and 8. When pH $> pK_a$, the log P versus pH plot is a straight line and $|H^+|$ is much lower than K_a; then the ratio $K_a/|H^+|$ is much higher than 1 and

$$\log P_{app} \sim \log P_{oct} + pK_a - pH \quad (15)$$

The slope is -1; the intercept at $\log P_{app} = 0$ occurs for a pH value equal to $\log P_{oct} + pK_a$. In Fig. 2, this corresponds to pH = 4, 7, and 10 for pK_a values equal to 2, 5, and 8, respectively, with $\log P_{oct} = 2$. The $P_{app} = 1$ point is interesting in CCC. For $P = 1$, the retention volume V_R is exactly equal to the apparatus volume V_T, whatever the stationary phase volume retained (Eq. 1).

2. Basic Solutes

Basic solutes can be protonated to give positively charged ions. In this case, the acidic character of the AH^+ cation is considered:

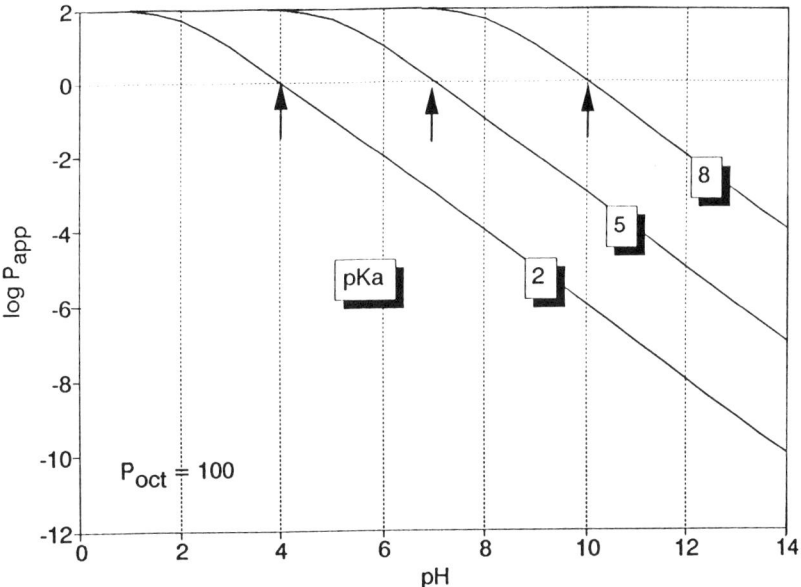

Figure 2 Observed partition coefficient P_{app} for an ionizable acid compound. At high pH values, the ion form is located in the water phase, producing a very low P_{app} value. Arrows mark where P_{app} equals 1 ($\log P_{app} = 0$), which is obtained for a pH value equal to $pK_a + \log P_{oct}$ (see text and Eqs. 13–15). At low pH values, $P_{app} = P_{oct}$.

$$A + H^+ \Leftrightarrow AH^+ \quad \text{and} \quad K_a = \frac{|A||H^+|}{|AH^+|} \tag{16}$$

With a development similar to Eqs. 12–14, we can form

$$\log P_{app} \sim \log P_{oct} - pK_a + pH \tag{17}$$

In this case, the plot of $\log P_{oct}$ versus pH is linear for pH $<$ pK_a, with a slope equal to 1 and, for $P_{app} = 1$, the pH value corresponds to the difference $pK_a - \log P_{oct}$.

In the case of other reactions, such as the dimerization of the solute in the organic phase, the equilibrium constant is used to establish the P_{app} equation, which depends on the solute concentration [7].

E. Effect of Temperature

Equation 5 expresses the standard free energy of transfer, ΔG_{tr}, of a solute A in the partitioning process between solvent 1 and solvent 2:

$$\Delta \mu = \Delta G_{tr} = RT \ln P \tag{18}$$

in which P is expressed as a mole-fraction ratio. Assuming the standard molar enthalpy is constant in a limited temperature range, the plot of ln P versus $1/T$ should produce a straight line with slope $\Delta G_{tr}/R$.

Unfortunately, it is not that simple. The mutual solubility of the two solvents is temperature-dependent. At the critical solution temperature, the biphasic system becomes monophasic [6]. As a general rule, it is possible to consider that the effect of temperature on the P value is not great if the solvents are not very miscible and the temperature change is not dramatic. An average change of 0.009 log P unit per degree, either positive or negative, was found for a variety of biphasic systems, including the octanol–water system [8].

III. THE OCTANOL–WATER PARTITION COEFFICIENT

A. Hydrophobicity Scale and Quantitative Structure–Activity Relationship

The use of liquid–liquid partition coefficients to predict solute solubilities and pharmaceutical activity was initiated at the end of the last century by Meyer and Overton (cited in [8]). Hansch [9] proposed the octanol–water partition coefficient P_{oct} as a reference parameter for hydrophobic bonding in biochemical and pharmacological systems. The polarity difference between octanol and water is close to the one between the aqueous medium and the living cell membranes. The P_{oct} parameter is used to predict the biological effect of organic chemicals. The Food and Drug Administration (FDA) and the Environmental Protection Agency (EPA) use P_{oct} values to estimate the tendency of an organic chemical to bioconcentrate into living cells [10]. A new drug cannot be accepted by FDA and EPA without the P_{oct} parameter [11].

As a hydrophobic scale, the P_{oct} or log P_{oct} allows one to estimate the water solubility of a compound. Yalkowski [12] has shown that the water solubility of a nonionizable compound, c^w, could be related to its P_{oct} coefficient through the empirical equation

$$\log c^w = -1.05 \log P_{oct} - 0.012\theta_m + 0.87 \tag{19}$$
$$n = 167, \quad r = 0.995, \quad \sigma = 0.242$$

where θ_m is the melting point of the compound in degrees Celsius. Taft and coworkers [13] based their water solubility predictions on intermolecular interactions such as cavity formation, nonspecific polar and polarizability effects, and specific donor–acceptor interactions. For non-hydrogen-bond-donating solutes and water as the solvent, their equation at 25°C is

$$\log c^w = 0.65 - 0.5 \log P_{oct} + 3.5\beta - \frac{V_s}{50} \tag{20}$$
$$n = 102, \quad r = 0.989, \quad \sigma = 0.17$$

where β is a solvatochromic parameter scaling for hydrogen-bond accepting ability (= basicity) of the solute, V_S is the solute molar volume (i.e., molecular weight divided by density). This approach was questioned by Yakolwski [14]. It makes sense, however, to link solute water solubility and octanol–water partition coefficients. The choice of the octanol–water system as a reference for hydrophobicity may not be perfect, but this issue is not the topic of this chapter.

B. Theoretical P_{oct} Evaluation. The Rekker–Hansch Fragmental Constants

In 1935 Hammett [15] showed it was possible to evaluate the dissociation constant of a substituted benzoic acid from the benzoic acid constant and a set of σ values, the Hammett constants, related to the substituent. By analogy, Hansch et al. [16] proposed to describe lipophilicity as follows:

$$\log \frac{P(SX)}{P(SH)} = \pi(X) \quad (21)$$

where $P(SX)$ and $P(SH)$ represent the partition coefficients of the solutes SX and SH, respectively. $\pi(X)$ is the hydrophobic substituent constant, that is, the contribution of substituent X to the lipophilicity of structure SH when X replaces an H atom in SH [7]. The $\log P_{oct}$ value of a given molecule can be estimated using the Rekker–Hansch π constants listed in Table 2 for every substituent:

$$\log P_{oct} = \Sigma \pi(X) + M \Sigma \kappa \quad (22)$$

M was called the "Magic" constant by Rekker [7], and κ is a correction factor for ring joining, aromatic conjugation, intramolecular interaction (e.g., folding), or electronegative proximity effect.

The $\log P_{oct}$ value of some molecules (Fig. 3) is evaluated to illustrate the use of Eq. 22 and Table 2. *Biphenyl* can be split in two C_6H_5 moieties (2 × 1.886) with an aromatic conjugation (1.00), which gives a $\log P_{oct}$ value of 4.04 (2 × 1.886 + 1.00 × 0.268 = 4.04). The experimental $\log P_{oct}$ value of biphenyl is exactly 4.04 [8]. *8-hydroxy quinoline* is a chelating agent whose molecule can be split in a pyridine ring (C_5H_4N) minus an H atom (0.526 − 0.175); then there is two-thirds of a benzene ring (2/3 × 1.886), and OH group (−0.343), and 1.00 times 0.268 for fused aromatic rings, which makes 1.533 for the estimated $\log P_{oct}$ value. The experimental $\log P_{oct}$ value is 2.00 [8] with a −0.466 deviation. For the isomer *2-hydroxy quinoline*, the theoretical $\log P_{oct}$ is different. It is a pyridine ring minus two H atoms (0.526 − 2 × 0.175) plus two-thirds of a benzene ring (2/3 × 1.886), an OH aromatic group (−0.343) and 1.00 × 0.286 for fused aromatic rings, which makes 1.376. The measured value is 1.26, that is, a 0.116 deviation. *Quinoline* alone has a calculated $\log P_{oct}$ value of 0.526 − 0.175 + 2/3 × 1.886 + 1.00 × 0.268 = 1.876; the measured value is 2.03 (a −0.154 deviation) [8]. These

Table 2 The Rekker–Hansch Constants for Selected Substituents Bonded on an Alkyl (sp^3 Hybridized) Carbon or on an Aryl (sp^2 Hybridized) Carbon

Fragment	On alkyl	On aryl	Fragment	On alkyl	On aryl
–Br	0.270	1.131	=CH–	0.235	0.235
–Cl	0.061	0.922	(O)–OH	—	–0.38
–F	–0.462	0.399	–COOH	–0.954	–0.093
–H	0.175	0.175	–O–CO–NH–	–1.943	–0.795
–H (neg)	0.462	0.462	–CH$_2$–	0.530	0.530
–I	0.587	1.448	–CO–NH$_2$	–1.970	–1.102
			–O–CO–NH$_2$	–1.481	—
=N–	–2.160	–1.012			
–NH–	–1.825	–0.964	–CH$_3$	0.702	0.702
–NH$_2$	–1.428	–0.854	C–C≡CH	0.730	0.730
–NO$_2$	–0.939	–0.78	–CH=CNO$_2$–	—	0.220
–O–	–1.581	–0.433	–CH=CH–NO$_2$	—	0.395
–OH	–1.491	–0.343	–CH=CH$_2$	0.935	0.935
–S–	–0.51	0.11	–O–CH$_2$–COOH	–1.155	–0.581
–S–S–	0.37	—	–CH=CH–COO–	—	0.042
–SH	0	0.62	C$_3$H$_3$N$_2$	–0.119	—
–SO–	–2.75	–2.05	–CH=CH–CONH–	—	–1.1
–SO$_2$–	—	–1.87	C$_5$H$_4$N	0.526	0.526
–SO$_2$N=	—	–2.454	C$_6$H$_3$	1.431	1.431
–SO$_2$NH–	—	–1.992	C$_6$H$_4$	1.688	1.688
–SO$_2$NH$_2$	—	–1.53	C$_6$H$_5$	1.886	1.886
C (quat.)	0.15	0.15	Correction factors κ and M		
–CCl$_3$	1.79	—			
–CF$_3$	0.757	1.331	Proximity effect 1 C between 2 groups		3.0
–C≡N	–1.066	–0.205	Proximity effect 2 C between 2 groups		2.0
–C=N–	—	–1.88	Aryl–aryl conjugation		1.0
–N=C=O	–2.894	—	Intramolecular interaction		1.0
–CO–	–1.703	–0.842	Ring joining		1.0
–O–CO–	–1.292	–0.431	Magic constant, M		0.268

Source: Data from Refs. 7–9.

examples show the Hansch–Rekker approach allows one to estimate the lipophilicity of a given molecule, but it cannot replace experimental measurement.

C. Experimental P_{oct} Measurements

1. The Shake Flask Method

The most extensive and useful partition coefficient data were obtained by simply shaking a solute with the two immiscible octanol and water phases and then

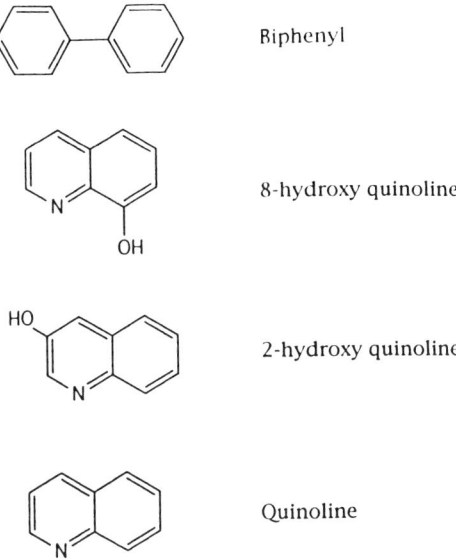

Figure 3 Molecular structures of the compounds used in text to illustrate the Rekker–Hansch fragmental approach of P_{oct} estimation.

analyzing the solute concentration in one or both phases [4,6,8]. For many solutes, repeated inversion (say ~100) of a 25-ml tube with ~0.01 M solute and the two phases establishes equilibrium in ~5 minutes. Very vigorous shaking can produce troublesome emulsions. The solute can be analyzed in only one phase, and the concentration in the other can be obtained by difference [8]. If the theoretical estimation of P_{oct} gives a very high value, it is wise to use a small volume of octanol shaken with a large volume of water to optimize the P_{oct} measurement error. The phase analysis is most often done by gas chromatography, liquid chromatography, or UV-visible spectroscopy. This method gives reliable results over the wide 10^{-4}–10^{+4} P_{oct} range. It requires highly pure solutes, however, and is very sensitive to the smallest contamination.

2. Reversed Phase Liquid Chromatography

Reversed phase liquid chromatography (RPLC) is used to estimate values of $\log P_{oct}$ from the corresponding $\log k'$ values. k' is the capacity factor, directly related to the retention time of the solute of interest. Good correlations are generally found between $\log k'$ and $\log P_{oct}$ for structurally similar compounds [17]. Unfortunately, the correlations are much poorer with dissimilar compounds [18]. Trace amounts of octanol were added in the mobile phase to enhance $\log k'$–

$\log P_{oct}$ correlations with a wide variety of solutes [19,20]. The P_{oct} range is 1–$10^{+5.5}$. The advantages of the RPLC method are its relative simplicity and that it does not need highly pure solutes. At the moment, the correlation remains the main drawback.

3. Countercurrent Chromatography

Equation 1 clearly shows the liquid–liquid partition coefficient is the only solute physicochemical parameter responsible for retention. The partition coefficient can be obtained from the solute retention volume:

$$P = \frac{V_R - V_M}{V_S} = 1 + \frac{V_R - V_t}{V_S} \tag{23}$$

If octanol is the stationary phase and water the mobile phase, the descending mode or head to tail direction is used, and P_{oct} values are obtained from the retention times without any assumptions.

All CCC techniques were used for P, P_{oct}, $\log P$, and $\log P_{oct}$ measurements: droplet CCC [21], toroidal-coil CCC [22], coil-planet CCC [23], and, of course, centrifugal partition chromatography (CPC) [24–35]. The next part of this chapter is dedicated to $\log P$ measurements using CCC, especially CPC.

IV. THE USE OF CCC FOR PARTITION COEFFICIENT DETERMINATION

A. Direct Measurements

The decisive advantage of CCC in P and P_{oct} measurement is that there is no correlation. The P value obtained from Eq. 23 is the liquid–liquid partition coefficient without any assumption. Correlations of the P or $\log P$ values obtained with the same liquid system by the shake flask method and by CCC should produce straight lines with a slope of unity and a nil intercept. This was actually obtained by several authors [22,23,27–29,33]. The validity and solidity of the method was assessed by Gluck and Martin [29] for P_{oct} coefficients.

1. Partition Coefficient Range

Figure 4 shows a 3-D plot of the retention volume *versus* the partition coefficient and the stationary phase volume retained inside the CCC apparatus. It shows the retention volume increases linearly with the P coefficient. The duration of the experiment and/or the ability to detect the solute diluted in large volumes of mobile phase will be the limiting factors for high P value determination. With an apparatus of 125 ml internal volume, a 6 ml/min flow rate, and a 12 hour maximum duration of the experiment, the retention volume is 4300 ml (4.3 L). If the stationary phase retention volume is 60 ml (48% retention ratio), the maximum

LIQUID–LIQUID PARTITION COEFFICIENTS

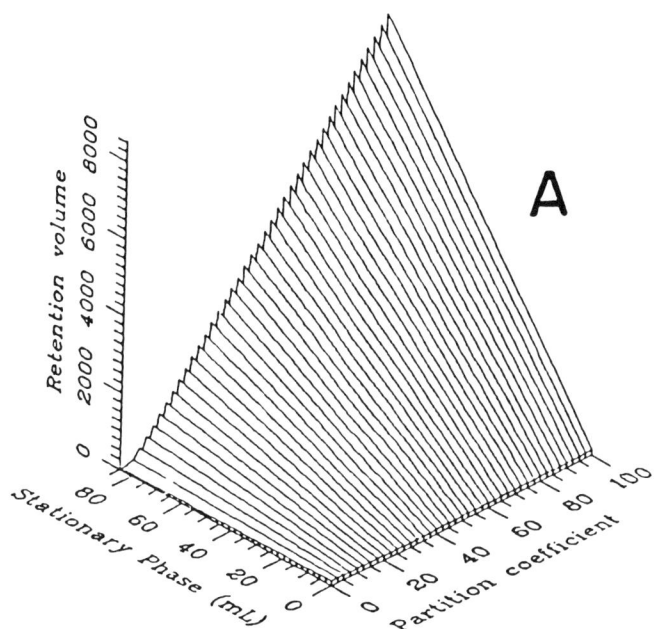

Figure 4 (A) 3-D plot of Eq. 1. The retention volume increases linearly with the P value and with the V_S volume. (B) Isoretention curves corresponding to the 3-D plot in (A). The retention volume can be decreased by reducing the P value, that is, changing the liquid system, or reducing the V_S volume, that is, sacrificing resolution.

measurable P value is 71 (Eq. 23), or $\log P = 1.85$. A P_{oct} value of 71 corresponds to a solute with moderate lipophilicity. It is desirable to be able to obtain much higher P_{oct} values.

The first way to increase the measurable range is to reduce V_s, the stationary phase (octanol) volume retained in the apparatus. If the stationary phase volume is 20 ml (16% retention ratio) instead of 60 ml, the maximum P value becomes 210 instead of 71. To allow the volume ratio of the two phases to be chosen at will, the CCC apparatus can be filled using two pumps, one for each phase. The apparatus is first filled by the water mobile phase; next, the desired volume of octanol is pumped in with the rotor set at the desired rotation speed; finally, the mobile phase flow rate is resumed [23]. Using such a procedure, the P_{oct} range that can be obtained directly by CCC is 0.05–200, or a $\log P_{oct}$ range $-1.3-+2.3$ [23,27]. Figure 5 shows the direct measurement of the P_{oct} value ranging from 4.4 ($\log P_{oct} = 0.64$) to 167 ($\log P_{oct} = 2.23$) of seven solutes injected together in a CPC machine loaded with six cartridges ($V_T = 125$ ml) containing only 22 ml of octanol [27].

Another way to directly measure the P_{oct} value of very lipophilic solutes is to

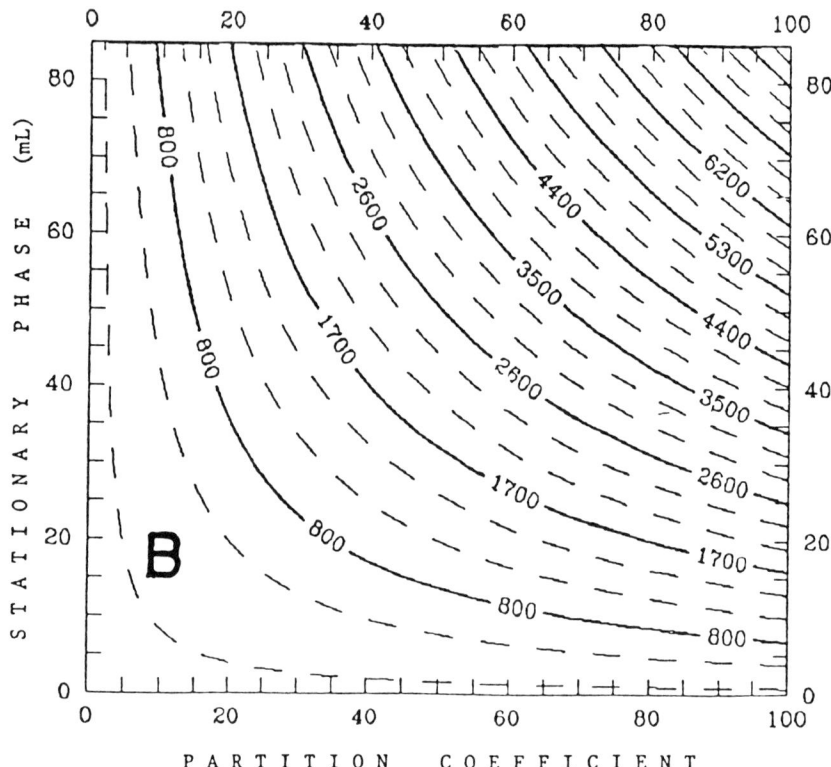

reverse the role of the phases, that is, to use octanol as the mobile phase in the ascending mode and water as the stationary phase. The problem is that the error can become very high.

2. Accuracy and Error Margins

In most experiments, it was found that the retention volumes can be determined with an experimental error margin of 1% with a minimum error volume of 0.5 ml [23,26,29,34]. The error margin can be estimated as follows [26]. Differentiating Eq. 23, one obtains

$$\frac{dP}{dV_R} = \frac{1}{V_S} \quad (24)$$

Using Eqs. 1, 23, and 24, the relative error is expressed by

$$\frac{dP}{P} = \frac{dV_R}{V_R - V_M} \quad (25)$$

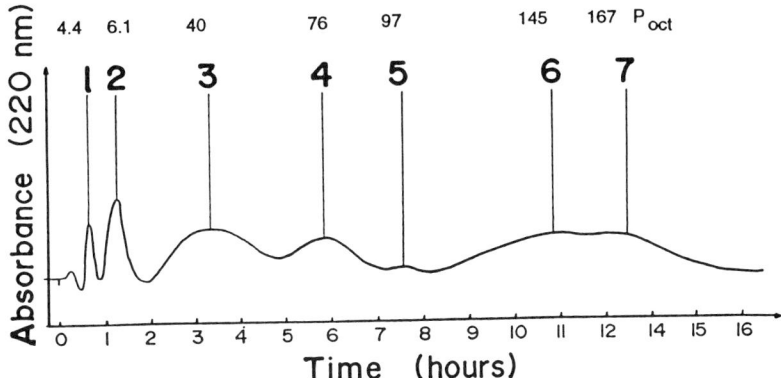

Figure 5 P_{oct} direct determination of seven compounds. CPC LLN machine with six cartridges, $V_T = 125$ ml, 5 ml/min water, pH 4, $21.7 < V_{oct} < 22.7$ ml, 100 µl injection volume of about 2 mg of each solute; **1**: benzamide; **2**: 2-acetoxy benzoic acid; **3**: acetophenone; **4**: benzoic acid; **5**: 2-chlorobenzoic acid; **6**: 2-chlorophenol; **7**: 2-chloronitrobenzene. The small figures are the P_{oct} values. Reprinted with permission from *J. Liq. Chromatogr.*, **11**: 1451 (1988), Figure 2.

The differentials (d) can be replaced by finite differences (Δ):

$$\frac{\Delta P}{P} = \frac{\Delta V_R}{V_R - V_M} \tag{26}$$

Equation 26 shows that the relative error on P increases dramatically as the solute retention decreases. For example, we can use typical volume values of a six-cartridge Sanki CPC unit: 125 ml, 50 ml, and 75 ml as the total internal volume V_T, the dead or mobile phase volume V_M, and the stationary phase volume V_S, respectively. If the retention volume of a solute is 50.6 ml, the partition coefficient is 0.008 (Eq. 23). Taking into account ΔV_R, the experimental error margin of 0.5 ml on the retention volume, the partition coefficient is somewhere between 0.0013 and 0.0143. As indicated by Eq. 26, the relative error is as high as 80%. In the case of reversed phases, octanol is the mobile phase and the calculation is the same, but the obtained P value corresponds to $1/P_{oct}$. A 50.6 ml octanol volume of retention produces a 0.008 P value, that is, $P_{oct} = 125$. The relative error on P is 80%; however, the relative error on P_{oct} is 560% because the P_{oct} value is somewhere between 70 and 750. Of course, such a low accuracy is not acceptable.

If a relative error of 10% is considered as acceptable for P_{oct}, with 0.5 ml as the minimum ΔV_R value, Eq. 26 indicates that the retention volume V_R must be at least 5 ml higher than the dead volume V_0, *whatever the apparatus volume V_T is*. Equation 23 gives the minimum value of P, with $V_R - V_M = 5$ ml:

$$P = \frac{5}{V_S} \tag{27}$$

With the 75 ml V_S volume, the minimum P value is 0.067 with a 10% relative error. The minimum P value is contained between 0.06 and 0.073. If octanol is the mobile phase, the maximum P_{oct} value measurable with a 10% error is $1/0.067 = 15$. This clearly shows that there is no gain in using octanol as the mobile phase to measure a high P_{oct} value. The accuracy drops very rapidly as the retention volume becomes low.

3. Stationary Phase Volume Determination

Equations 23–27 use V_S, the stationary phase volume retained inside the CCC apparatus. This a measured parameter. The complete differentiated form of Eq. 23 is

$$dP = \left(\frac{\partial P}{\partial V_R}\right)_{V_S} dV_R + \left(\frac{\partial P}{\partial V_S}\right)_{V_R} dV_S \tag{28}$$

which can be simply expressed, using Eqs. 23, 25, and 28,

$$\frac{\Delta P}{P} = \frac{\Delta V_R}{PV_S} + \frac{1-P}{P}\frac{\Delta V_S}{V_S} \tag{29}$$

Equation 29 shows that, at constant ΔV_S error, say 0.5 ml, the error on P is low if the stationary phase volume V_S is high. Also, the error on P is lower with high P values than with low p values. This is another point in favor of the use of a water mobile phase to measure high P_{oct} values, which are low $1/P_{oct}$ values with the octanol mobile phase. In this case, it was demonstrated that a low stationary phase (octanol) volume was desirable to reduce the retention time (Fig. 4). Then, V_S should be known precisely with a minimum ΔV_S.

There are several ways to measure V_S. The simplest way is to measure with a graduated cylinder the stationary phase volume displaced by the mobile phase during the CCC apparatus equilibration. This volume corresponds to V_M. Since all the internal apparatus volume is filled by liquids, V_S is simply $V_T - V_M$. Another way is to use a "tracer" solute [26,34]. The tracer molecule can be a solute whose partition coefficient, P_{trac}, in the liquid system is known. Its retention volume V'_R gives V_S:

$$V_S = \frac{V'_R - V_T}{P_{trac} - 1} \tag{30}$$

The tracer solute may be an unretained compound; then

$$V_R' = V_M \quad \text{and} \quad V_S = V_T - V'_R \tag{31}$$

Equation 31 corresponds to Eq. 30 with $P_{trac} = 0$ [26].

LIQUID–LIQUID PARTITION COEFFICIENTS

The interest in the direct measurement of P_{oct} values is that there is no assumption and no correlation [29]. The method, however, cannot be used to determine the P_{oct} value of very lipophilic solutes. The P_{oct} range is only from 0.01 to 200, or a $\log P_{oct}$ range from -2 to $+2.3$. To extend this range, CCC was used with liquid systems differing from a pure octanol and water system. Water–(40% octanol–60% hexane) [27], water–(20% octanol–80% hexane), acetonitrile–hexane [24], hexanol–water, cyclohexane–water [28], chloroform–water, or chloroform–methanol–water [21] liquid systems were used with CCC. This does extend the P_{oct} range up to $P_{oct} = 30{,}000$ ($\log P_{oct} = 4.5$) at the cost of the main advantage of the direct measurement: correlations are needed. Equations such as

$$\log P_{oct} = a \log P_{solvent} + b \qquad (32)$$

are needed to obtain P_{oct} from the measured $P_{solvent}$ value. The problems linked to correlations are apparent: the correlations are excellent within a family of compounds; they become poor if solutes belonging to different families are studied together [21,23,24,27,28]. In this case, RPLC is faster and easier to operate than CCC.

B. Dual-Mode Centrifugal Partition Chromatography: The Back-Flushing Technique

1. Method Description

Working together, Gluck and Martin [30] and Menges et al. [31] proposed the dual-mode, or back-flushing, technique to extend the P_{oct} range measurable by CPC. The idea is simple: Solutes with very high P_{oct} values move very slowly in the octanol phase; they need too long a time to emerge outside the apparatus. To force them out of the CCC apparatus, the role of the aqueous and octanol phases and their flowing direction are reversed after some reasonable flowing time in the normal direction.

Defining v, the volumic speed, in cm^3/s, as the apparatus volume the solute traveled by time unit, the mobile aqueous phase flow rate F and the solute retention volume are related by

$$v = \frac{FV_T}{V_R} \qquad (33)$$

At time t after the injection, the solute has moved by a volume V_t toward the outlet of the apparatus:

$$V_t = vt = \frac{V_T V_{aq}}{V_R} \qquad (34)$$

where V_{aq} is the aqueous mobile phase volume pushed into the CCC machine during time t, $V_{aq} = vt$. The ratio V_t/V_T ($= V_{aq}/V_R$) corresponds to the apparatus volume percentage traveled by the solute at time t. When this ratio equals unity, the solute is detected in the detector. If at time t the phase role and direction are reversed, the mobile phase becomes the octanol phase moving in the ascending mode. Equations 33 and 34 are valid for the octanol flow rate, octanol mobile phase volume, and octanol solute retention volume (i.e., the P value is $1/P_{oct}$). To leave the top of the apparatus, the solute should cover the volume V_t (Eq. 34) with the octanol parameters. Then

$$\frac{V_T V_{oct}}{V_{R,oct}} = \frac{V_T V_{aq}}{V_R} \tag{35}$$

with

$$V_{R,oct} = V_T + \left(\frac{1}{P_{oct}} - 1\right)(V_T - V_S) \tag{36}$$

and the following very simple equation can be formed:

$$P_{oct} = \frac{V_{aq}}{V_{oct}} \tag{37}$$

Table 3 lists the P_{oct} values obtained with the dual-mode methods. Figure 6 illustrates the method with the P_{oct} determination of naphthalene, with 4.8 L of

Table 3 P_{oct} Measurements Using the Dual-Mode Back-Flushing Method

Solute	V_{aq} (ml)	V_{oct} (ml)	P_{oct}	t_R (hours)	$\log P_{oct}$ measured	literature
Phenol	157	4.15	37.8	0.75	1.58	1.47
Benzoic acid	178	2.36	75.4	0.85	1.88	1.87
Benzene	617	2.92	211	2.2	2.32	2.14
2-Chlorophenol	487	2.71	180	2.2	2.25	2.17
Toluene	2,627	5.48	480	9	2.68	2.71
	1,072	2.59	414	4.6	2.62	2.71
Ethyl benzoate	1,169	1.96	596	5	2.78	2.64
	2,133	2.66	802	7.3	2.90	2.64
1-Naphthol	2,924	2.16	1,354	12.3	3.13	2.98
Naphthalene	4,795	1.58	3,035	20	3.48	3.23
Biphenyl	19,595	1.01	19,400	65	4.29	3.80

V_{aq} is the aqueous volume pushed in the descending mode, V_{oct} is the octanol volume pushed in the ascending mode, t_R is the experiment duration in hours. Apparatus: Sanki CPC LLN units loaded with six cartridges, 2400 channels.
Source: Data from Refs. 29 and 31.

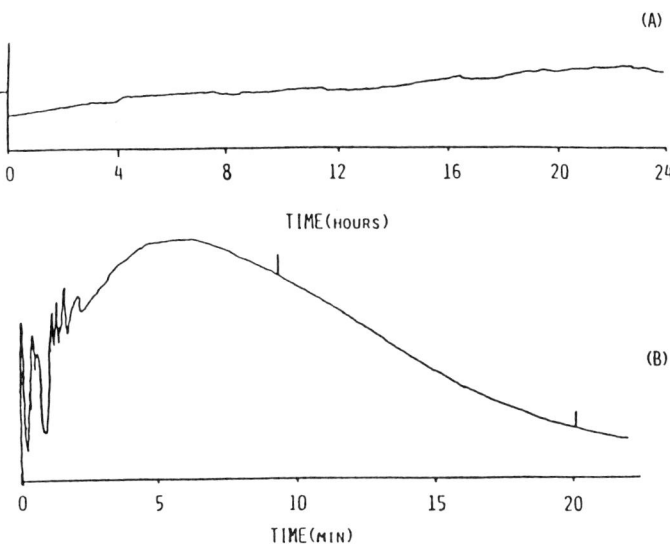

Figure 6 Dual-mode determination of naphthalene P_{oct}. (A) the normal mode 24 hours flush at 4 ml/min descending water flow rate; (B) the phase switching and reverse mode with a 0.3 ml/min octanol flow rate ascending. V_{aq} = 4795 ml, V_{oct} = 1.58 ml, P_{oct} = 3090. Reprinted with permission from *J. Liq. Chromatogr.*, **13:** 3072 (1990), Figure 3.

aqueous V_{aq} in the descending mode and only 1.6 ml of octanol V_{oct} in the ascending mode. P_{oct} is 4800/1.6 = 3000, or $\log P_{oct}$ = 3.48. The highest measured P_{oct} value was 19,000 ($\log P_{oct}$ = 4.3) for biphenyl, but the error was 0.5 log units [30]. The accepted P_{oct} value for biphenyl is 6300 ($\log P_{oct}$ = 3.8).

2. *Error Estimation*

The problem of this method is that the octanol and the water phase volume must rigorously not change during the phase switching. This is very difficult to fulfill. Menges et al. [31] used a three 6-port valve system: one valve to inject, the second valve for mode switching, and a third valve to isolate the whole system when the input line and the pump are rinsed by octanol before mode switching. Figure 6 shows the UV signal after the back-flushing. Perturbations are obvious; they can easily induce a significant error in the V_{oct} measurement. Gluck and Martin [30] used an 8-port valve and two pumps in an uninterrupted process that reduced the perturbations at the phase change. Differentiating Eq. 37, the error is expressed by

$$\frac{dP_{oct}}{P_{oct}} = \frac{\Delta V_{aq}}{V_{aq}} - \frac{\Delta V_{oct}}{V_{oct}} \tag{38}$$

We want to measure high P_{oct} values, which means that V_{aq} is far higher than V_{oct}. For example, a P_{oct} value of 1000 ($\log P_{oct} = 3$) needs an aqueous V_{aq} volume 1000 times higher than the octanol V_{oct} volume. If the V_{aq} volume is 1 L (3.4 hours at 5 ml/min), the V_{oct} volume is only 1 ml. The first term of the error equation (Eq. 38) is always small. The second term is critical. A ΔV_{oct} error of 1 ml is easy to get if some octanol remains as dead volume in any tube or, more likely, if the peak maximum is difficult to locate with the octanol phase. With $V_{oct} = 1$ ml, the dP_{oct}/P_{oct} value is 1 (100%), and the actual P_{oct} value could be 2000 instead of 1000. The V_{oct} error is reduced using a very low flow rate for the octanol phase (0.1–0.3 ml/min). Gluck [30] recommends that the minimum V_{oct} volume be 4 ml. If the maximum reasonable V_{aq} volume is estimated to be 20 L (55 hours at 6 ml/min) of aqueous phase in the descending mode, with the 4 ml octanol volume in the ascending mode, the maximum measurable P_{oct} value is 5000, or $\log P_{oct} = 3.7$. The biphenyl P_{oct} value was obtained with 19.6 L of aqueous phase in the descending mode and only 1.01 ml of octanol in the ascending mode [30], which may explain the 0.5 $\log P_{oct}$ error on this solute. We note that this error with such a highly lipophilic solute is within the error margin obtained using most other methods. More information on the dual-mode back-flushing technique can be found in Chapter 8.

C. Cocurrent Chromatography: The Moving Stationary Phase

The dual-mode method uses the unique property of CCC: the stationary phase is a liquid. Berthod [35] developed another method to determine high P values with CCC using the liquid property of the stationary phase: the cocurrent CCC method. If a lipophilic solute stays too long inside the CCC apparatus, why not push it out, pushing slowly the liquid stationary phase in the same direction as the mobile phase?

1. Theoretical Treatment

Defining F_{oct} and F_{aq} as the octanol "stationary phase" flow rate and the aqueous "mobile phase" flow rate, respectively, Berthod [35] demonstrated that the retention time t_R of a solute was expressed by

$$t_R = \frac{V_{aq} + P_{oct}V_{oct}}{F_{aq} + P_{oct}F_{oct}} \tag{39}$$

The global retention volume corresponds to the liquid flow rate ($F_{aq} + F_{oct}$) multiplied by the retention time expressed by Eq. 39:

$$V_R = (F_{aq} + F_{oct})\left(\frac{V_{aq} + P_{oct}V_{oct}}{F_{aq} + P_{oct}F_{oct}}\right) \tag{40}$$

The retention volume V_R of lipophilic solutes decreases dramatically with F_{oct}. For example, with a CPC apparatus of 125 ml (CPC unit with six cartridges), $V_{oct} =$

40 ml and $V_{aq} = 85$ ml and an aqueous flow rate of 6 ml/min, a $P_{oct} = 1000$ solute has a retention volume of 40 L and a retention time of 4 days and 15 hours. With the cocurrent method, a second pump is used to push slowly the octanol phase with a F_{oct} flow rate of only 0.1 ml/min; Eq. 39 shows that the retention volume drops to 2.3 L and the retention time drops to 6 hours and 20 minutes, a 94% reduction in experiment duration.

Figure 7 shows the changes of the retention volume *versus* its log P_{oct} value. The experimental conditions were a V_T volume of 49 ml (a CPC unit with only two cartridges) with $V_{oct} = 23$ ml and $V_{aq} = 26$ ml, and an aqueous and octanol flow rate of 8 and 0.02 ml/min, respectively. Three areas can be seen in the sigmoid curve: Area I, where the hydrophile solutes are carried by the aqueous phase; Area II, where intermediate polarity solutes are carried by both liquid phases; and Area III, where the lipophilic solutes are carried by the octanol phase. The maximum retention volume change with log P_{oct} corresponds to the selectivity maximum. At this point, the second derivative of Eq. 40 is nil. The first derivative is [35]

$$\frac{dV_R}{d\log P_{oct}} = 2.3 P_{oct}(F_{aq} + F_{oct})\left[\frac{V_{oct}F_{aq} - V_{aq}F_{oct}}{(F_{aq} + P_{oct}F_{oct})^2}\right] \quad (41)$$

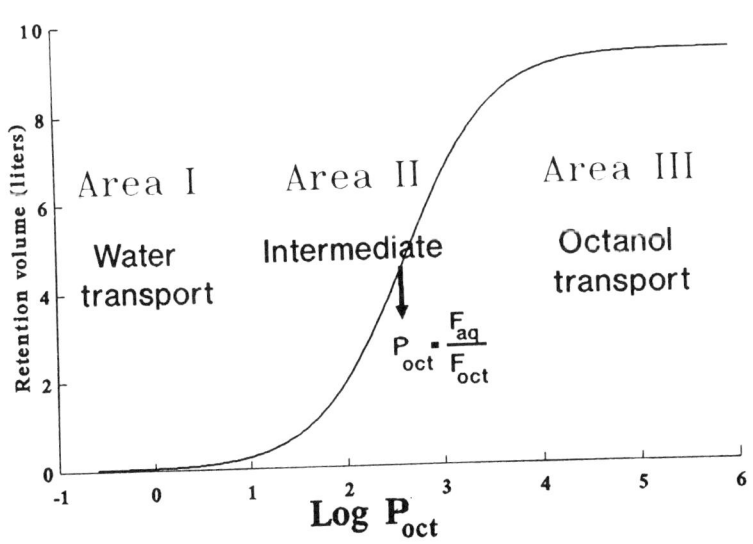

Figure 7 The moving stationary phase, or cocurrent, method for P_{oct} measurements. The arrow corresponds to the selectivity maximum, $P_{oct} = 400$, log $P_{oct} = 2.60$, $V_T = 49$ ml, $V_{oct} = 23$ ml, $F_{aq} = 8$ ml/min, $F_{oct} = 0.02$ ml/min. Reprinted with permission from *J. Liq. Chromatogr.*, **15**: 2771 (1992), Figure 1.

Differentiating Eq. 41 gives the second derivative of Eq. 40:

$$\frac{d^2V_R}{d(\log P_{oct})^2} = 2.3 \frac{dV_R}{d\log P_{oct}} \left(\frac{F_{aq} - P_{oct}F_{oct}}{F_{aq} + P_{oct}F_{oct}} \right) \quad (42)$$

The second derivative becomes nil for the same conditions as the first derivative: $P_{oct} = 0$, $P_{oct} \to \infty$, and $V_{oct}F_{aq} = V_{aq}F_{oct}$. The last term of Eq. 42 becomes nil at the interesting point $P_{oct} = F_{aq}/F_{oct}$, which corresponds to the selectivity maximum. For the Fig. 7 experimental conditions, the ratio F_{aq}/F_{oct} is $8/0.02 = 400$, $\log P_{oct} = 2.6$.

2. Experimental Description

Figure 8 shows the experimental setup used to obtain the P_{oct} value listed in Table 4. The role of pump 1 for the aqueous phase and pump 2 for the octanol phase are straightforward. A third pump was used to alleviate the detection problem. Two nonmiscible phases leave the CPC unit. A clarifying agent (2-propanol) was added to homogenize the eluent and to make continuous UV detection possible. Figure 9A shows the actual chromatogram of the phenanthrene P_{oct} determination ($P_{oct} = 20,000$, $\log P_{oct} = 4.3$). The P_{oct} value is easily computed from Eq. 39 or 40 using the solute retention time t_R:

$$P_{oct} = \frac{t_R F_{aq} - V_{aq}}{V_{oct} - t_R F_{oct}} \quad (43)$$

In all experiments, a low UV signal noise was observed before injection and at the beginning of the chromatogram. The UV signal noise increased as the

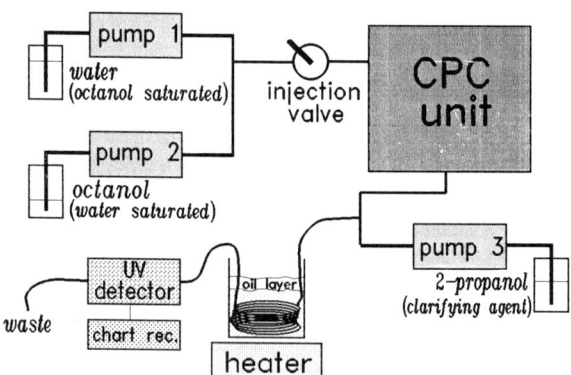

Figure 8 The cocurrent CCC setup. The CPC unit must be a CCC machine retaining firmly the liquid phases. Reprinted with permission from *J. Liq. Chromatogr.*, **15**: 2772 (1992), Figure 2.

LIQUID–LIQUID PARTITION COEFFICIENTS

Table 4 P_{oct} Measurements Using the Cocurrent Method

Solute	F_{oct} (ml/min)	F_{aq} (ml/min)	V_{oct} (ml)	V_R (ml)	t_R (hours)	P_{oct}	$\log P_{oct}$ CPC	$\log P_{oct}$ literature
Phthalimide	0.02	8	25	287 ± 15	0.60	11	1.04 ± 0.03	1.15
p-Hydroxybenzoic acid	0.02	8	25	653 ± 20	1.36	27	1.43 ± 0.02	1.58
Benzoic acid	0.01	4	29.4	1490 ± 30	6.20	57	1.76 ± 0.01	1.87
Benzene	0.01	4	29.7	2530 ± 30	10.51	107	2.03 ± 0.006	2.14 ± 0.02
O-Chlorophenol	0.02	8	25	2720 ± 30	5.63	147	2.17 ± 0.01	2.17 ± 0.02
Toluene	0.01	9	20.2	6520 ± 50	12.08	500	2.70 ± 0.01	2.71 ± 0.02
1-Naphthol	0.02	8	25	7270 ± 50	15.10	1,050	3.02 ± 0.02	2.98 ± 0.04
Naphthalene	0.01	9	20.2	15500 ± 150	28.63	5,100	3.70 ± 0.03	3.23 ± 0.21
Biphenyl[a]	0.02	8	23.0	8700 ± 100	18.10	6,700	3.8 ± 0.1	3.8 ± 0.2
Phenanthrene	0.02	9	22.2	9790 ± 100	18.08	20,000	4.3 ± 0.2	4.4 ± 0.3

[a]Experiment done at only 800 rpm rotation speed.
Apparatus CPC Sanki LLN, V_T = 49 ml, rotation speed = 1000 rpm, two cartridges, 800 channels, descending mode, V_{oct} is the volume of octanol retained inside the two cartridges.
Sources: Data from Ref. 32; log P_{oct} literature values were obtained from the Pomona College of Medicinal Chemistry Log P_{oct} Database.

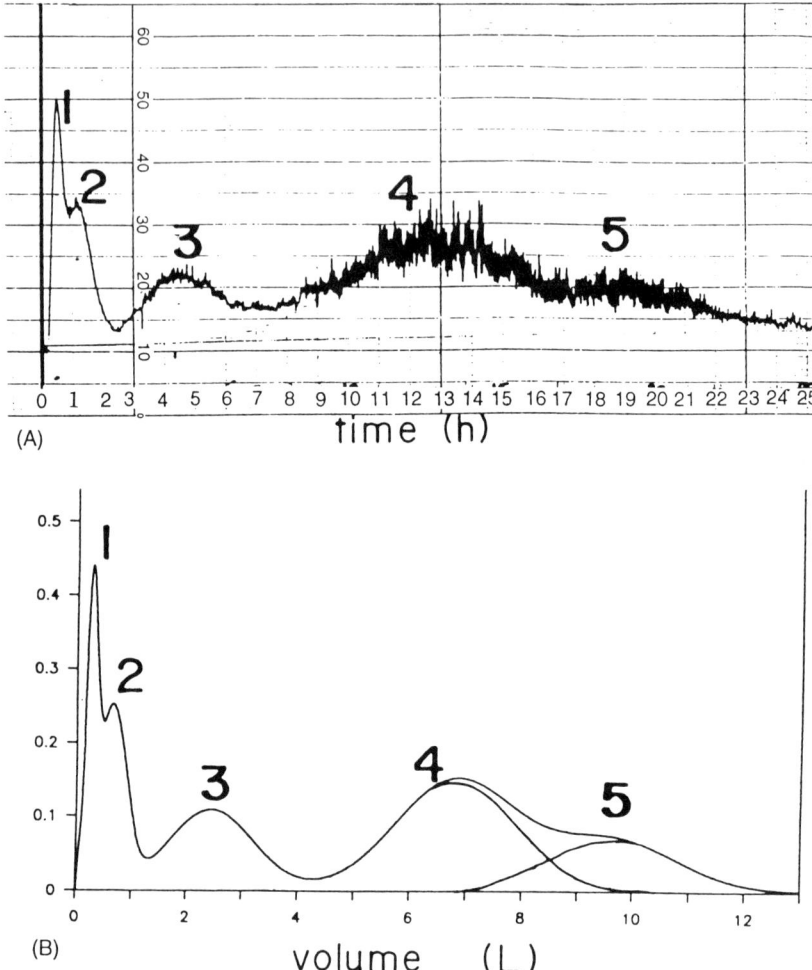

Figure 9 Actual cocurrent chromatogram (A) and the corresponding smoothed chromatogram (B) used for the retention volume measurements. Injected volume: 0.3 ml; **1**: phthalimide (0.9 mg inj.); **2**: p-hydroxybenzoic acid (3.1 mg); **3**: o-chlorophenol (4 mg); **4**: 1-naphthol (2.2 mg); **5**: phenanthrene (1 mg). F_{aq} = 9 ml/min, F_{oct} = 0.02 ml/min, V_{oct} = 26.3 ml, detection UV, 210 nm, 0.16 aufs, postcolumn addition of 2-propanol at 2.7 ml/min. Reprinted with permission from *J. Liq. Chromatogr.*, **15:** 2775 (1992), Figure 5.

chromatogram developed if solutes were present. This noise occurs because the solutes are not evenly distributed in the liquid phase. As the chromatogram develops, the solutes are more and more lipophilic, that is, more and more located in the octanol phase. The octanol phase comes out of the CPC machine in little droplets with some bursts due to pump pulsations or rotor vibrations. The postcolumn addition of 2-propanol and heat (Fig. 8) homogenizes the liquid phase. There are, however, local octanol concentration variations not UV detected when no solute or hydrophilic solutes are present. The local octanol concentration variations are UV detected when a lipophilic and UV absorbing solute is located in the octanol phase. Modern data handling electronic devices alleviate this problem (Fig. 9B) [35].

The interest in the method is that there is no abrupt change—it is continuous. The octanol volume retained in the CPC machine is very stable, more stable than with other methods because there is a constant input of octanol. The octanol volume changes due to dissolution, as noted in the direct method [27,29], or to phase reversal in the back-flushing method [30,31] do not exist with the cocurrent CCC method. The V_{oct} volume was determined using a test solute (2-chlorophenol, P_{oct} = 147 in Ref. 33). Another very important effect that was experimentally observed is the increased peak efficiency due to the octanol flow rate. Figure 6 shows a peak with about 4 plates for the dual-mode back-flushing method when octanol was the mobile phase. Peak efficiencies of 10 plates or less (80 channels per plate or more) were obtained in the direct method (Fig. 5) [27,29,31,32]. Peak efficiencies 20 times higher were obtained in cocurrent CCC [32,33]. Octanol flowing in the back of the channels pushes the solutes, restraining peak tailing [32].

3. Method Precision and P_{oct} Range

The first derivative (Eq. 41) can be used to estimate the method error. Figure 10 is the plot of $d(\log P_{oct})/dV_R$ for six different F_{oct} octanol flow rate values and F_{aq} = 8 ml/min. It shows the error is maximal for solutes in the Area I region (hydrophile solutes) and in the Area III region (lipophilic solutes). The error is minimal for solutes transported by both the octanol and the aqueous phase (Area II region). For $\log P_{oct}$ values higher than 4, a ±0.2 log unit error can be accepted. Equation 41 gives the error in log units per ml. Table 4 lists the phenanthrene retention volume as 9790 ml ± 100 ml. With F_{aq} = 9 ml/min and F_{oct} = 0.02 ml/min, Eq. 41 gives $d(\log K_{oct})/dV_R$ = 0.0019 log units per ml. The error in the $\log P_{oct}$ value of phenanthrene is 100 × 0.0019 = 0.19 log units (Table 4). One should note that the result, 4.3 ± 0.2, for the $\log P_{oct}$ phenanthrene value means the P_{oct} value of phenanthrene is somewhere between 12,500 ($\log P_{oct}$ = 4.1) and 32,000 ($\log P_{oct}$ = 4.5). With a minimum value of 0.01 ml/min for F_{oct} and a maximum value of 9 ml/min for F_{aq}, the maximum measurable P_{oct} value is 40,000, $\log P_{oct}$ = 4.6, in 38 hours and with 20 L of octanol-saturated aqueous phase and 22 ml of water-

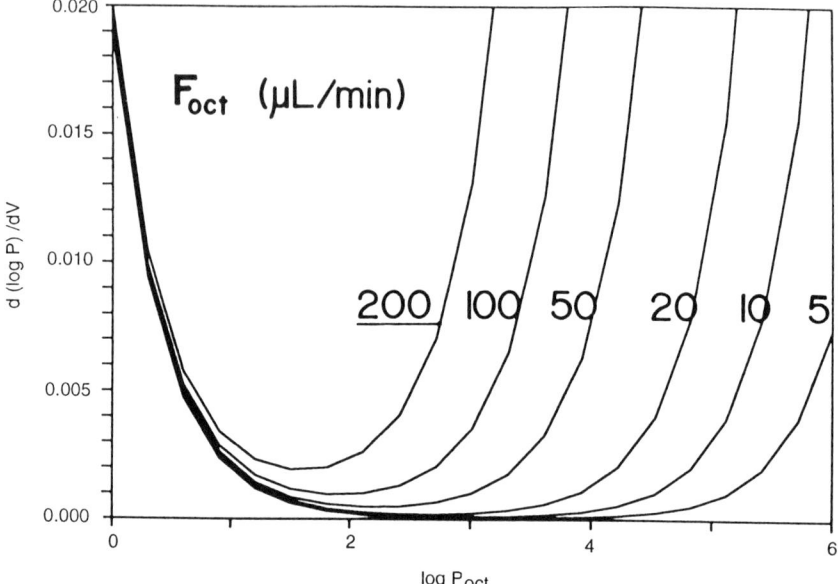

Figure 10 Theoretical curves of the error function plotted versus $\log P_{oct}$ for different octanol flow rates. $F_{aq} = 8$ ml/min. The octanol volume was kept constant: $V_{oct} = 23$ ml. In experimental measurements, it was observed that the V_{oct} volume increased with the F_{oct} flow rate at constant F_{aq} flow rate (Table 4). Reprinted with permission from *J. Liq. Chromatogr.*, 15: 2779 (1992), Figure 6.

saturated octanol phase [33]. The error on the 4.6 value of $\log P_{oct}$ is ± 0.2 log units if the error on the 20 L retention is ± 200 ml.

In the case of an unknown P_{oct} value, a rapid screening can be done at both high aqueous flow rate (8–9 ml/min) and high octanol flow rate (0.1–0.2 ml/min). This will give a low selectivity but a high efficiency (a sharp peak). The retention volume and/or retention time of the solute gives a rough estimation of its P_{oct} value (Eq. 43). Then the octanol flow rate can be reduced to a value slightly higher than F_{aq}/P_{oct} to be close to the selectivity maximum (Eq. 42). A second P_{oct} determination is done with this new flow rate setting to get a higher precision optimizing the duration of the experiment [32,33,35].

D. Coil Planet Centrifuge Machines Versus Centrifugal Partition Chromatographs

High-speed CCC is mainly performed with two types of commercially available devices: The coil-planet centrifuge machines developed by Ito and coworkers [36] and the centrifugal partition chromatographs developed by Sanki [37]. We note that both devices would use the CPC acronym. In this book, the CPC acronym is dedicated to centrifugal partition chromatographs. The coil-planet centrifuge machine is denoted HSCCC, high-speed countercurrent chromatograph [38]. A comparison of the two systems showed that the HSCCC machines gave a much higher efficiency than the CPC machines in direct P_{oct} measurements [32]. The HSCCC machines, however, cannot retain the octanol phase tightly enough to perform both the dual-mode method or the cocurrent method. Only the CPC machines can be used for high P_{oct} measurements with the dual-mode method and/or the cocurrent method [32]. If we take in account the noise generated by the HSCCC machines, the almost silent CPC units are definitely more adapted for P_{oct} determinations than the very noisy HSCCC machines.

To conclude this chapter, Table 5 compares the different methods used to determine the P_{oct} coefficients that were presented. The log P_{oct} range and the main advantages and drawbacks are listed. The methods are used for P_{oct} determinations; they could be used for liquid–liquid partition coefficient determinations in other liquid systems [34]. The ability of the CPC machines to retain strongly a liquid stationary phase (especially octanol) makes them the apparatus of choice if P_{oct} measurements are the main purpose of the work.

Table 5 Comparison of the Different P_{oct} Measurement Methods

Method	log P_{oct} range	Advantages	Drawbacks
Shake flask	−4 to +4	Officially recognized method	Requires highly pure solutes
HPLC	0 to +5.5	Fast, does not need highly pure solutes, possible automation	Poor correlations with dissimilar compounds
CCC direct	−2 to +2.3	No correlation, no need for pure solutes, possible automation	Does not work for lipophilic solutes, time consuming
CCC dual-mode[a]	+1 to +4	No correlation, no need for pure solutes, recycling possible	Possible error when switching modes, very long experiment duration, low efficiency
CCC cocurrent[a]	−1 to +4.6	No correlation, continuous method, high efficiency, adjustable experiment duration	No recycling possibility, detection problems

[a]Both the dual-mode and cocurrent method can be performed only with hydrostatic units capable of tightly holding the octanol phase.

ACKNOWLEDGMENT

The author thanks the continuous support of the Centre National de la Recherche Scientifique, UA 435.

REFERENCES

1. V. Golg, K. L. Loenig, A. D. McNaught, and P. Sehmi, *Compendium of Chemical Technology*, IUPAC 87B Recommendations, Blackwell Scientific Publications, Oxford (1987).
2. J. H. Hildebrand, J. M. Prausnitz, and R. L. Scott, *Regular and Related Solutions*, Van Nostrand Reinhold, Princeton (1970).
3. L. R. Snyder, *Principle of Adsorption Chromatography*, Marcel Dekker, New York (1968).
4. C. Reichardt, *Solvents and Solvents Effects in Organic Chemistry*, VCH Publishers, Weinheim (1988).
5. D. W. J. Grant and T. Higushi, *Solubility Behavior of Organic Compounds*, Wiley Interscience, New York (1990).
6. J. Rydberg, C. Musikas, and G. R. Choppin, *Principles and Practices of Solvent Extraction*, Marcel Dekker, New York (1992).
7. R. F. Rekker, *The Hydrophobic Fragmental Constant*, Elsevier, Amsterdam (1977).
8. A. Leo, C. Hansch, and D. Elkins, *Chem. Rev.*, **71**: 525 (1971).
9. C. Hansch, *Accounts Chem. Res.*, **2**: 232 (1969).
10. Rule and Regulations, *Federal Register*, Vol. 50, **188**: 39252 (Sept. 1985).
11. *Code of Federal Regulation*, **40**: CFR 769.1550 (1990).
12. S. H. Yalkowski and S. H. Valvani, *J. Pharm. Sci.*, **69**: 912 (1980).
13. R. W. Taft, M. H. Abraham, G. R. Famini, R. M. Doherty, and M. J. Kamlet, *J. Pharm. Sci.*, **74**: 807 (1985).
14. S. H. Yalkowski, R. Pinal, and S. Banerjee, *J. Pharm. Sci.*, **77**: 74 (1988).
15. L. P. Hamett, *Chem. Rev.*, **17**: 125 (1935).
16. C. Hansch, P. P. Maloney, T. Fujita, and R. M. Muir, *Nature*, **194**: 180 (1962).
17. S. E. Krikorian, T. A. Chorn, and J. W. King, *Quant. Struct.-Act. Relat.*, **6**: 65 (1987).
18. K. Miyake and H. Terada, *J. Chromatogr.*, **240**: 9 (1982).
19. J. E. Garst and W. C. Wilson, *J. Pharm. Sci.*, **73**: 1616 (1984).
20. D. J. Minick, D. A. Brent, and J. Frenz, *J. Chromatogr.*, **461**: 177 (1989).
21. F. Gago, J. A. Builla, and J. Elguero, *J. Chromatogr.*, **360**: 247 (1986).
22. R. S. Tsai, N. El Tayar, B. Testa, and Y. Ito, *J. Chromatogr.*, **538**: 119 (1991).
23. P. Vallat, N. El Tayar, B. Testa, I. Slacanin, A. Martson, and K. Hostettmann, *J. Chromatogr.*, **504**: 411 (1990).
24. H. Terada, Y. Kosuge, N. Nakaya, W. Murayama, Y. Nunogaki, and K. I. Nunogaki, *Chem. Pharm. Bull.*, **35**: 5010 (1984).
25. H. Terada, Y. Kosuge, N. Nakaya, W. Murayama, Y. Nunogaki, and K. I. Nunogaki, *J. Chromatogr.*, **400**: 343 (1987).
26. A. Berthod and D. W. Armstrong, *J. Liq. Chromatogr.*, **11**: 1187 (1988).

27. A. Berthod, Y. I. Han, and D. W. Armstrong, *J. Liq. Chromatogr.*, **11**: 1441 (1988).
28. N. El Tayar, A. Martson, A. Bechalany, K. Hostettmann, and B. Testa, *J. Chromatogr.*, **469**: 91 (1989).
29. S. J. Gluck and E. J. Martin, *J. Liq. Chromatogr.*, **13**: 2529 (1990).
30. S. J. Gluck and E. J. Martin, *J. Liq. Chromatogr.*, **13**: 3559 (1990).
31. R. A. Menges, G. L. Bertrand, and D. W. Armstrong, *J. Liq. Chromatogr.*, **13**: 3061 (1990).
32. A. Berthod and V. Dalaine, *Analusis*, **20**: 325 (1992).
33. A. Berthod, R. A. Menges, and D. W. Armstrong, *J. Liq. Chromatogr.*, **15**: 2769 (1992).
34. A. Berthod and M. Bully, *Anal. Chem.*, **63**: 2508 (1991).
35. A. Berthod, *Analusis*, **18**: 352 (1990).
36. Y. Ito, *CRC Crit. Rev. Anal. Chem.*, **17**: 65 (1986).
37. W. D. Conway, *Countercurrent Chromatography*, VCH Publishers, New York (1990).
38. A. P. Foucault, *Anal. Chem.*, **63**: 569A (1991).

8
Centrifugal Partition Chromatography for the Determination of Octanol–Water Partition Coefficients

Steven J. Gluck
Dow Chemical, Midland, Michigan

Eric Martin
Chiron Corporation, Emeryville, California

Marguerite Healy Benko
DowElanco, Indianapolis, Indiana

I. INTRODUCTION TO K_{ow} MEASUREMENTS

Octanol–water partition coefficients (K_{ow}) have been established as the most significant quantitative physical property correlated with biological activity [1,2]. In this capacity, they have found extensive use in drug and pesticide design as a parameter for quantitative structure activity relationships (QSAR) [3,4]. In addition, $\log K_{ow}$ is used to predict bioconcentration factors in aquatic organisms, water solubility, and soil adsorption coefficients [5]. Because of these relationships between K_{ow} and environmental parameters, state and federal agencies in the United States require the accurate determination of K_{ow} by prescribed methodologies for product registration [6]. These procedures, commonly referred to as the shake flask and generator-column methods, are sensitive to impurities and thus require a selective detection scheme such as high-performance liquid chromatography (HPLC). They are expensive because of the effort required to assure an accurate measurement and the development of HPLC procedures for each compound. They are not amenable to automation for a set of unrelated compounds. Hence, use of the accurate government-approved methodology is not practical for routine K_{ow} determination for QSAR purposes.

Alternative means have therefore been developed for estimating K_{ow}.

Additive molecular fragment approaches have been developed, which have the advantage that they can be used on compounds for which no sample is available [5, 7,8]. Extrathermodynamic relationships have also been found that correlate $\log K_{ow}$ with the log of reversed phase liquid chromatography capacity factors using empirically fitted but theoretically unexplained constants. (There are more than 100 publications that attempt to correlate HPLC capacity factors with K_{ow}.) Besides ease of automation, HPLC has the important advantage of insensitivity to impurities and small sample requirements. Both of these approaches work best for estimating K_{ow} of structural homologs, however, and work poorly for structurally unrelated compounds [9–16]. In general, these approaches do not have the accuracy desired for quantitative work because they only mimic true octanol–water partitioning. They are not direct measurements of K_{ow}.

Centrifugal partition chromatography (CPC) using octanol and water as the two phases is a useful alternative for providing octanol–water partition coefficients [17–23]. Despite some confusing references [24,25], CPC has been demonstrated to have an operating range of $\log K_{ow}$ up to 4.3. In CPC, the retention mechanism is the partitioning of a solute between octanol and water. It offers the automation advantages, small sample size, and insensitivity to impurities of the HPLC procedures with the potential accuracy of the shake flask method. Three approaches to determining K_{ow} by CPC are described in this chapter.

II. NORMAL MODE

That octanol and water are the CPC chromatographic phases does not in itself guarantee a direct measurement. The system must be proven to be in equilibrium and the fundamental chromatographic relationships on which the partition coefficient calculations are based must be shown to hold. In a feasibility study, Berthod and Armstrong [18] used CPC to successfully determine K_{ow} for 17 compounds out of a group of 55 compounds with $\log K_{ow}$ values up to 5, selected and supplied by Gluck and Martin. The method was claimed to be direct because K_{ow} was correctly determined from V_r and the octanol stationary phase (V_o) and aqueous mobile phase V_w) volumes using chromatographic theory. This claim was, however, ambiguous. An assumed, unverified relationship was used to determine V_o, and then the same relationship was used to determine K_{ow} of other compounds. Although they clearly showed a linear, empirical relationship between K_{ow} and V_r, this procedure could have been merely another extrathermodynamic chromatographic correlation where "V_o" and "V_w" were theoretically meaningless fitted constants with no relationship to the actual phase volumes. Any linear function would have given the appearance of a direct measurement. In that case, applying the method to diverse compounds outside of the training set would be far less secure. The assumed relationship must be assessed to confirm whether this procedure is indeed a direct method. If proven to be direct, it can be assumed to give accurate K_{ow} determinations for structurally diverse compounds.

It is useful to ascertain what procedures are required to obtain a desired precision. Replicate runs of a sample do not suffice, because the result is also dependent on the precision of the determination of system parameters such as V_o and V_w. A propagation of errors analysis is needed to determine precisions over the range of application [20].

A. Experimental Procedures

1. Apparatus

The system consisted of a CPC, Model CPC-LLN rotor (Sanki Laboratories, Mount Laurel, New Jersey). Mobile phase flowed from a 10-L reservoir through the Sanki Laboratories Model LBP-V pump into a Hitachi Model 655A-40 autosampler. Analyte partitioning took place in the centrifugal rotor, which was thermostated at 20°C. Six rotor cartridges were used, corresponding to 2400 individual extraction stages of approximately 50 μl each. Compounds were detected with a Kratos Model 757 variable-wavelength UV absorbance detector equipped with a preparatory, short pathlength, flowcell. The UV detector was followed with a Molytek Thermalpulse flow rate monitor. The absorbance and the flow rate outputs were monitored with the PE Nelson ACCESS*CHROM chromatography data system (Fig. 1).

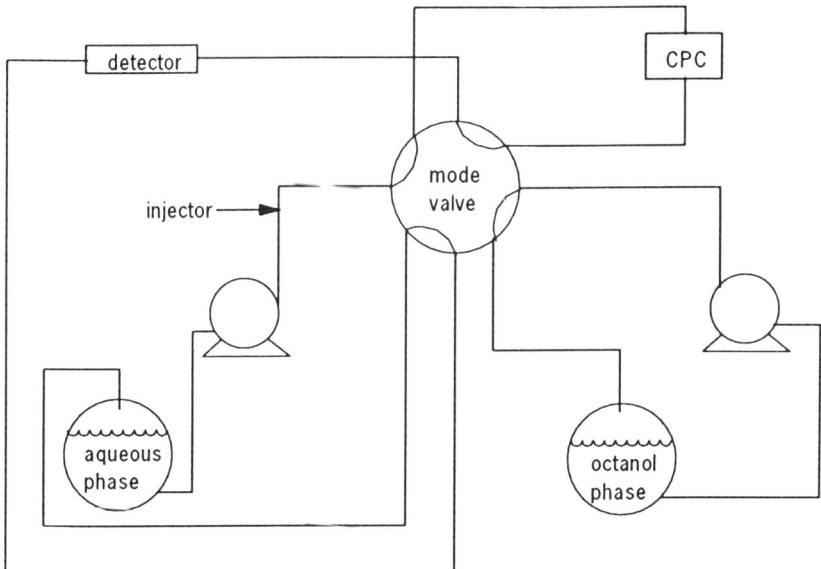

Figure 1 Block diagram of CPC system used for normal and dual–mode K_{ow} determinations.

2. Filling the Cartridges with Stationary Phase

Isopropanol was used to flush the system before 150–250 ml of water-saturated octanol was pumped through the system to completely fill the cartridges with octanol. While the rotor was spinning at 300 rpm, octanol-saturated water was pumped in the reverse-ascending mode as fast as possible without overpressurizing the system, until the first drop of water appeared at the detector exit as indicated by both the appearance of an emulsion and a large shift in the absorbance output. Pumping in this direction is supposed to empty the system of stationary phase, but in practice some remains, thus providing a smaller fraction of stationary phase and correspondingly shorter retention times. This operation has been referred to as "underloading" [19]. The mode valve was then switched to the reverse-descending position, and the water flow continued for 20–30 minutes at 8–20 ml/min. Under these high flow rates and low rpm conditions, more of the stationary phase is removed. Finally, the rotor speed was increased to 700 rpm and the flow rate was decreased to 4 ml/min. The baseline would usually stabilize within 20 minutes. Typical system pressure was 20 kg/cm^2.

3. Retention Volume Determination

The V_r of an analyte was determined by its retention time and either the integrated flow rate of the eluent with respect to time or the average flow rate multiplied by the time. Extensive comparisons between both procedures showed that the simple method of multiplying the average flow rate by retention time gave better reproducibility than integration of the flowmeter output in the retention volume measurements. The flowmeter was not well suited to the measurement of saturated solutions.

Analyte retention time was determined manually as the peak maximum using the postrun plotting capabilities of the PE Nelson ACCESS*CHROM software on a VAX-based data system. Data were collected at a 1-point-per-second rate. Run times varied from 25 minutes to 16 hours, depending on the K_{ow} of the analyte.

4. Standards and Analytes

The standard mixture consisted of 5% phenol, 8% benzyl alcohol, 1% benzamide, and 2% dimethyl formamide in octanol. The detector wavelength was 255 nm. The analytes consisted of a mixture of acetone, acetanilide, acetophenone, and 2-chlorophenol in octanol. These compounds were detected at 227 nm. Repetitive 100-μl injections of the standards were performed by the autosampler. Data collection was started at the time of injection by the autosampler. In this mode, the system can run for weeks without maintenance. Literature values for all compounds were selected as Log P Star values from the Pomona College Medicinal Chemistry database [26]. Typical sample masses injected ranged from 0.5 to 5 mg.

B. Theory

The determination of K_{ow} from V_r is based on the fundamental chromatographic relationship

$$V_r = V_o P + V_w \tag{1}$$

where P is the partition coefficient [17]. If octanol is the stationary phase and water is the mobile phase, then P is K_{ow}.

Knowing the system parameters V_w and V_o, K_{ow} is obtained as a linear function of the only compound-dependent variable, V_r:

$$K_{ow} = \frac{V_r - V_w}{V_o} \tag{2}$$

Several approaches have been reported for determining V_w and V_o. Terada [16] determined V_w by measuring the amount of stationary phase displaced by mobile phase into a graduated cylinder during start-up. The V_o was calculated from the difference between V_w and the independently known instrument constant V_t:

$$V_o = V_t - V_w \tag{3}$$

P was then calculated from Eq. 1 (Terada did not use the octanol–water solvent system).

Berthod and Armstrong [17,18] determined V_o and V_w from V_r for a standard compound with a known K_{ow}, using Eqs. 2 and 3 along with the independently determined value for V_t.

Our approach is to determine V_o and V_w from a set of four compounds of known K_{ow} covering a wide range of values. From Eq. 1, the slope of a plot of V_r versus literature K_{ow} is V_o, and the intercept is V_w. Since the uncertainty in V_r increases with increasing K_{ow}, weighted regression is used to determine the slope and intercept [27]. The weights are reciprocal variances from 33 identical injections of the calibration mixture (see Section II. C). System calibration may be repeated several times for improved accuracy. After establishing V_o and V_w, Eq. 2 is used again to determine K_{ow} of the unknowns. Since Eq. 3 is not used in our procedure, it serves as an internal check for consistency in V_w and V_o. Furthermore, separate experiments described below show that this slope and intercept correspond with independently determined values for V_o and V_w.

The precision of the K_{ow} determination for an unknown compound can be assessed by analyzing the propagation of random error. The uncertainty in determining $\log K_{ow}$ from Eq. 2 is given by [28]

$$\lambda(\log K_{ow}) = 0.4343 \sqrt{\frac{\lambda^2 V_r + \lambda^2 V_w}{(V_r - V_w)^2} + \left(\frac{\lambda V_o}{V_o}\right)^2} \tag{4}$$

where λ indicates the 95% confidence interval.

Estimates for the standard deviations in $V_o(sV_o)$ and $V_w(sV_w)$ are calculated from the standard deviations of the slope and intercept from the weighted regressions of the 33 identical injections of the standard mixture described above. λV_w and λV_o for a new experiment are then calculated from those standard deviations along with t for the total number of observations in the calibration runs for the current experiment minus 2 (two degrees of freedom are required to determine a line):

$$\lambda V_w = t \cdot sV_w \tag{5}$$

$$\lambda V_w = t \cdot sV_w \tag{6}$$

The 95% confidence interval for the mean value of $V_r(\lambda V_r)$ is calculated from the standard error of the mean for V_r of the compound to be tested:

$$\lambda V_r = \frac{t \cdot sV_r}{\sqrt{n}} \tag{7}$$

where n is the number of replicate injections of the test sample, and t is for 95% confidence at $n - 1$ degrees of freedom. This requires an estimate for the standard deviation of $V_r(sV_r)$ for the unknown. An empirical estimate of sV_r as a function of K_{ow} is described below.

C. Results and Discussion

This apparatus and procedure were designed for the unattended, automatic measurement of standards and samples. Hence, the eluent is recycled and a data system has been employed to collect raw data for manual postrun manipulations. Ideally, the data system could determine the V_r of each compound automatically, but the high viscosity of the octanol stationary phase resulted in broad peaks for the more lipophilic compounds. This low efficiency leads to poor signal-to-noise ratios for these compounds, and the chromatography data algorithms have trouble picking the peak maximum. In addition, small amounts of stationary phase occasionally collect slowly and release rapidly in the detector flowcell, resulting in significant baseline shifts. These shifts, combined with wide peaks and occasional low signal-to-noise ratios, mandate a manual determination of V_r.

1. System Calibration and Characterization

The system's precision was studied by injecting the set of four calibration compounds 33 consecutive times, once every 334 minutes. A typical chromatogram is shown in Fig. 2. The means and standard deviations from those values of V_r are in Table 1. The V_o was determined for each individual injection by the weighted regression procedure (see below) and was fitted versus the cumulative volume pumped. The slope of this linear regression fit showed the volume of stationary phase increased by 0.6 ml (2.6%) over the course of the week-long experiment.

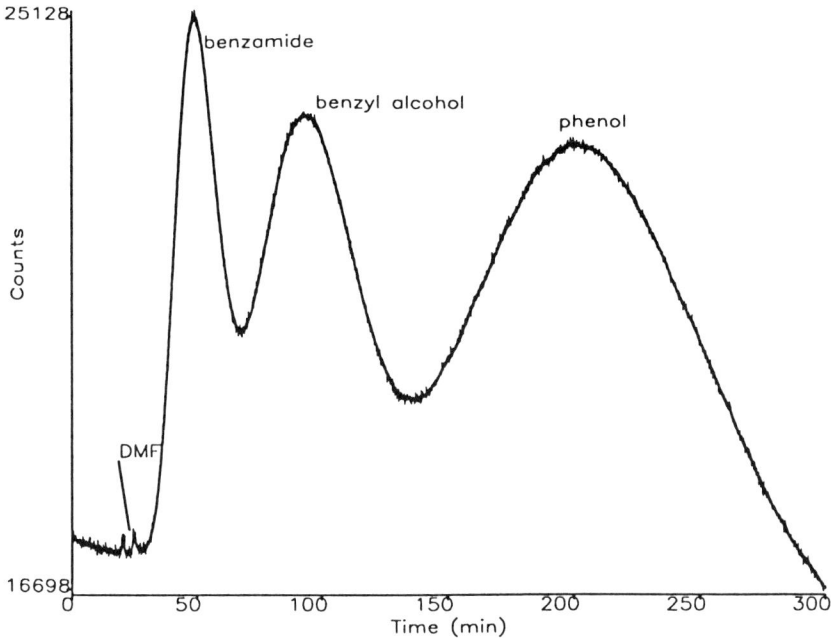

Figure 2 Typical chromatogram of the calibration mixture.

This very slight increase of 3.3 µl/hour is actually less than the expected increase of 15 µl/hour due to 84 µl of octanol added with each injection. This indicates a slow bleed rate of 12 µl/hour, only 7% of that in a previous report [17,19]. This difference might be due to a lower operating pressure, better octanol presaturation of the mobile phase from recycling the solvent, or temperature-related phenomena. Since the autoinjector reruns the standard mixture every 2 or 3 days, these small volume changes can easily be monitored, and the instrument can be operated for weeks without recharging the centrifuge. This is an important reason for

Table 1 Means and Standard Deviations of the Retention Volumes[a] for 33 Replicate Injections of a Standard Mixture

Parameter	Dimethyl formamide	Benzamide	Benzyl alcohol	Phenol
Mean	97.2	194	370	788
Standard deviation	0.5	1.4	4.6	14.3
Literature K_{ow}	0.0977	4.36	12.6	29.5

[a]All retention volumes are given in milliliters (ml).

determining V_o and V_w in situ by the multiple-point, weighted regression procedure [20] rather than using volumetric [16] or single-point [17,18] methods when the instrument is loaded or unloaded.

2. Weighted Least Squares Regression

A plot of V_r versus K_{ow} for the combined data from 33 injections of the calibration set (Fig. 3) shows the basic linear relationship of Eq. 1. The scatter in V_r increases with K_{ow}. Since ordinary least squares regression assumes that error in the y values are constant, this variation of sV_r with K_{ow} dictates the use of weighted regression.

3. Test Samples

Four compounds, acetone, acetanilide, acetophenone, and 2-chlorophenol, were run as a set of unknowns. Before these compounds were injected, the system was calibrated by four injections of the standard mixture. Table 2 contains V_r, literature $\log K_{ow}$, and CPC $\log K_{ow}$ for each test compound. All the CPC K_{ow} values agree well within the targeted ±0.1 log units of the literature K_{ow} values.

4. Demonstration of a Direct Method

The validation of this procedure as a direct determination of K_{ow}, that is, one with no theoretically unexplained fitted parameters, is a key aspect of this study. Indirect procedures such as HPLC, Terada's CPC method [16], and Tayar's CPC method [29], which rely on empirical correlations between octanol–water parti-

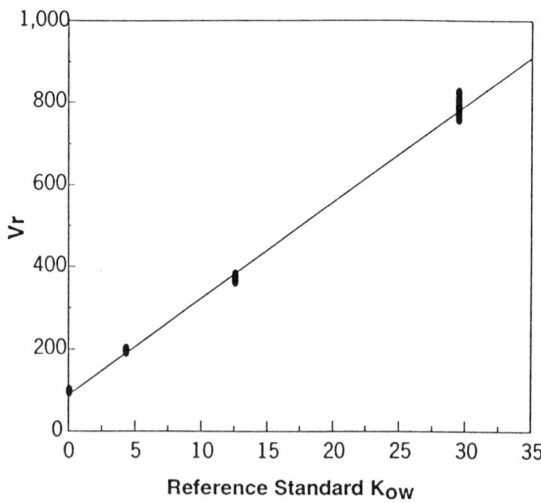

Figure 3 Retention volume versus K_{ow} for 33 injections of the calibration mixture.

Table 2 Test Compounds Determined by CPC Using the Weighted Regression Procedure

Compound	$\log K_{ow}$ literature	CPC	Residual	V_r (ml)
Acetone	−0.24	−0.31	0.07	107.0
Acetanilide	1.16	1.16	0.00	394.5
Acetophenone	1.58	1.59	0.01	890.8
2-Chlorophenol	2.15	2.11	0.04	2716

The V_o was 20.24 ± 0.40 ml, V_w was 97.57 ± 0.20 ml, and V_t was 117.8 ± 0.45 ml, as determined by four injections of the calibration mixture.

tioning and other partitioning phenomena can yield erroneous results, particularly when applied to structurally diverse compounds.

The fundamental relationship (Eq. 1) assumes that the system is in equilibrium. If the system were poorly mixed, short-circuit fluid flow or channeling could occur [30], which would decrease the effective V_w and/or V_o. Increased mixing or lower flow rates would then be expected to increase V_r.

Three experiments were performed to confirm that the method gives direct values of K_{ow}. In the first experiment, the aqueous phase flow rate was varied from 0.4918 to 7.578 ml/min for running the calibration set. In each case, the V_o, V_w, and V_t were identical, independent of flow rate. Indeed, after the experiment, the V_o and V_w determined by flushing the apparatus were the same. A similar experiment was performed at constant flow rate (4.5 ml/min) but over a range of rotor rpm gave V_r that were independent of the centrifugal force induced by the spinning rate. In the third experiment, V_o and V_w were experimentally set at different values but their sum, V_t, was always the same [21].

Together, these experiments demonstrate that the system is in equilibrium throughout the practical range of operating conditions. Furthermore, the slope and intercept of the calibration line relating V_r to K_{ow} do indeed correspond to the physical system parameters described in Eq. 1, as determined by independent analytical means. Indeed, as a direct method, it is unnecessary to do a correlation study; normal mode CPC-derived K_{ow} values are the same as shake flask-derived K_{ow} values.

5. Propagation of Errors

A goal of this project was to develop a CPC procedure to determine $\log K_{ow} \pm 0.1$ at 95% confidence throughout the range of 0−+2.5. Values below zero are best measured in normal-ascending mode (with octanol as the mobile phase) and might be expected to have uncertainties comparable with those of their positive counterparts. Propagation-of-error analysis helps to clarify the experimental protocol required to achieve this precision. The relevant equations (Eqs. 4–7) were presented in the Section II. B. With these equations, the expected uncertainty in measured K_{ow} can be calculated as a function of actual K_{ow} throughout the desired range.

Experimental values for sV_w, sV_o, and sV_r (as a function of V_r) are required to solve Eqs. 5–7. From the weighted regressions for 33 replicate injections of the standard mixture, the standard deviation of the slope and intercept (sV_o and sV_w) were found to be 0.098 and 0.99, respectively. Plotting sV_r versus the mean of V_r for each of the four calibration standards revealed a simple linear correlation:

$$sV_r = 0.0206V_r - 2.295$$
$$n = 4, \quad r^2 = 0.988, \quad s = 0.836, \quad F = 169.7 \quad (p = 0.004) \qquad (8)$$

Thus, Eqs. 5–8 can be used to calculate the expected random experimental error in a CPC $\log K_{ow}$ determination as a function of actual $\log K_{ow}$ of an unknown. Examination of Eq. 4 shows that for compounds with very low K_{ow}, which are barely retained, so $V_r - V_w$ is small, the relative uncertainty in K_{ow} will be large. This limits the low range of K_{ow} that can be measured with water as the mobile phase. Equations 7 and 8 show that sV_r and therefore λV_r increase linearly with V_r (and therefore K_{ow}). Thus, both the numerator and the denominator of Eq. 4 increase at the same rate when V_r is much larger than V_w, and the relative uncertainty in K_{ow} approaches a constant value for lipophilic compounds. There is a minimum at intermediate K_{ow} values where λV_r is small but V_r is still substantially greater than the dead volume, V_w. All uncertainties depend on the t values and therefore on the number of replications of each part of the experiment. Figure 4 shows the expected error in CPC-determined $\log K_{ow}$ as a function of actual $\log K_{ow}$ for an experiment with a single injection of the four-compound calibration mixture and duplicate injections of each unknown. This procedure is seen to be sufficient to achieve the desired precision throughout the $\log K_{ow}$ range of -0.5 to 2.5.

D. Conclusions

An automated CPC technique was developed for determining $\log K_{ow}$ in the range from -0.5 to 2.5. The procedure was a direct measurement of K_{ow} with no empirically fitted parameters. A multipoint in situ calibration employing weighted regression was more accurate than the single-point method for determining the phase volumes [16–18], which is critical for accurate K_{ow} determination. Propagation of error showed that a single four-component calibration followed by duplicate injections of the unknown provides, at 95% confidence, uncertainty of less than $0.1 \log K_{ow}$ units over the accessible range.

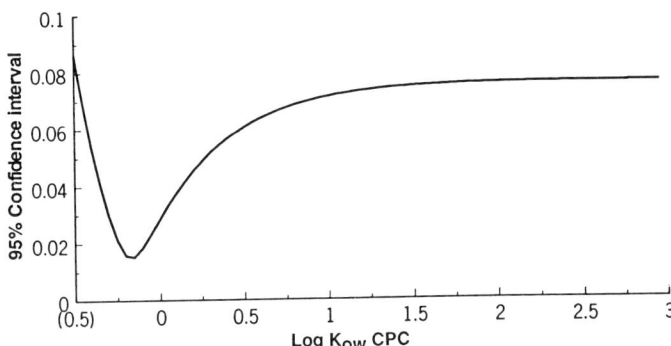

Figure 4 Error in $\log K_{ow}$ at 95% confidence ($\lambda \log K_{ow}$) versus $\log K_{ow}$ determined.

The main practical shortcoming of this procedure is the limitation to $\log K_{ow}$ values less than 2.5. Therefore, the development of the dual-mode and cocurrent methods have followed as described in the next sections.

III. DUAL-MODE METHOD

A. Introduction

A solution to the problem of limited range for normal mode CPC–determined values of K_{ow} is to apply a common operation used in CPC, called dual-mode CPC [19–22,31–33]. After a predetermined time of pumping in the descending mode, the mode is switched to ascending and the compounds of interest elute in the octanol. The advantage of dual-mode CPC is that the analyte only has to move a small distance through the CPC for a partition coefficient to be determined. Therefore, the determination range of partition coefficients can be extended. In addition, since the analyte remains largely in a small volume of the octanol phase, it is easily detectable.

B. Theory

From Eq. 1, K_{ow} is shown to be fully explained by the basic chromatographic theory relating the V_r to K_{ow}, V_w, and V_o. Therefore, the fractional distance (X, where $0 \leq X \leq 1$) an analyte migrates through the CPC at a volume of water pumped in the descending mode (V_d) is

$$X = \frac{V_d}{K_{ow}V_o + V_w} \qquad (9)$$

When the mode is switched from descending to ascending, the analyte moves in the reverse direction, and from Eqs. 1 and 9 the retention volume in the ascending mode solvent, octanol (V_{ra}), is

$$V_{ra} = \frac{XV_w}{XV_o + K_{ow}} \qquad (10)$$

K_{ow} can thus be deduced from Eqs. 9 and 10 to be independent of every parameter except V_d and V_{ra}:

$$K_{ow} = \frac{V_d}{V_{ra}} \qquad (11)$$

This derivation assumes that V_o and V_w remain constant. Indeed, V_o and V_w do change, but at a rate of 0.20 µl/min at a 4.0 ml/min aqueous flow rate and 700 rpm [20]. Bleed is sequential, however; that is, the stationary phase from the last chamber leaves the rotor first. In practice, the analyte never comes into contact with the bled phase that has left the rotor.

C. Experimental Procedures

1. Apparatus and Method

The system is set up as shown in Fig. 1. For safety concerns and convenience, solvents were presaturated in a hood using a thermostated, jacketed 3-L mixing apparatus. This consisted of a round bottom, three-neck, 3-L flask with a paddle-type electric stirrer and a Teflon stopcock affixed to the bottom of the flask. The phases were separated and kept in similar thermostated flasks below the mixing apparatus (Fig. 5). This apparatus minimized the handling of solvents and assured well-saturated, temperature-controlled phases.

Typically, a 5-µl aliquot of 1–50% analyte in octanol is injected from the Hitachi model 655A-40 autosampler into the octanol-saturated water stream flowing at 4.89 ml/min from the Sanki model LBP-V pump. The flow passes through the eight-port mode valve (Valco Instruments) and into the CPC in the descending mode at 600 rpm. The aqueous phase is recycled. After a predetermined descending flow volume, the mode valve is switched via the data acquisition/control system (Perkin Elmer/Nelson Analytical ACCESS-CHROM) to the ascending mode. In this mode, water-saturated octanol is pumped at 0.317 ml/min from a Waters M-45 pump up through the CPC and out through the Linear Instruments Model 204 detector with the semiprep flowcell. V_d is measured from the time of sample injection to the time of mode switching. V_{ra} is measured from the abrupt change in apparent absorbance when the octanol phase entered the flowcell to the manually determined peak maximum of the replotted detector output on a graphics terminal. Measuring V_{ra} from the abrupt change instead of the time of the valve switch is important to reduce a bias that can be introduced due to

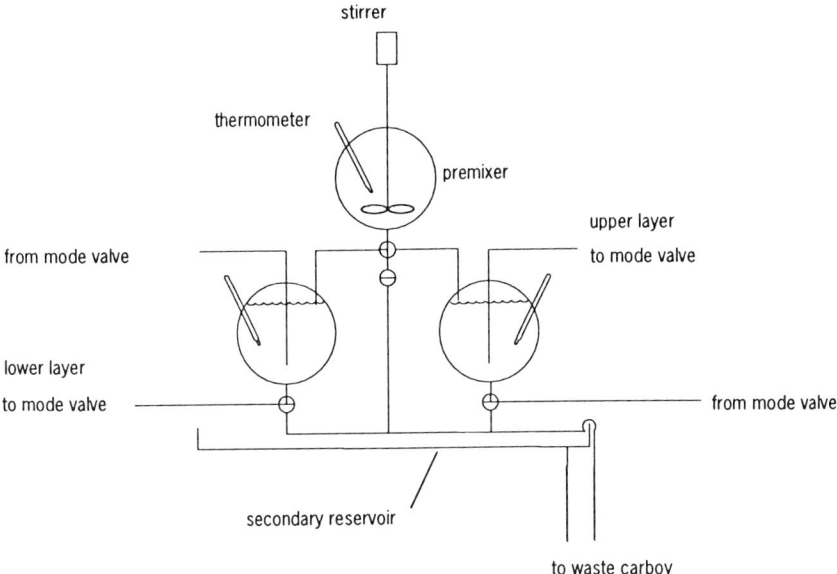

Figure 5 Solvent handling apparatus.

the leftover aqueous phase between the mode valve and the detector and aqueous phase that has displaced octanol from the beginning chambers due to bleed. Suggestions that error could be introduced from bleed in this procedure neglected the way V_{ra} should be measured [23]. Figure 6 is a typical replotted output from the detector.

2. Test Compounds

The quality of the comparison between literature values of K_{ow} and those determined by this methodology ultimately depend on the variation in the literature values. The results of an OECD Laboratory Intercomparison Testing of the shake flask method for determining K_{ow} gave an average range of reported values of $\log K_{ow}$ of 1.0 for six compounds spanning K_{ow} of -1.2 to 5.6 [34], and error was not correlated with $\log K_{ow}$. Although this one study cannot be used to assign an error to literature partition coefficients in general, it highlights the potential for error in reference compound $\log K_{ow}$ values. Table 3 contains the compounds used to test the dual-mode CPC methodology. Included in the table are the number of corroborating literature values for $\log K_{ow}$ and the range of those values. These data were selected from the Pomona College Medicinal Chemistry database and the values of $\log K_{ow}$ are the most recent Log P star values from that database as of July 1993 [26].

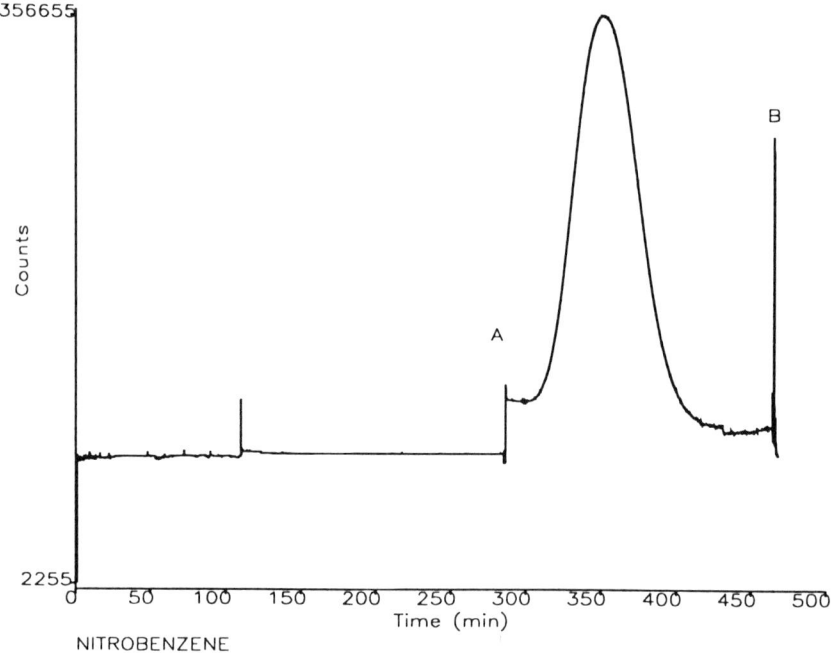

Figure 6 Typical dual–mode detector output. "A" marks the switch from descending mode to ascending mode; "B" marks the switch from ascending mode to descending mode.

D. Results and Discussion

1. Pump Stability

For accurate V_d and V_{ra} measurements, the pumping flow rate accuracy and precision were assessed. The aqueous flow rate was 4.890 ml/min with a standard deviation of 0.0058 ml/min ($n = 9$), checked periodically over an 85-day period. The octanol flow rate was 0.317 ml/min with a standard deviation of 0.0022 ($n = 7$) over a 55-day period. The flow rates were determined by timing volume displacement into volumetric flasks during the course of an analysis. Propagating this random variation into the K_{ow} results in only 1.7% relative error at 95% confidence.

2. Selection of V_d

If the V_d selected is too small, then the V_{ra} cannot be determined because the peak merely decreases from an initial maximum. For the same compound, as the V_d increases, the peak goes from a monotonically decreasing profile to a Gaussian profile. If V_d is too large, then the later eluting portion of the peak appears

Table 3 Compounds Used in This Study

Compound	$\log K_{ow}$	Range	Number of References
Phenol	1.46	1.46–1.75	12
2-Ethoxyphenol	1.68	1.68	1
Benzene	2.13	2.03–2.34	15
2-Nitrotoluene	2.30	2.30–2.36	2
Ethyl benzoate	2.64	2.20–2.64	2
Chlorobenzene	2.84	2.46–2.89	6
Nitrobenzene	1.85	1.79–1.85	6
Anisole	2.11	1.98–2.11	5
Toluene	2.73	2.11–2.80	7
Biphenyl	4.09	3.76–4.17	8
Ethylbenzene	2.64	2.20–2.64	2
o-Xylene	3.12	2.77–3.12	2
Bromobenzene	2.99	2.98–3.01	3
1-Napthol	2.84	2.31–2.98	3

truncated because some of the component is eluted in the descending mode. An extreme of this situation is that the component completely elutes in the descending mode and is not even observable in the ascending mode. Gaussian-shaped peaks are not expected in chromatography if there are less than about 24 theoretical plates [35]. A compound with a high partition coefficient for the chosen V_d would not have many theoretical exchanges between the phases and would thus not

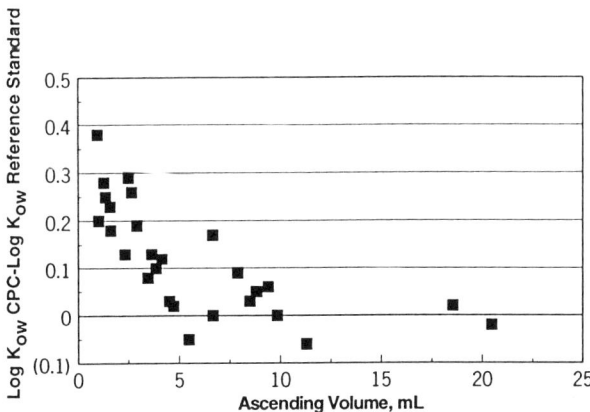

Figure 7 Determined error in $\log K_{ow}$ versus V_{ra}.

appear symmetrical. The peak maximum in these nonsymmetrical peaks is not an accurate measure of V_{ra}. Ideally, the peak centroid should be used. It was difficult to assess the peak centroid because of the large error in determining the baseline of broad, unsymmetrical peaks. Figure 7 and Table 4 show the error in $\log K_{ow}$ CPC versus the V_{ra}. At V_{ra} greater than 8 ml, the bias appears to be minimal relative to the random portion of the error. Thus, we estimate K_{ow}, either by a fragment method or by a screening CPC run. We then select a V_d that will result in a V_{ra} estimated to be large enough to minimize the bias in practical applications.

Table 4   Error in $\log K_{ow}$ Determined Versus V_{ra} for All Compounds Studied

Compound	V_d	V_{ra}	$\log K_{ow}$ determined	literature	Error
Phenol	77.27	1.581	1.69	1.46	0.23
Phenol	157.1	4.154	1.58	1.46	0.12
Phenol	313.5	9.394	1.52	1.46	0.06
2-Ethoxyphenol	235.3	2.506	1.97	1.68	0.29
2-Ethoxyphenol	470.0	6.653	1.85	1.68	0.17
Benzene	308.7	0.949	2.51	2.13	0.38
Benzene	616.7	2.922	2.32	2.13	0.19
Benzene	1,233	8.493	2.16	2.13	0.03
2-Nitrotoluene	489.6	1.276	2.58	2.30	0.28
2-Nitrotoluene	987.6	3.638	2.43	2.30	0.13
2-Nitrotoluene	1,957	7.883	2.39	2.30	0.09
Ethyl benzoate	1,067	1.361	2.89	2.64	0.25
Ethyl benzoate	2,133	2.659	2.90	2.64	0.26
Ethyl benzoate	4,263	8.801	2.69	2.64	0.05
Chlorobenzene	1,688	1.615	3.02	2.89	0.13
Chlorobenzene	3,380	4.717	2.86	2.89	−0.03
Nitrobenzene	342.9	4.525	1.88	1.85	0.03
Nitrobenzene	690.1	9.852	1.85	1.85	0.00
Nitrobenzene	1,389	20.48	1.83	1.85	−0.02
Anisole	631.4	3.863	2.21	2.11	0.10
Anisole	1.262	8.833	2.16	2.11	0.05
Anisole	2.524	18.55	2.13	2.11	0.02
Toluene	2,627	5.477	2.68	2.73	−0.05
Toluene	5,248	11.30	2.67	2.73	−0.06
Biphenyl	19,595	1.010	4.29	4.09	0.20
o-Xylene	4,103	2.333	3.25	3.12	0.13
Bromobenzene	4,103	3.464	3.07	2.99	0.08
1-Napthol	4,671	6.677	2.84	2.84	0.00

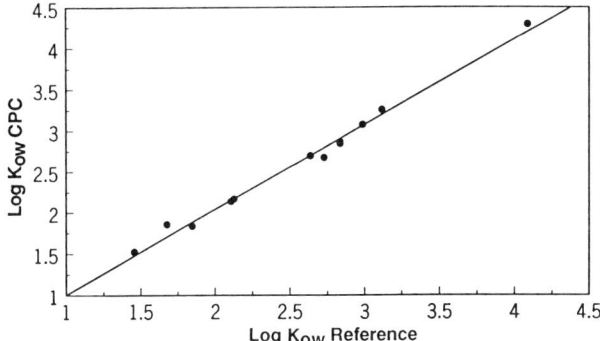

Figure 8 Correlation of literature K_{ow} with dual–mode CPC–determined K_{ow} at maximum V_{ra}.

3. Results from Maximum V_d

Most papers that present alternative procedures for determining K_{ow} show a correlation table or plot of K_{ow} determined versus K_{ow} from literature. In the case of this methodology, an analogy is to plot all the compounds used in this study at maximum V_d, in the cases where several values of V_d were chosen for the same compound, versus $\log K_{ow}$ (Fig. 8). The determination range evaluated in this technique was up to $\log K_{ow}$ literature of 4.09 (K_{ow} = 12,300), and the determined K_{ow} values agreed well with the literature values.

E. Conclusions

Dual-mode CPC is effective in determining values of K_{ow} up to 12,300. The dual-mode CPC system complements the normal elution mode of CPC for determining K_{ow} with a demonstrated range of 29–12,300. ($\log K_{ow}$ range 1.5–4.1). The procedure is fundamentally derived without requiring the use of extrathermodynamic parameters. The technique has a determination bias dependent on the distance the material traveled through the CPC system as reflected in V_{ra}. For the best accuracy, V_{ra} should be at least 8 ml. Below 8 ml, the bias could perhaps be subtracted, and this would lead to even faster analysis times.

IV. COCURRENT MODE

A. Introduction

This operation, first demonstrated by Sutherland [36] in the toroidal-coil centrifuge, and later adapted to K_{ow} determinations by other workers [23,37], relies on the simultaneous pumping of a ratio of a small flow of octanol and a larger flow of water to elute strongly retained compounds ($\log K_{ow} > 3$). A third solvent such as isopropanol or acetonitrile is added after the CPC rotor to clarify the octanol and water stream before the detector. The intent is to minimize optical noise due to the difference in the refractive indices of octanol and water.

B. Theory

From Sutherland [36], the V_r of a solute is a function of phase ratio in the rotor (β), the flow rate ratio (α), V_t and K_{ow}:

$$V_r = \left(\frac{K_{ow}(1 - \beta) + \beta}{K_{ow}(1 - \alpha) + \alpha} \right) V_t \tag{12}$$

Equation 12 can be rearranged to solve for K_{ow}:

$$K_{ow} = \frac{V_t \beta - V_r \alpha}{V_r - V_r \alpha - V_t + V_t \beta} \tag{13}$$

V_t may be determined by injecting a solute at constant flow rate in the system while it contains only a single phase. α is set by the pumps for each phase, and β is determined by injecting a nonretained solute or using compound(s) of known K_{ow}.

C. Conclusions

This method may hold promise in some high $\log K_{ow}$ applications. In the feasibility study of Berthod et al. [23], short-term noise was filtered to give an elution profile expected from several injected components. The source of the noise could have been poor mixing of the cosolvent, a signal caused by the solubilization of large, discrete segments of the separate phases, or a temperature-related influence on refractive index. The main drawback of the procedure was the large amount of effluent generated, which cannot be recycled due to the addition of the clarifying solvent.

V. PRACTICAL OPERATING SUGGESTIONS

1. Modify the system to use 0.5-mm bore stainless steel tubing wherever possible for safety and to minimize unwanted leaks.
2. Replace small-diameter tubing in analytical-style detector flowcells to minimize excess pressure.
3. Minimize excess tubing lengths outside of the rotor by closely arranging all critical fluid connections.
4. Use heat exchanger fluid-jacketed 3- or 4-L reservoirs for mixing the phases at constant temperature. Keep the phases at the same temperature as the rotor. This careful control of temperature along with our efforts to minimize

pressure buildup could be the reason other workers report significantly more bleed than we observe [18,20,24,25] Figure 5 shows such an apparatus.

5. Carefully calibrate both the octanol and water pumps over the flow rate ranges used.
6. Operate at a phase ratio of approximately 1 volume octanol stationary phase to 5 volumes water mobile phase.
7. Use an injection volume up to 1 ml for a 120-ml rotor. Poorly soluble compounds maybe dissolved in dimethyl formamide to improve detectability without significantly affecting large values of K_{ow}.
8. Buffers may be used throughout the applicable pH range; however, appropriate protocol requires measuring the pH of the aqueous layer *after* equilibrating with the octanol.

REFERENCES

1. A. Leo, C. Hansch, and C. Church, *J. Med. Chem.*, *12*: 767–771 (1969).
2. P. C. Taylor, "Hydrophobic Properties of Drugs," *Comprehensive Medicinal Chemistry: The Rational Design, Mechanistic Study and Therapeutic Application of Chemical Compounds* (J. C. Deardon, ed.), Pergamon Press, New York, pp. 241–294 (1990).
3. Y. C. Martin, *Quantitative Drug Design*, Marcel Dekker, New York (1978).
4. R. Franke, *Theoretical Drug Design Methods*, Elsevier, Amsterdam (1984).
5. W. J. Lyman, W. F. Reehl, and D. H. Rosenblatt, eds., *Handbook of Chemical Property Estimation Methods*, McGraw-Hill, New York, Chapters 1, 2, and 4 (1982).
6. Rules and Regulations, *Federal Register*, Vol. 50, 188: 39252 (Sept. 1985).
7. R. F. Rekker, *The Hydrophobic Fragmental Constant—Its Derivation and Application—A Means of Characterizing Membrane Systems*, Elsevier, Amsterdam (1977).
8. A. Leo, P. Y. C. Jow, C. Silipo, and C. Hansch, *J. Med. Chem.*, *18*: 865–870 (1975).
9. H. Terada, *Quant. Struct.-Act. Relat.*, *5*: 81–88 (1986).
10. G. D. Veith, N. M. Austin, and R. T. Morris, *Water Res.*, *13*: 43 (1979).
11. L. P. Brukhard, D. W. Kuehl, and G. D. Vieth, *Chemosphere*, *14*: 1551–1560 (1985).
12. J. J. Sabatka, D. J. Minick, T. K. Shumaker, G. L. Hodgson, and D. A. Brent, *J. Chromatogr.*, *384*: 349–356 (1987).
13. P. M. Sherblom and R. P. Eganhouse, *J. Chromatogr.*, *454*: 37–50 (1988).
14. D. Minick, D. A. Brent, and J. Frenz, *J. Chromatogr.*, *461*: 177–199 (1989).
15. M. J. Kamlet, M. H. Abraham, P. W. Carr, R. M. Doherty, and R. W. Taft, *J. Chem. Soc., Perkin Trans. II*, 15: 2087–2091 (1988).
16. H. Terada, Y. Kosuge, W. Murayama, N. Nakaya, Y. Nunogaki, and K. Nunogaki, *J. Chromatogr.*, *400*: 343–351 (1987).
17. A. Berthod and D. W. Armstrong, *J. Liq. Chromatogr.*, *11*(3): 547–566 (1988).
18. A. Berthod and D. W. Armstrong, *J. Liq. Chromatogr.*, *11*(6): 1187–1204 (1988).
19. A. Berthod, D. W. Armstrong, and Y. I. Han, *J. Liq. Chromatogr.*, *11*(7): 1441–1456 (1988).
20. S. J. Gluck and E. J. Martin, *J. Liq. Chromatogr.*, *13*(13): 2529–2551 (1990).

21. S. J. Gluck and E. J. Martin, *J. Liq. Chromatogr.*, *13*: 3559 (1990).
22. R. Menges, G. Bertrand, and D. Armstrong, *J. Liq. Chromatogr.*, *13*(15): 3061–3077 (1990).
23. A. Berthod, R. Menges, and D. Armstrong, *J. Liq. Chromatogr.*, *15*(15/16): 2769–2785 (1992).
24. N. El Tayar, R.-S. Tsai, P. Vallat, C. Altomare, and B. Testa, *J. Chromatogr.*, *556*: 181–194 (1991).
25. R.-S. Tsai, N. El Tayar, B. Testa, and Y. Ito, *J. Chromatogr.*, *538*: 119–193 (1991).
26. Pomona MedChem Software, Version 3.52, Pomona College, Pomona, California.
27. N. R. Draper and H. Smith, *Applied Regression Analysis*, 2nd ed., John Wiley, New York, pp. 108–116 (1981).
28. D. P. Shoemaker, C. W. Garland, and J. I. Steinfeld, *Experiments in Physical Chemistry*, 3rd ed., McGraw-Hill, Inc., New York, pp. 51–55 (1974).
29. N. E. Tayar, A. Marston, A. Bechalany, K. Hostettmann, and B. Testa, *J. Chromatogr.*, *469*: 91–99 (1989).
30. O. Levenspiel, *Chemical Reaction Engineering*, 2nd ed., Wiley, New York, p. 254 (1972).
31. R. C. Bruening, E. M. Oltz, J. Furukawa, K. Nakanishi, and K. Kustin, *J. Natural Prod.*, *49*(2): 193–204 (1986).
32. J. Cazes, *Amer. Biotechnol. Lab.*, *6*(2A): 20 (1988).
33. Sanki Laboratories, *Instruction Manual for Centrifugal Partition Chromatograph Model LLN* (1987).
34. *Partition Coefficient in Octanol Water*, OECD Guideline #107 (May 12, 1981).
35. D. G. Peters, J. M. Hayes, and G. M. Hieftje, *Chemical Separations and Measurements*, W. B. Saunders Co., Philadelphia, pp. 496–499 (1974).
36. J. A. Sutherland, *J. Liq. Chromatogr.*, *7*(2): 363–384 (1984).
37. S. Gluck, personal correspondence with D. Armstrong (June 29, 1989).

9
Mutual Separation of Lanthanoid Elements by Centrifugal Partition Chromatography

Kenichi Akiba
Institute for Advanced Materials Processing, Tohoku University, Sendai, Japan

I. INTRODUCTION

The lanthanoid group comprises 15 elements, from lanthanum ($Z = 57$) through lutetium ($Z = 71$), in the same place on the periodic table. They are inner transition elements characterized by progressive filling of $4f$ electrons from the $[Xe]4f^0$ to $[Xe]4f^{14}$ configuration. An increase in effective nuclear charge with increasing atomic number attends a successive shrinkage in atomic and ionic size. This has been known as lanthanoid contraction [1]. As Fig. 1 shows, ionic radii of trivalent lanthanoids gradually decrease from 1.061 Å for La^{3+} to 0.848 Å for Lu^{3+} in six coordination [2]. All the trivalent lanthanoid ions closely resemble each other in their physical and chemical properties, in particular, to aqueous solutions. Furthermore, yttrium and scandium, situated above lanthanum in Group 3A, have a similar M^{3+} ion. These lanthanoids, yttrium, and scandium are sometimes classed as rare earths because of their similar properties and general occurrence in lanthanoid minerals. Their separation has been a large problem, in which one is confronted differentiation based on only slight differences in the chemical properties of the individual elements.

This chapter deals with liquid–liquid extraction and its application to centrifugal partition chromatography (CPC) in order to accomplish the mutual separation of adjacent lanthanoids.

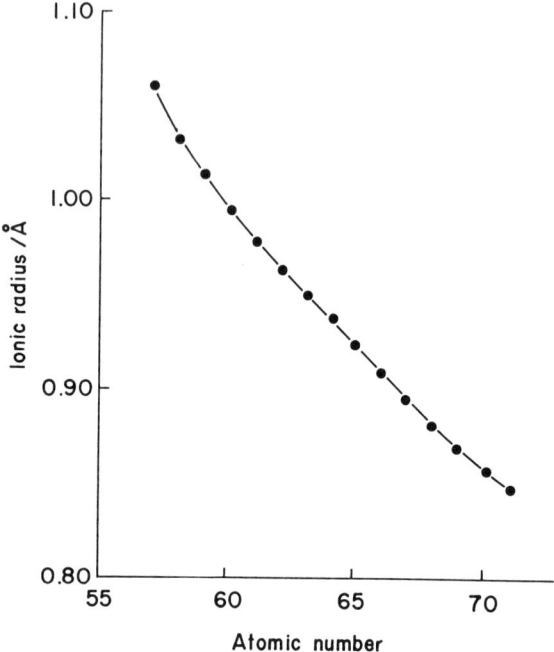

Figure 1 Ionic radii of trivalent lanthanoids.

II. SEPARATION OF LANTHANOIDS

The mutual separation of lanthanoid ions has usually been carried out through multiple processes. Classical methods of fractional crystallization and fractional precipitation of inorganic salts resulted in incomplete isolation. At present, more satisfactory methods such as ion exchange and solvent extraction have been widely employed, not only in analytical but in preparative separations of lanthanoids.

A. Ion Exchange Method

Ion exchangers have rather poor selectivity for lanthanoids, and selectivity is usually achieved through complex formation with suitable ligands. In ion exchange processes, mixed lanthanoids are adsorbed onto cation exchange columns, and hence, the lanthanoids are chromatographically eluted with a solution containing complexing agents such as ethylenediaminetetraacetic acid. The lanthanoid complexes are separated following their stability constant sequence [3,4]. Though the ion exchange method is a powerful separation technique, its application to the separation of macro amounts of elements has been limited. Because it requires large amounts of ion exchange resins, the solubilities of reagents and their complexes are rather low, and this method still requires long separation time [5].

B. Solvent Extraction Method

Lanthanoid separation by solvent extraction depends upon differences in the distribution ratios of individual lanthanoids between organic and aqueous phases. Solvent extraction of lanthanoids has been extensively investigated by using a wide variety of extracting reagents [6]. Organic esters of phosphoric acid, including tributyl phosphate (TBP) and *bis*(2-ethylhexyl)phosphoric acid (DEHPA), have been most commonly used, owing to their high extractability and relatively large separation factors [7–11], for example, the average value of 2.5 with DEHPA between successive lanthanoids [9]. Phosphonic acids (RO)RPO(OH) and phosphinic acids $R_2PO(OH)$ also have been accessible for the extraction of lanthanoids [12–17]. Since separation factors are not sufficiently large to achieve complete isolation of each lanthanoid by batch extraction, multiple processes such as batteries of mixer-settlers and countercurrent extraction are necessary for quantitative separation of this series of elements [18,19].

III. EXTRACTION EQUILIBRIUM OF LANTHANOIDS

A. Acid Organophosphorus Extractants

Organophosphorus compounds have some particular advantages not only in solvent extraction, but in centrifugal partition chromatography, because of their good loading and stripping behavior, low solubility in the aqueous phase, and relatively large separation factor for lanthanoids. Typical acidic organophosphorus extractants are shown below:

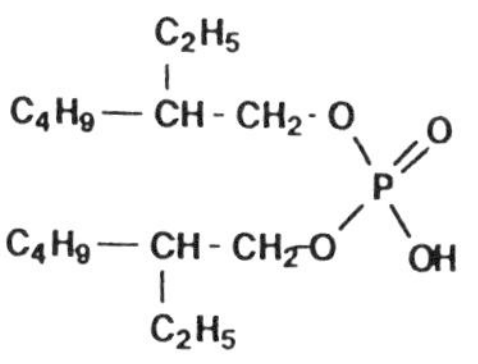

DEHPA 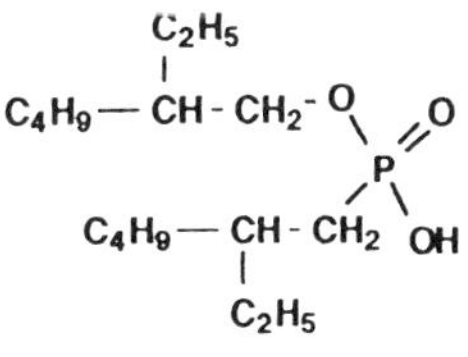   EHPA 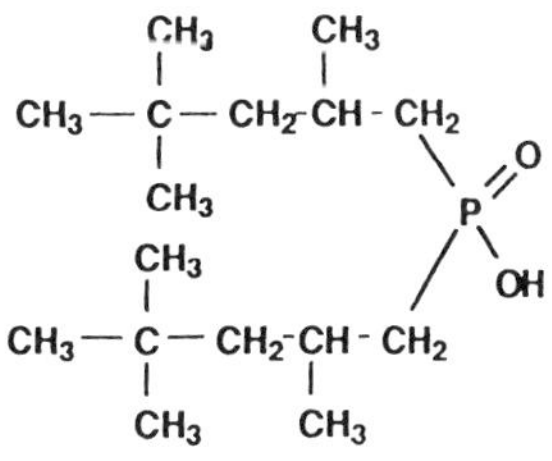   DTMPPA

DEHPA: bis(2-ethylhexyl)phosphoric acid
EHPA: 2-ethylhexyl 2-ethylhexylphosphonic acid
DTMPPA: bis(2,4,4-trimethylpentyl)phosphinic acid

The extractability of lanthanoids decreases in the order DEHPA > EHPA > DTMPPA, in accordance with their decreasing order of acid dissociation constant [20]. The extractant DEHPA, having high extractability, requires high acidity for stripping, while the extractants such as EHPA or DTMPPA are rather preferable for stripping lanthanoids with a dilute acid solution. It is convenient to employ these acidic extractants as a stationary phase in CPC, since the retention volume will be optimized by controlling the pH of a mobile phase. In practice, lanthanoid separation has been accomplished by CPC employing DEHPA [21–23], EHPA [14, 24], and DTMPPA [25,26] as the stationary phase.

B. Extraction Equilibrium

Equilibrium of liquid–liquid extraction is fundamental for preparing the suitable conditions of stationary and mobile phases. In lanthanoid separation, distribution ratios and their differences between individual lanthanoids are of great importance. Distribution ratios between the aqueous and organic phases of a series of lanthanoids are shown as a function of pH in Fig. 2 for EHPA and in Fig. 3 for

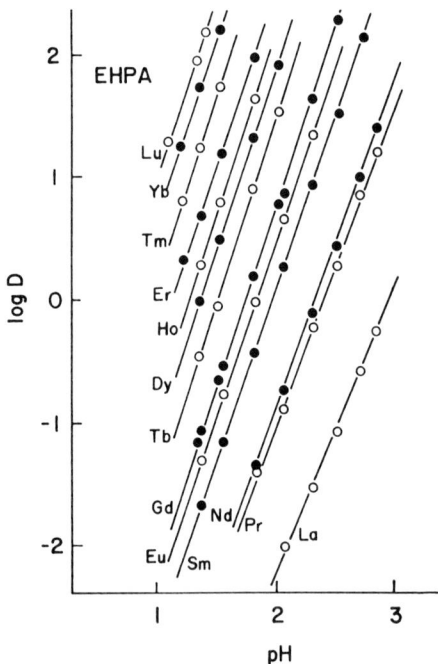

Figure 2 Distribution ratios of lanthanoids as a function of pH. 0.02 M (EHPA)$_2$ in kerosene; 0.1 M (H,Na) Cl$_2$CHCOO; 25°C. (From Ref. 14.)

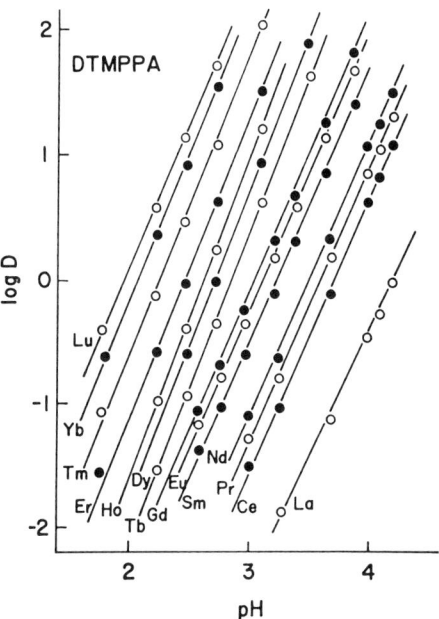

Figure 3 Distribution ratios of lanthanoids as a function of pH. 0.02 M (DTMPPA)$_2$ in kerosene; 0.1 M (H,Na)Cl$_2$CHCOO; 25°C. (From Ref. 26.)

DTMPPA. Plots of log D versus pH gave a series of parallel lines with slopes close to 3. The third-power dependence of the distribution ratio on reagent concentration was also confirmed. Overall equilibrium for the extraction of a trivalent lanthanoid M^{3+} can be expressed by the following equation:

$$M_{aq}^{3+} + 3(HA)_{2,org} \rightleftarrows M(HA_2)_{3,org} + 3H_{aq}^+ \qquad (1)$$

where (HA)$_2$ stands for the dimer of an organophosphorus acid, EHPA or DTMPPA. The extraction constant (K_{ex}) for this two-phase reaction is given by

$$K_{ex} = \frac{[M(HA_2)_3]_{org}[H^+]_{aq}^3}{[M^{3+}]_{aq}[(HA)_2]_{org}^3} \qquad (2)$$

The values of K_{ex}, evaluated from the distribution data with EHPA and DTMPPA, are plotted against the atomic number of lanthanoids in Fig. 4. The K_{ex} values increase from lanthanum to lutetium, indicating that cations with small ionic radii form stable extractable complexes. The K_{ex} values with EHPA are about two orders of magnitude larger than those with DTMPPA, while they vary with atomic number in a similar manner. The K_{ex} values are not a linear function of Z, but they

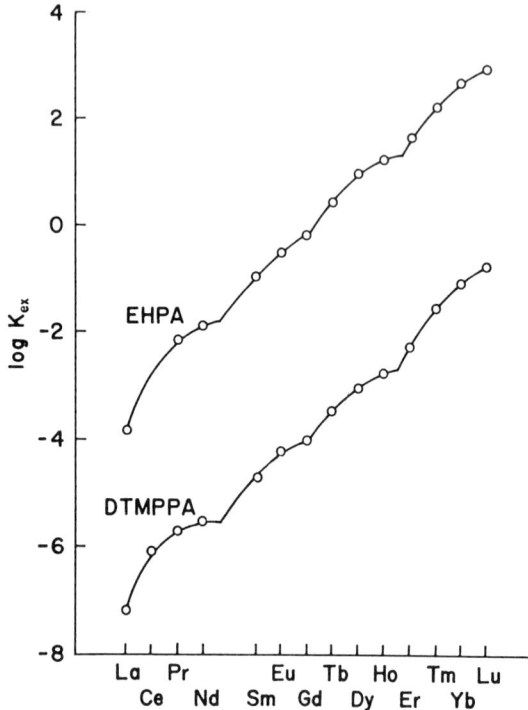

Figure 4 Variation of extraction constants with atomic number.

are divided into four smooth curves, each consisting of four lanthanoid elements. These regularities are well known as the "tetrad" effect, relating to the behavior of lanthanoids and actinoids in many extraction systems [27–30]; there is a special stability associated with half-filled and filled subshells [28,31,32].

The separation factor (α_D) in liquid–liquid extraction is given by the ratio of the distribution ratios of components 1 and 2:

$$\alpha_D = \frac{D_2}{D_1} = \frac{K_{ex2}}{K_{ex1}} \qquad (3)$$

The values of α_D evaluated from the data obtained with EHPA and DTMPPA are listed in Table 1, together with those of $\log K_{ex}$. In accordance with the regularities in K_{ex} values, the α_D values are commonly large for the pairs placed in the former part of each tetrad curve, while those for the Pr–Nd, Eu–Gd, Dy–Ho, and Yb–Lu pairs in the latter part of the curves are small, and hence their mutual separation will be particularly difficult.

Table 1 Extraction Constants and Separation Factors

Lanthanoid	EHPA[a] logK_{ex}	α_D	DTMPPA[b] logK_{ex}	α_D
La	−3.80 ⎫		−7.23	13.8
Ce	⎬ 47.9	−6.09	2.29
Pr	−2.12 ⎭		−5.73	1.51
		1.95		
Nd	−1.83		−5.55	7.41
		8.51		
Sm	−0.90		−4.68	3.02
		3.02		
Eu	−0.42		−4.20	1.51
		1.86		
Gd	−0.15		−4.02	3.98
		5.18		
Tb	0.56		−3.42	2.63
		2.70		
Dy	0.99		−3.00	1.62
		1.98		
Ho	1.29		−2.79	3.47
		2.50		
Er	1.69		−2.25	5.25
		3.62		
Tm	2.25		−1.53	3.24
		3.06		
Yb	2.74		−1.02	2.14
		1.67		
Lu	2.96		−0.69	

[a] 0.02 M (EHPA)$_2$ in kerosene.
[b] 0.02 M (DTMPPA)$_2$ in kerosene.
0.1 M (H,Na)Cl$_2$CHCOO, 25°C.

IV. CENTRIFUGAL PARTITION CHROMATOGRAPHY

A. Application to CPC

Centrifugal partition chromatography (CPC) employing a stationary phase without any solid support [33] offers great possibilities for chromatographic separation, because it affords a wide selection of suitable stationary and mobile phases according to their requirements on the basis of liquid–liquid extraction behavior [34].

1. Procedures

An apparatus for centrifugal partition chromatography was utilized by arranging partition cartridges in a rotor [33], where Type 30E and 250W cartridges are composed of 54 and 400 partition channels, respectively. A kerosene solution containing EHPA or DTMPPA was employed as the stationary phase [14,26]. The sample solution (1 cm^3) usually containing 10^{-3} M (M = mol dm^{-3}) of each lanthanoid was injected. The lanthanoid ions were eluted with a mobile phase buffered with 0.1 M (H,Na) Cl$_2$CHCOO to an appropriate pH. The separated lanthanoids were detected by measuring the absorbance of colored complexes at 650 nm through a post column reaction with Arsenazo III [35]. Chromatographic behavior has been examined in order to accomplish the mutual separation of adjacent lanthanoids [14,26].

2. Retention Volume

In chromatography, the retention volume (V_R), indicating relative affinity for the stationary phase, is related to the distribution ratio of the desired component:

$$V_R = V_m + DV_s \qquad (4)$$

where V_s and V_m are the volumes of the stationary and mobile phases, respectively. The retention volume obtained for europium was confirmed to be approximately proportional to the distribution ratio, as shown in Fig. 5. The D value of an individual lanthanoid will be optimized for adequate separation by varying pH and/or extractant concentration.

3. Chromatographic Parameters

The number (N) of theoretical plates is related to the retention volume and the bandwidth (W):

$$N = 16\left(\frac{V_R}{W}\right)^2 \qquad (5)$$

The separation factor (α_C) in chromatography is evaluated from the retention volumes of individual components 1 and 2:

$$\alpha_C = \frac{V_{R2} - V_0}{V_{R1} - V_0} \qquad (6)$$

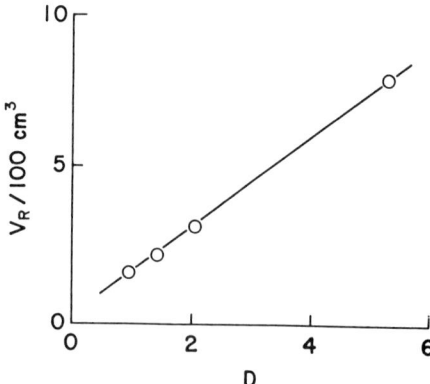

Figure 5 Relation between the retention volume and the distribution ratio for europium. Stationary phase: 0.01 M (EHPA)$_2$ in kerosene/cartridges (30E × 6); mobile phase: 0.1 M (H,Na)Cl$_2$CHCOO; 25°C. (From Ref. 24.)

where V_0 is the elution volume for an unretained component. The degree of separation of adjacent peaks is defined as resolution (Rs):

$$\mathrm{Rs} = 2\frac{V_{R2} - V_{R1}}{W_1 + W_2} \qquad (7)$$

B. Selection of Mobile Phase

Effects of the mobile phase on chromatographic behavior have been observed by employing the stationary phase of EHPA in kerosene loaded on six cartridges (30E) [14]. Figure 6a illustrates a chromatogram obtained with an aqueous mobile phase for a binary mixture of Sm and Gd. Peaks with appreciable tailing may be attributed to slow equilibration of liquid–liquid extraction including complex formation. Figures 6b and 6c illustrate chromatograms produced by mixed mobile phases of 10 and 20% v/v ethanol (EtOH) in water, respectively; ethanol additives appreciably improved peak resolution.

1. Water-Miscible Organic Solvents

Several organic solvents, miscible with water, have been tested as components of the mobile phase. Effects of water-miscible organic solvents are summarized in Table 2 for the separation of Sm and Gd. An acetonitrile (AN)–water mixture gave

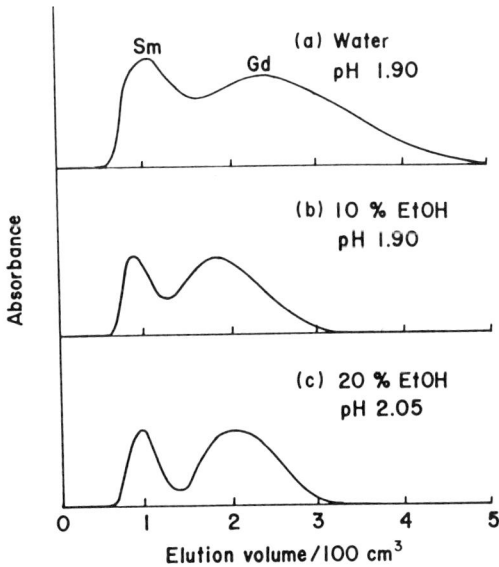

Figure 6 Elution curves of a binary mixture of Sm and Gd. Stationary phase: 0.02 M (EHPA)$_2$ in kerosene/(30E × 6); mobile phase: 0.1 M (H,Na)Cl$_2$CHCOO; 900 rpm; 25°C. (From Ref. 14.)

Table 2 Effect of 20% v/v Miscible Organic Solvents–Water on the Separation of Sm and Gd

Solvent	Peak shape	N_{Sm}	N_{Gd}	α_C	Rs
Acetonitrile	No peak				
Methanol	Tailing	29.0	17.5	3.0	0.83
Ethylene glycol	Symmetry	31.2	23.6	3.8	1.15
Ethanol	Symmetry	31.7	21.7	2.9	0.95
Dimethyl sulfoxide	Symmetry	28.1	23.3	3.6	1.14
N,N-Dimethyl formamide	Asymmetry	19.5	11.9	3.4	0.75
Water alone	Asymmetry	27.1	15.5	2.9	0.82

Six cartridges (30E), 0.02 M (EHPA)$_2$, 900 rpm, 55°C.
Source: Ref. 14.

nondistinct peaks and was found to be undesirable for the mobile phase. Mixtures with N,N-dimethyl formamide (DMF) and methanol (MeOH) yielded asymmetric peaks with small N values owing to pronounced tailing, and hence these mobile phases result in rather low resolution. Symmetrical peaks have been obtained by the use of mixtures with ethylene glycol (EG), EtOH, or dimethyl sulfoxide (DMSO). The N values for individual peaks and the α_C values appreciably increased compared with the case of water alone.

2. Effects of Mobile Phase Composition

Effects of the composition of the mobile phase on elution behavior have been examined by varying the content of water-miscible solvents [24]. High content, larger than 30% v/v of EG, MeOH, and EtOH, brought about a partial loss of stationary phase owing to an increase in mutual solubility between the stationary and mobile phases. Figure 7 shows variations in resolution of the peaks of Sm and Gd for EG–water mixed mobile phases. The Rs value increased with pH value and also with ethylene glycol content. Thus, if necessary, the mixture of 20% EG–water appears to be desirable as a mobile phase.

While little is known about kinetics of separation processes in CPC, the influence of miscible organic additives may be related to following factors: (1) a lowering of surface tension causes a decrease in drop sizes; (2) an increase in distribution of extractant into the aqueous phase accelerates a complex formation; and (3) a decrease in water activity reduces hydration of metal ions. These associated factors may enhance the extraction rate and improve peak resolution.

C. Selection of Stationary Phase

One of principal advantages of CPC is that a wide variety of stationary phases is available, based on solvent extraction behavior. Chromatographic behavior has

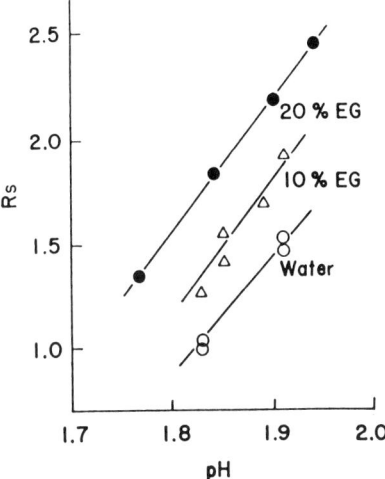

Figure 7 Peak resolution for Sm–Gd as a function of pH. (From Ref. 24.)

been examined by using six cartridges (250W) holding DTMPPA in different kinds of organic solvents [26]. Figure 8 illustrates typical chromatograms through the stationary phase of a carbon tetrachloride solution. Sharp peaks without pronounced tailing were obtained, and their resolution was improved with increasing pH value of the mobile phase. Similar results were also obtained for a toluene solvent. Resolution between Tm and Yb decreased in the order kerosene > CCl_4 > toluene. A hexane solution of DTMPPA resulted in significant tailing and poor separation of a binary mixture of Tm and Yb. Thus, kerosene is the most promising solvent for the stationary phase.

D. Operating Parameters

Since chromatographic separations are usually sensitive to operating conditions, effects of operating variables have been examined by employing the stationary phase of DTMPPA in kerosene held in six cartridges (250W) and the aqueous mobile phase without organic additives [26].

1. Effects of Temperature

Figure 9 illustrates a typical sharp chromatogram obtained at 70°C together with that at 25°C. Separation is greatly improved at a high temperature. Variations in individual chromatographic parameters are presented in Fig. 10. Since rising temperature enhances the retention volume, these values were normalized at a definite value of V_R for Tm. Similar progressive increases in V_R for Tm and Yb had

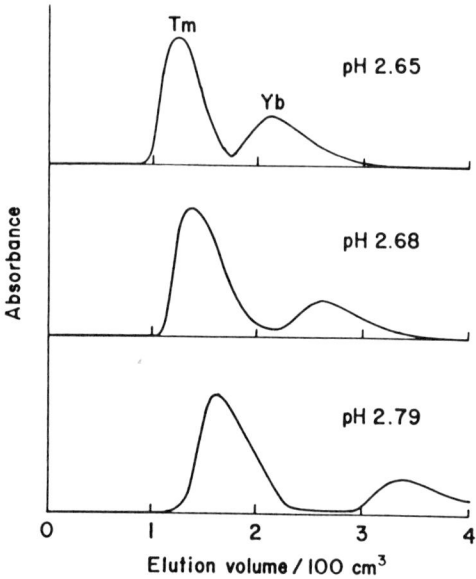

Figure 8 Chromatograms for the binary mixture Tm–Yb with CCl_4 solution. Stationary phase: 0.02 M $(DTMPPA)_2$ in CCl_4/cartridges (250W × 6); 600 rpm; 25°C.

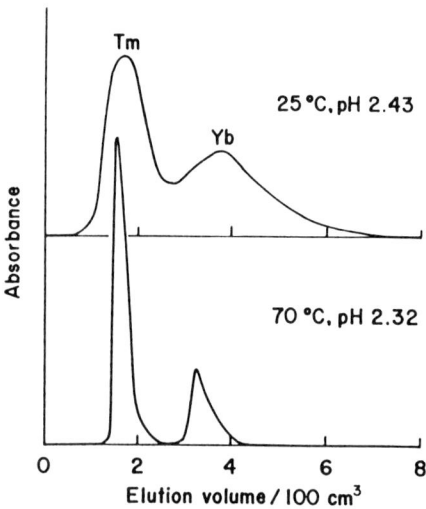

Figure 9 Chromatograms at different temperatures. Stationary phase: 0.02 M $(DTMPPA)_2$ in kerosene/(250W × 6); 900 rpm.

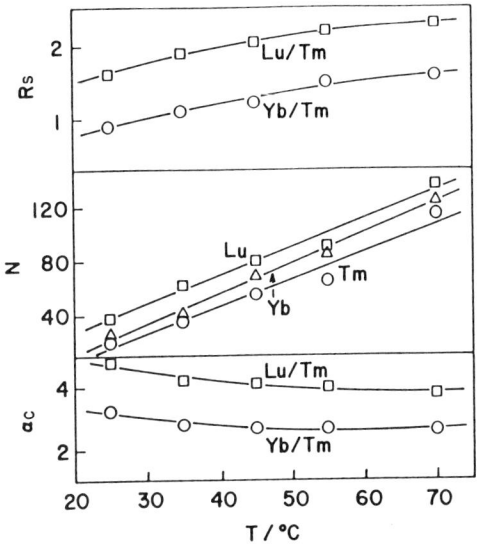

Figure 10 Effects of temperature on chromatographic parameters. (From Ref. 26.)

only a small effect on the separation factor, which decreased slightly with a rise in temperature from 25 to 70°C. On the other hand, the N values of each element increased almost linearly with temperature; resolution was thereby enhanced to a certain extent.

2. Effects of Rotational Speed

Figure 11 shows effects of rotational speed on chromatographic parameters. The N values increased linearly with rotating speed, the remaining α_C values being constant. The number of microdroplets of the mobile phase probably increased at high rotation, and this would accelerate the rate of distribution. Therefore, the Rs value was enhanced almost linearly with the N value. Thus, a high speed of rotation is preferable for improving resolution, as long as pressure is maintained less than a limited value of 60 kg cm^{-2} for this system.

3. Effects of Flow Rate

Effects of different flow rates of the mobile phase are shown in Fig. 12. The retention volume was almost unaltered at any flow rate, and the α_C value remained nearly constant, indicating normal retention behavior, while the bandwidth was considerably broadened at a high flow rate and the value of N greatly decreased. The flow rate of the mobile phase should be sufficiently low to attain distribution equilibria between the stationary and mobile phases. Thus, the separation greatly

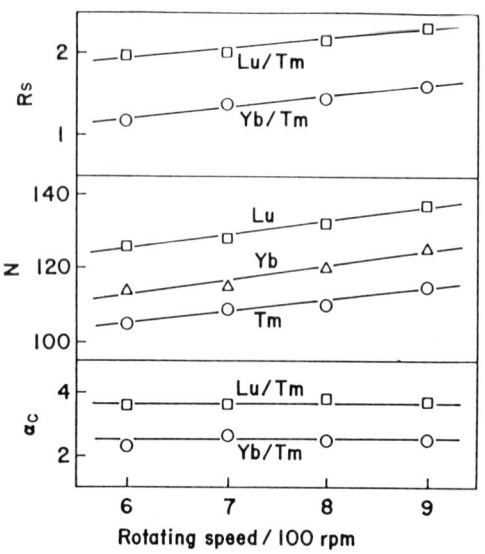

Figure 11 Effects of rotational speed on chromatographic parameters at 70°C. (From Ref. 26.)

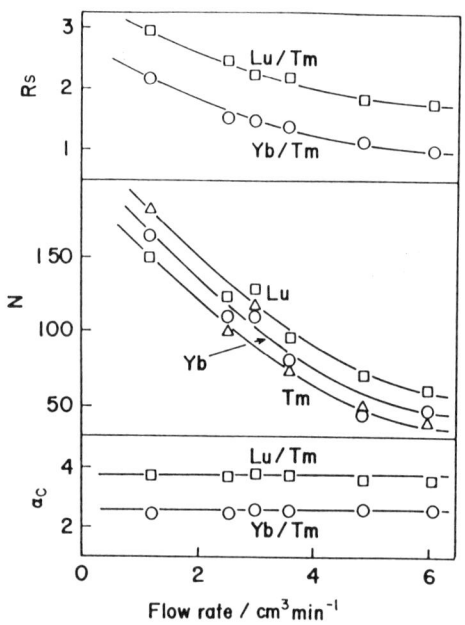

Figure 12 Effects of flow rate of mobile phase on chromatographic parameters. (From Ref. 26.)

depends on the N value, and a relatively high Rs value, above 2, was obtained at low flow rate, around 1 cm^3 min^{-1}.

V. MUTUAL SEPARATION OF LANTHANOIDS BY CPC

A. Separation with EHPA

1. Lighter Lanthanoids

Chromatographic separations of a mixture of light and middle lanthanoids have been performed by CPC using 12 cartridges (30E) holding the stationary phase of EHPA [14,16]. It is important to control pH of the mobile phase in order to optimize the distribution ratio of each component for adequate separation. When D is small, resolution will be poor owing to low retention on the stationary phase, while in the case of large D a long time will be taken for elution. Figure 13

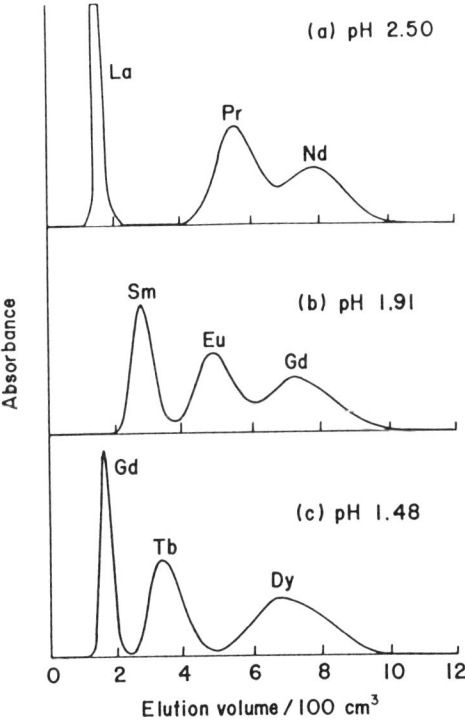

Figure 13 Separation of ternary mixtures of light and middle lanthanoids by the use of 12 cartridges of 30E. Stationary phase: 0.02 M (EHPA)$_2$ in kerosene; mobile phase: 0.1 M (H,Na)Cl$_2$CHCOO in 20% v/v EG–water; 3 cm^3 min^{-1}; 800 rpm; 55°C. (From Ref. 14.)

illustrates typical chromatograms for the separation of ternary mixtures of lanthanoids. Light lanthanoids were weakly retained and were preferentially eluted with a mild acid solution. Praseodymium was completely separated from lanthanum with the mobile phase of pH 2.50, as expected from the large α_D (=31.6) value, while the peaks of Pr and Nd with low α_D (=1.66) were incompletely resolved (Fig. 13a). Figure 13b illustrates a chromatogram for a mixture of Sm, Eu, and Gd, one of the most difficult mixtures for complete separation. The pH of the mobile phase was adjusted to 1.91 in a optimum range for the Eu–Gd pair of low α_D value. The quantitative separation of adjacent lanthanoids, Sm and Eu, has been accomplished with good resolution, Rs = 1.14. The separation of Eu and Gd with the minimum α_D (= 1.70) was incomplete as α_C = 1.64 and Rs = 0.72. Figure 13c illustrates a chromatogram for the mixture Gd–Tb–Dy eluted with the mobile phase of pH 1.48. Each pair is quantitatively separated as expected based on its relatively large separation factor, that is, α_D = 5.75 for Gd–Tb and 3.09 for Tb–Dy.

The separation parameters evaluated from the chromatograms are listed in Table 3. Most of the lanthanoid pairs with large α_D (>2) have been separated with fairly good resolution Rs > 1, while the pairs of small α_D (<2), such as Pr–Nd and Eu–Gd, have been incompletely resolved with low Rs less than 1. The chromato-

Table 3 Separation Parameters for CPC Chromatogram with EHPA Loaded on 12 Cartridges (30E) at 55°C

Lanthanoid	N	α_C	Rs
La	169.0		
Pr	85.8	9.86	2.70
Nd	86.2	1.51	0.77
Sm	91.8		
Eu	64.0	2.23	1.14
Gd	49.4	1.64	0.72
Gd	108.2		
Tb	51.4	4.71	1.33
Dy	41.6	2.70	1.17
Ho	67.6		
Er	27.0	2.19	0.68
Tm	19.5	2.99	0.93

Source: Ref. 14.

graphic separation factor α_C values are close to the α_D values; therefore, separation processes in CPC seem to essentially correspond to those in liquid–liquid extraction provided the distribution ratios for each element are close to optimum. The resulting peaks were well separated; however, resolution was still low owing to band broadening; the values of N were around one-tenth compared with the total number (54 × 12) of channels of 12 cartridges.

Similar chromatographic separations have also been performed by adopting the stationary phase of DEHPA diluted in heptane, toluene, and chloroform, and the mobile phase of dilute hydrochloric acid [21–23].

2. Heavier Lanthanoids

The heavy lanthanoid elements such as Tm and Lu are more strongly retained, and it is sometimes difficult to elute metal ions from the stationary phase even with a mobile phase of relatively low pH. Even if the stationary phase is repalced with a lower concentration (5 × 10^{-3} M) of EHPA, resulting peaks for mixtures of heavier lanthanoids are considerably broadened and only partially resolved; this is probably due to a decrease in extraction rate with increasing atomic number [36].

B. Separation with DTMPPA

As mentioned above, the separation of heavier lanthanoids was rather difficult by the EHPA stationary phase loaded on the cartridges (Type 30E). An enhancement in lanthanoid separation has been attempted using six cartridges (Type 250W) having more precise channels at a higher temperature of 70°C and a low flow rate of the mobile phase [26]. The kerosene solution of DTMPPA was employed as the stationary phase, while the aqueous mobile phase was 0.1 M (H,Na)CCl$_2$CHCOO without any water-miscible organic solvents. The chromatographic separation of middle to heavy lanthanoids has also been performed at an appropriate pH. The mutual separation of Gd–Tb–Dy have been sufficiently accomplished, that is, Rs > 1, owing to their large α_D values. Figure 14a illustrates a chromatogram obtained at pH 2.62 and low flow rate of 0.68 cm^3/min^{-1} for a mixture of Dy–Ho–Er. In spite of sharp peaks for Dy and Ho, their separation was incomplete, that is, Rs = 0.74, reflecting the tetrad effect. This is the most difficult separable pair with low α_D value, while the separation of Ho and Er was accomplished as Rs = 1.75. A chromatogram for a mixture of Ho–Er–Tm with the mobile phase of pH 2.56 is shown in Fig. 14b. The chromatographic peaks have higher N values compared with the case of 30E cartridges. The peaks for ER and Tm were completely separated, as Rs = 2.76.

Figure 15 illustrates a chromatogram for a ternary mixture of Tm, Yb, and Lu. Each lanthanoid yielded individual peaks with sufficiently high N values, that is, 150 for Tm, 160 for Yb, and 170 for Lu. These N values were still rather low compared with the total number of channels, that is, 400 × 6 = 2400. The

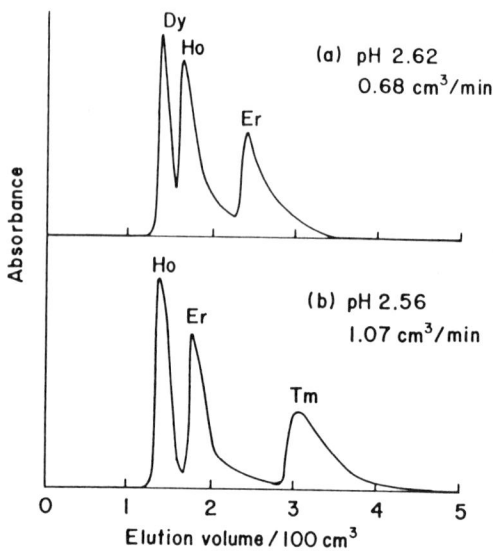

Figure 14 Separation of ternary mixtures of middle and heavy lanthanoids by the use of six cartridges of 250W. Stationary phase: 0.02 M $(DTMPPA)_2$ in kerosene; mobile phase: 0.1 M $(H,Na)Cl_2CHCOO$; 900rpm; 70°C.

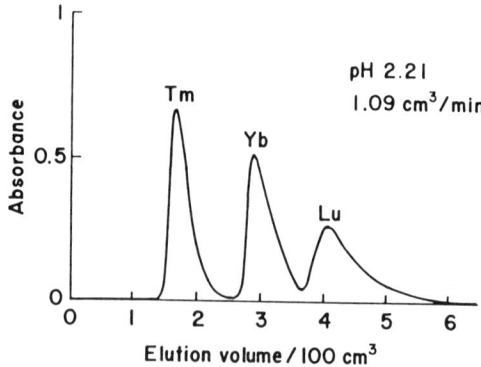

Figure 15 Separation of Tm, Yb, and Lu. Conditions are same as those in Fig. 14. (From Ref. 26.)

quantitative separation of adjacent lanthanoids has been thus achieved with sufficient resolution, 1.64 for Tm–Yb and 1.02 for Yb–Lu.

The chromatographic parameters obtained with DTMPPA loaded on 250W cartridges are summarized in Table 4. The α_C values were rather smaller than the α_D values, while they varied in a similar way only with a few exceptions; complete separation Rs > 1 has been accomplished for most of adjacent lanthanoids. As is seen for the mixture of Gd–Tb–Dy, the α_C values with DTMPPA were only slightly lower than those with EHPA; however, the mutual separation of heavy lanthanoids has been achieved by optimizing conditions of the mobile phase in addition to operating conditions: high temperature of 70°C and low flow rate of around 1 cm^3 min^{-1}. Further, the separation of a mixture of light and heavy lanthanoids has been performed in a single CPC run with the stationary phase of DTMPPA in heptane by gradient pH elution [25].

C. Separation with TBP

By employing a neutral organophosphorus extractant TBP as a stationary phase, CPC has been applied to the separation of actinoids and lanthanoids [37]. In order to enhance the distribution ratio and the separation factor, a high concentration of lithium nitrate solution was adopted as the mobile phase, based on the salting-out effect in liquid–liquid extraction [38]; the distribution ratios of lanthanoids into TBP are large only as a result of high salt concentration. Figure 16 illustrates a

Table 4 Separation Parameters for CPC Chromatogram with DTMPPA Loaded on Six Cartridges (250W) at 70°C

Lanthanoid	N	α_C	Rs
Gd	265		
Tb	429	2.13	1.69
Dy	286	1.84	1.79
Dy	603		
Ho	278	1.29	0.74
Er	365	1.76	1.75
Ho	485		
Er	256	1.81	0.96
Tm	251	2.76	2.76
Tm	150		
Yb	160	2.34	1.64
Lu	170	1.52	1.02

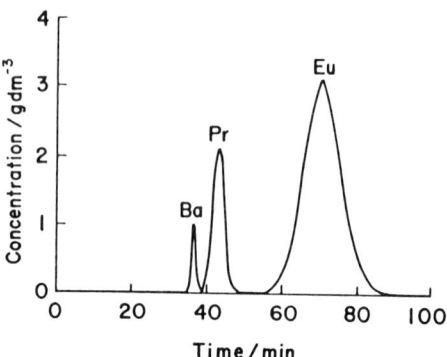

Figure 16 Separation of Pr and Eu (Ba corrects the retention time) by the use of six cartridges of 250W loading TBP. Stationary phase: 30% TBP–28% n-dodecane–42% CCl_4; mobile phase: 5 M $LiNO_3$–1 M HNO_3; 1200 rpm; 25°C. (From Ref. 37.)

typical chromatogram for the separation of Ba, Pr, and Eu, where Ba was used for correcting the retention time. The composition of the stationary phase was optimized not only for large separation factor but also for an appropriate specific gravity by the use of carbon tetrachloride. These individual peaks were satisfactorily separated as Rs = 2.0 owing to improved large separation factor of 5.1 and relatively large N values, that is, 490 for Pr and 180 for Eu. These severe conditions of high concentrations of salt and acid, however, are usually undesirable to protect devices from corrosion and for stable operations.

REFERENCES

1. H. E. Kremers, *Rare Metals Handbook*, 2nd ed. (C. A. Hampel, ed.), Reinhold Publishers, London, p. 393 (1961).
2. J. E. Powell, *Handbook on the Physics and Chemistry of Rare Earths*, Vol. 3 (K. A. Gschneidner, Jr., and L.R. Eyring, eds.), North-Holland Pub., Amsterdam, p. 81 (1979).
3. D. B. James, J. E. Powell, and F. H. Spedding, *J. Inorg. Nucl. Chem.*, *19*: 133 (1961).
4. M. M. Zeligman, *Anal. Chem.*, *37*: 524 (1965).
5. F. H. Spedding, E. I. Fulmer, T. A. Butler, E. M. Gladrow, M. Gobush, P. E. Porter, J. E. Powell, and J. M. Wright, *J. Am. Chem. Soc.*, *69*: 2812 (1947).
6. B. Weaver, *Ion Exchange and Solvent Extraction 6* (J. A. Marinsky and Y. Marcus, eds.), Marcel Dekker, New York, p. 189 (1974).
7. D. F. Peppard, J. P. Faris, P. R. Gray, and G. W. Mason, *J. Phys. Chem.*, *57*: 294 (1953).
8. D. F. Peppard, W. J. Driscoll, R. J. Sironen, and S. McCarty, *J. Inorg. Nucl. Chem.*, *4*: 326 (1957).

9. D. F. Peppard, G. W. Mason, J. L. Maier, and W. J. Driscoll, *J. Inorg. Nucl. Chem.*, *4*: 334 (1957).
10. T. B. Pierce and P. F. Peck, *Analyst*, *88*: 217 (1963).
11. Z. Kolarik, S. Drazanova, and V. Chotivka, *J. Inorg. Nucl. Chem.*, *33*: 1125 (1971).
12. E. O. Otu and A. D. Westland, *Solvent Extr. Ion Exch.*, *8*: 759 (1990).
13. D. F. Peppard, G. W. Mason, and I. Hucher, *J. Inorg. Nucl. Chem.*, *18*: 245 (1961).
14. K. Akiba, S. Sawai, S. Nakamura, and W. Murayama, *J. Liq. Chromatogr.*, *11*: 2517 (1988).
15. K. Li and H. Freiser, *Solvent Extr. Ion Exch.*, *4*: 739 (1986).
16. Y. Komatu and H. Freiser, *Anal. Chim. Acta*, *277*: 397 (1989).
17. H. Matsuyama, T. Okamoto, Y. Miyake, and M. Teramoto, *J. Chem. Eng. Japan*, *22*: 627 (1989).
18. O. D. Michelsen and M. Smutz, *J. Inorg. Nucl. Chem.*, *33*: 265 (1971).
19. T. Goto, *J. Inorg. Nucl. Chem.*, *30*: 3305 (1968).
20. X. Fu, Z. Hu, Y. Liu, and J. A. Golding, *Solvent Extr. Ion Exch.*, *8*: 573 (1990).
21. T. Araki, T. Okazawa, Y. Kubo, H. Ando, and H. Asai, *J. Liq. Chromatogr.*, *11*: 267 (1988).
22. T. Araki, T. Okazawa, Y. Kubo, H. Asai, and H. Ando, *J. Liq. Chromatogr.*, *11*: 2437 (1988).
23. T. Araki, H. Asai, H. Ando, N. Tanaka, K. Kimata, K. Hosoya, and H. Narita, *J. Liq. Chromatogr.*, *13*: 3673 (1990).
24. K. Akiba, Y. Ishibashi, and S. Nakamura, *Solvent Extraction 1990* (T. Sekine, ed.), Proceedings of ISEC'90, Elsevier Science Publishers, Amsterdam, p. 1677 (1992).
25. S. Muralidharan, R. Cai, and H. Freiser, *J. Liq. Chromatogr.*, *13*: 3651 (1990).
26. Y. Suzuki, S. Nakamura, and K. Akiba, *Anal. Sci.*, *7*: Supplement, 249 (1991).
27. D. F. Peppard, G. W. Mason, and S. Lewey, *J. Inorg. Nucl. Chem.*, *31*: 2271 (1969).
28. B. Ceccaroli and J. Alstad, *J. Inorg. Nucl. Chem.*, *43*: 1881 (1981).
29. A. C. Preez and J. S. Preston, *Solvent Extr. Ion Exch.*, *10*: 207 (1992).
30. S. Nakamura, Y. Surakitbanharn, and K. Akiba, *Anal. Sci.*, *5*: 739 (1989).
31. L. J. Nugent, *J. Inorg. Nucl. Chem.*, *32*: 3485 (1970).
32. S. Siekierski and I. Fidelis, *Extraction Chromatography* (T. Braun and G. Ghersini, eds.), Elsevier Scientific Publishers, Amsterdam, p. 226 (1975).
33. W. Murayama, T. Kobayashi, Y. Kosuge, H. Yano, Y. Nunogaki, and K. Nunogaki, *J. Chromatogr.*, *239*: 643 (1982).
34. T. Araki, Y. Kubo, T. Toda, M. Takata, T. Yamashita, Y. Murayama, and Y. Nunogaki, *Analyst*, *110*: 913 (1985).
35. D. J. Barkley, M. Blanchette, R. M. Cassidy, and S. Elchuk, *Anal. Chem.*, *58*: 2222 (1986).
36. S. P. Novikov and B. F. Myasoedov, *Solvent Extr. Ion Exch.*, *3*: 267 (1985).
37. S. Usuda, H. Abe, S. Tachimori, H. Takeishi, and W. Murayama, *Solvent Extraction 1990* (T. Sekine, ed.), Proceedings of ISEC'90, Elsevier Science Publishers, Amsterdam, p. 717 (1992).
38. D. Scargill, K. Alcock, J. M. Fletcher, E. Hesford, and H. A. C. McKay, *J. Inorg. Nucl. Chem.*, *4*: 304 (1957).

10
Separator-Aided Centrifugal Partition Chromatography

Takeo Araki
Kyoto Institute of Technology, Matsugasaki, Kyoto, Japan

I. INTRODUCTION

Separator- (or carrier-) aided CPC is a technique of centrifugal partition chromatography (CPC). The original form of CPC is a chromatography that is constituted with mutually immiscible liquid stationary phase (usually a solvent) and liquid mobile phase, that is, a continuous liquid–liquid extraction technique. As far as a simple partition mechanism such as solubility acts in CPC, a set of sample species having almost the same solubility cannot be separated by this liquid–liquid chromatography. In 1985 this author reported that alkali metal picrates can be readily separated in CPC by addition of a crown ether compound as carrier (separator) to the stationary chloroform phase (this work appears in Section II).

The basic concept of separator-aided CPC is quite simple: A separator (or carrier) compound (Y) interacts with sample species (S) to form complexes with different equilibrium constants. When the complex state ($[S\text{---}Y]$) in the stationary phase is reversible to a certain extent, the $[S\text{---}Y]$ dissociates to the Y and S species depending on the instability of the complexes. This situation can be shown as

$$S + Y \rightarrow [S\text{----}Y] \rightarrow Y + S \tag{1}$$

Thus, the use of a separator (or carrier) compound in CPC seems to greatly improve the potential of CPC because two types of mechanisms, that is, partition plus complexation or some other interactions, are involved.

A modern liquid-membrane transport (LMT) system also uses immiscible liquid phases. The same formula as Eq. 1 can be written in the LMT system where one chemical species (S) in a liquid phase is transported by the aid of a carrier compound (Y) into the other liquid phase, that is, a liquid membrane, by forming a complex ([S---Y]). Dissociation of the complex at the other side of the liquid-membrane surface completes transport of a substrate species S through a liquid membrane.

The CPC results obtained with the crown ether showed that the behaviors of the alkali picrates correlated with the corresponding extraction data. Since the LMT behaviors are known to correlate also with extraction data, a relation between CPC and LMT systems became indirectly clear. Hence, the transport data of a LMT system can be used for CPC and the separation data of CPC can be applied to LMT systems.

This chapter collects the work of this author on separator-aided CPC and the corresponding LMT systems for the reader's convenience in surveying these subjects.

Section II describes the CPC separation of alkali metal picrates with the aid of the crown ether. In Section III, direct comparison of CPC with LMT is made by using 2,4,6-trinitrophenol (picric acid) and p-nitrophenol samples. The next five sections refer to the behaviors of rare earth metal ions: Sections IV–VI for the CPC separation with di(2-ethylhexyl) phosphoric acid separator, and Sections VII and VIII for selective transport in LMT systems based on the corresponding CPC results. The final section discusses the significance of CPC as a multistage liquid-membrane transport technique, from the viewpoint of LMT technology.

For further extension of separator-aided CPC, the author and his collaborators are now developing solubilized inorganic solids as separators. Partial success in the solubilization of solid silica gel (Aerosil 200) in toluene has just been obtained by introducing octadecyldimethyl silyl (ODS) groups [1]. The octadecylated silica gel is chemically almost identical with that employed in high-performance liquid chromatography (HPLC). Hence, it is expected that a majority of the data of HPLC obtained with ODS columns will be examined by this ODS-silica gel-aided CPC. Further extension of this new technique will advance the CPC separations. Unfortunately, the last subject cannot be discussed here now.

II. CARRIER-AIDED CPC FOR PREPARATIVE-SCALE SEPARATIONS*

A. Introduction

Although high-performance liquid chromatography (HPLC) is a powerful separation technique, its use has been limited; for preparative-scale applications, it requires large amounts of expensive column materials, packing of the column is a

---

*Adapted from T. Araki, Y. Kubo, T. Toda, M. Takata, T. Yamashita, W. Murayama, and Y. Nunogaki, *The Analyst, 110*: 913–916 (1985), by the courtesy of The Royal Society of Chemistry.

complicated procedure, and it is difficult for users to change the column materials according to their own requirements. For many years, attempts have been made to overcome these problems, but so far there is no effective solution.

A new separation technique, centrifugal partition chromatography, was introduced by Murayama et al. [2] at Sanki Engineering Ltd., Japan. It was developed on the basis of the classical countercurrent distribution method. The countercurrent distribution method was improved by the use of a rotary seal joint device, which makes it possible to pump the mobile phase through the system while it is under centrifugation. The conventional countercurrent partition cells are replaced by microcells connected in series to form a cell cartridge. One cartridge is composed of 50 microcells, and it is possible to increase the number of cartridges when a higher partition efficiency is desired. Application of a centrifugal force to the cartridge shortens the phase separation time in each microcell. Hence, a high-speed multistage countercurrent procedure is achieved.

This chromatographic method is attractive from several points of view: (1) scaling up to an industrial-scale separation is possible; (2) the stationary phase in the cartridges can be changed by an extremely simple procedure, and any possible stationary phase can be used arbitrarily; (3) the stationary phase can be chosen from both water and organic solvents; when water is used as the mobile phase, an immiscible organic solvent can be used as the stationary phase, or vice versa, which gives the possibility of wide applicability; and (4) the centrifugal force to be applied depends on the viscosity and density of the liquids used, but no strong force is needed when a combination of common solvents is employed.

Murayama and coworkers [2,3] tested the method on the separation of fatty acids, amino acid derivatives, sugars, and saponins by employing simple solvent-to-solvent mechanisms.

With this background, we have attempted to improve the efficiency by introducing chemical mechanisms in the partitioning microcells. In this section, we report the effect of the addition of a carrier molecule to the stationary phase, which has a host–guest interaction ability with a specific species in the sample. As an example, a crown ether was used as the host molecule. The simple method was found to be effective when the conventional solvent-to-solvent partition mechanism results in failure, as in the separation of alkali metal picrates (guest molecules). Although the present results were obtained only in limited systems, the principle can be extended to a large variety of chemical interactions, and with suitable carriers the technique can be widely applied to a variety of samples.

B. Experimental Procedures

1. Materials

Analytical reagent–grade solvents were used after distillation. Alkali metal picrates were prepared by the reaction of alkali metal carbonates with picric acid in water and recrystallized from warm water. Dibenzo-18-crown-6 was purchased from Nacalai Tesque Ltd., and used after recrystallization.

2. *Apparatus*

A Model L-90 centrifugal partition chromatograph (Sanki Engineering Ltd.) was connected to a Uvilog-5III UV detector (Oyo-Bunko Kiki Co. Ltd.) and a pen recorder. Pumping of the mobile phase (water) was carried out by using a common liquid chromatographic pump, viz., a Model 5GK50 (Oriental Motor Co., Ltd.), with a Model 5SK2SGKA reaction motor.

3. *Chromatography*

Three separating cartridges (each containing 50 microcells) were connected in series to form a partitioning device. Dibenzo-18-crown-6 was dissolved in a chloroform-rich organic solvent, and the solution was used as stationary phase. Water was eluted as the mobile phase by pumping at the inlet of the flow system. The chromatogram was monitored at the end of the flow line by using a UV detector at 380 nm and was operated under centrifugation at 600–1000 rpm at room temperature. The operating conditions, that is, choice of solvent, amount of the crown ether, sample size, inlet pressure (flow rate), and number of rotations of the centrifugal disk, were optimized by repeating the experiments. The total volume of the stationary phase was determined to be 45 ml and the dead volume 10.8 ml, irrespective of the nature of the stationary phase.

C. Results and Discussion

Dibenzo-18-crown-6 (**I**) is almost insoluble in water but has limited solubility in chloroform. In addition, its characteristic complexation ability has been widely studied on a series of alkali metal cations [4]. In this respect, the crown ether **I** was chosen as a carrier compound in the stationary phase for the centrifugal partition chromatography of alkali metal picrates. The following stationary phases were tested: (1) **I** (0.06 g) in $CHCl_3$, 700 rpm; (2) **I** (0.64 and 0.81 g) in $CHCl_3$ + p-cresol (19 + 1 and 9 + 1), 700 and 1000 rpm; (3) **I** (1.6 g) in $CHCl_3$ + ethanol (9 + 1), 700 rpm; and (4) **I** (1.6 g) in $CHCl_3$ + butan-1-ol (9 + 1), 700 rpm. When the crown ether was not added, all alkali metal picrates used showed the same retention volumes (V_R) in each solvent system.

The effectiveness of the crown ether as the host–guest partitioning reagent depended largely on the nature of the organic solvent system. Stationary phase (1) was found to be impractical because the values of the alkali metal picrates were too small owing to the low solubility of the crown ether in $CHCl_3$. In order to increase the efficiency of contact of the sample ions with the carrier molecule, stationary phases (2)–(4) were examined.

Stationary phase (2) was better than (1) in view of the longer retention time due to the higher concentration of the crown ether used. Considerable separation efficiencies were observed with this stationary phase, but the acidic nature of p-cresol caused some mechanical problems in the metal parts of the flow and detection systems. With (3), the retention time also increased. Escape of the crown ether from the stationary phase was significant, however, giving rise to deposition of solidified **I** in the various parts of the chromatographic lines.

Stationary phase (4) gave satisfactory results for our purpose. The role of butan-1-ol was important for increasing the concentration of **I** and increasing the affinity of the aqueous mobile phase for the $CHCl_3$ phase. The former is the solubility effect, and the latter the surfactant effect. Escape of **I** from the stationary phase was extremely slight in the $CHCl_3$ + butan-1-ol systems.

The retention values for the individual samples, which were obtained with (4), are shown in Table 1. The results show that the crown ether is an effective carrier for changing the V_R values of each alkali metal picrate. Hence, the addition of a carrier molecule to the stationary phase is seen to be an effective and simple method for improving the separation efficiency of centrifugal partition chromatography.

When the V_R values were plotted against the values of equilibrium constants (K_{ex}), which were determined by the extraction method [5,6] for the aqueous picrate solutions with dibenzo-18-crown-6 in $CHCl_3$, a roughly linear correlation was obtained (Fig. 1). This result indicates that the separation mechanism in carrier-aided chromatography is almost parallel to the nonchromatography steady state complexation behavior. Hence, the extent of host–guest interaction in the chromatography, and therefore relative V_R values, can be predicted approximately from the static physicochemical equilibrium constant or complexation constant, which are determined by the extraction method in closely related two-phase systems. Conversely, carrier-aided CPC may be used as a physicochemical method

Table 1 Retention Values of Alkali Metal Picrates Using Dibenzo-18-Crown-6-Aided CPC[a]

Alkali metal picrates: M^+	V_R (ml)	V_R per g crown ether (ml)	V_R per g crown ether per microcell (ml × 100)
Li^+	14.0	8.60	5.73
Na^+	13.8	8.47	5.65
K^+	25.6	15.58	10.39
Rb^+	20.5	12.58	8.39
Cs^+	19.0	11.66	7.77

[a]Stationary phase: dibenzo-18-crown-6 (1.63 g) in $CHCl_3$–butan-1-ol (9 + 1). Mobile phase: water, 1 ml min^{-1}. Rotation speed: 700 rpm. Number of cartridges: 3 (150 microcells). Sample size: 1.5 ml (concentration 0.01 M). Room temperature.

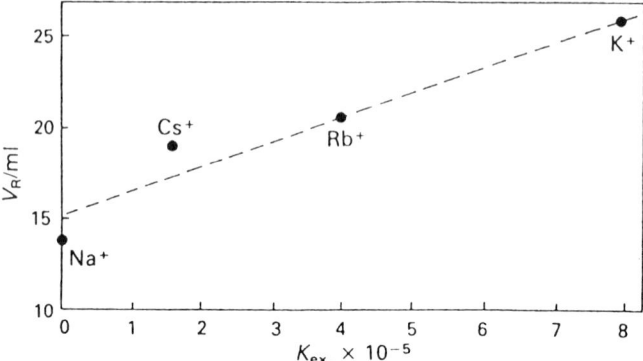

Figure 1 Correlation between V_R values in CPC and K_{ex} values [3,4] for alkali metal picrates in dibenzo-18-crown-6 systems. Correlation factor = 0.96.

for obtaining certain information about complexation behavior when the chromatography is carried out under suitable conditions. Simplicity in the procedures for exchanging the stationary phase, together with the mechanically controlled operation of chromatography (in which diffusion problems are minimized by applying centrifugal force), seems to be advantageous for such a physicochemical application.

The number of theoretical plates calculated from our chromatograms by using the equation $n = (4d/w)^2$ did not deviate much from the number of microcells used. If one knows the value of the chemical factor associated with a given stationary phase in one microcell, the approximate number of theoretical plates can be precalculated by multiplying the number of microcells by the value of the chemical factor. This effect applies because the theoretical model of the plate theory in partition chromatography is the same as the construction of the microcells in this apparatus [7], and the use of a centrifugal force eliminates the diffusional complications that occur in the partitioning cells.

The relative V_R values for the binary ion systems were determined (Table 2) from the chromatograms obtained with 150 microcells (a very small number of theoretical plates compared with that in HPLC). The best separation was observed for the pair of K^+/Na^+ ions; the $V_R(K^+)/V_R(Na^+)$ ratio was 1.99. This pair was separated with about 90% efficiency on the chromatogram (Fig. 2a). The observable limit of the separation on the chromatogram under the same conditions was found to be $V_R(M_1^+)/V_R(M_2^+) \approx 1.25$, as exemplified by the pair K^+/Rb^+ (Fig. 2b). Separation of the pairs Na^+/Rb^+, Na^+/Cs^+, and Li^+/Cs^+ exceeded this value, and incomplete separations were observed. The results for binary ion

Table 2 Separation Factors and $\log(K_{ex1}/K_{ex2})$ for Binary Alkali Metal Picrate Samples in Crown Ether–Aided Centrifugal Partition Chromatography[a]

Alkali metal picrate		$V_R(M_1^+)/V_R(M_2^+)$		
M_1^+	M_2^+	Observed	Calculated[b]	$\log(K_{ex1}/K_{ex2})$
Li+	Na+	1.00	1.01	—
K+	Na+	1.99	1.86	2.0
Rb+	Na+	1.52	1.49	1.7
Cs+	Na+	1.50	1.38	1.3
Cs+	Li+	1.46	1.36	—
K+	Rb+	1.26[c]	1.25	0.3

[a]Experimental conditions as in Table 1.
[b]Calculated from the values shown in Table 1.
[c]Estimated from the less separated chromatogram shown in Fig. 2b.

systems were almost consistent with expectation from the V_R values obtained in the single-ion experiments. When a mixture containing all five cations was used as the sample, group separation took place, that is, one group containing Li+ and Na+ ions and the other K+, Rb+, and Cs+ ions. As this type of chromatography can be applied to preparative-scale fractionation, this technique could be used to pretreat a complicated sample and to simplify further the separation procedures by other techniques.

Tables 1 and 2 give the values of V_R per gram of crown ether per microcell and the relationship of $V_R(M_1^+)/V_R(M_2^+)$ versus $\log(K_{ex1}/K_{ex2})$, respectively. These values will help to establish the optimum conditions under which a complete separation can be obtained without repeated trial and error measurements.

For the K+/Na+ binary system, each fraction was collected at the exit of the detector and rechromatographed under the same conditions. The samples were diluted during the first chromatography, but it is clear that each ion was purified in the first chromatographic process (Fig. 3).

Chromatographic experiments were carried out under the same conditions by using alkali metal benzoate and alkali metal p-nitrophenolate samples. The differences in V_R values were small, in spite of the long retention times. These results suggest that the benzoates and p-nitrophenolates behave as ion pairs rather than as free ions, and the interaction between sample and crown ether occurs mainly through the organic moiety of the samples. It appears to be important to consider the ionic dissociation constants of the salts.

The $CHCl_3$ solution used as the stationary phase was recovered simply by

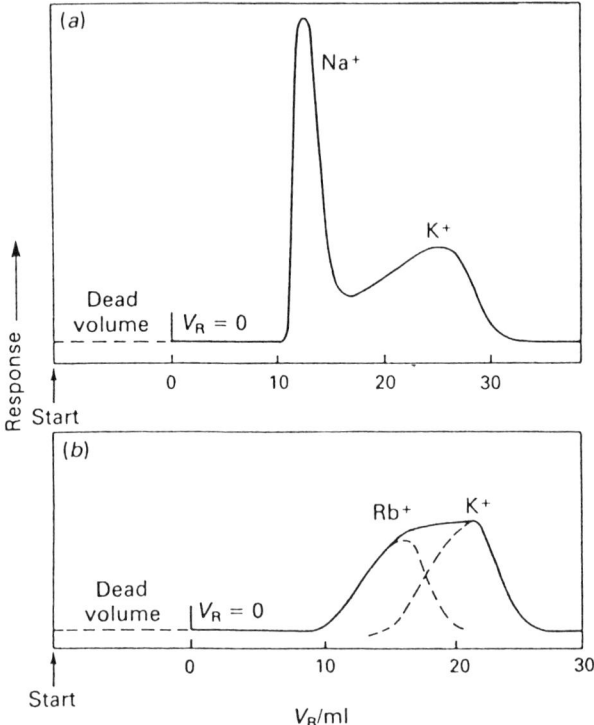

Figure 2 Separation of binary picrate samples in CPC with 150 microcells using dibenzo-18-crown-6 as a carrier compound in the stationary phase. (a) K and Na picrates in a 1:1 molar ratio; (b) K and Rb picrates in a 1:1 molar ratio.

pumping water in the absence of a centrifugal force. From this $CHCl_3$ solution it is possible to recover the crown ether almost quantitatively by separating the $CHCl_3$ phase from water, followed by drying and evaporating the solvent in vacuo. Recrystallization of the recovered crown ether gave the pure compound, which could be used again or taken for other purposes. This feature is also characteristic of the separating reagent in other chromatographic techniques.

These results have demonstrated briefly what happens in the chromatographic technique described. These involve several significant advantages together with some problems and limitations and also an outline of the physicochemical background. These observations will provide a basis for the future development of centrifugal partition chromatography, which will be important as a

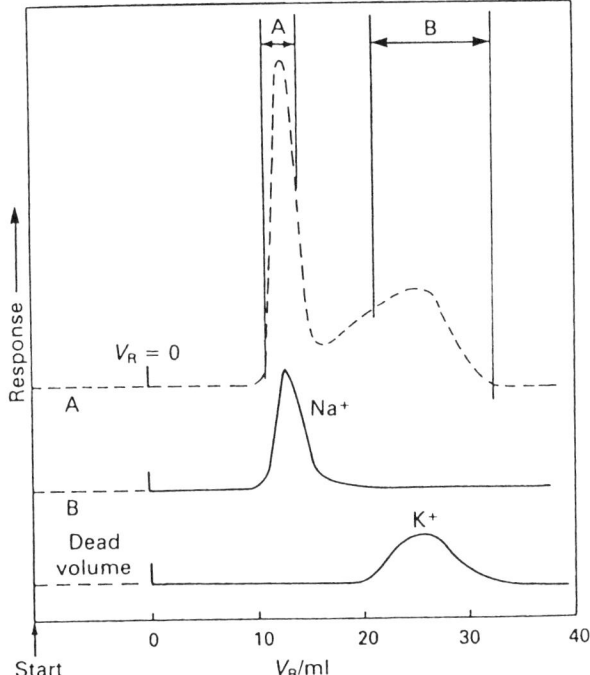

Figure 3 Rechromatography of picrate fractions collected. A: Na picrate fraction; B: K picrate fraction. The first chromatogram, shown as a dashed line, is the same as the chromatogram in Fig. 2a, in which fractions A and B are indicated as ranges.

complementary tool to overcome the defects of various other popular chromatographic techniques.

III. SELECTIVE LIQUID-MEMBRANE TRANSPORT OF NITROPHENOLS BY A SIMPLE AND COSTLESS CARRIER. AN APPLICATION OF THE RESULTS OF CPC*

Artificial liquid-membrane transport (LMT) [8] has been obtaining an important position [9] in chemistry in view of its dual significance. First, it is an experimental technique of biomimetic molecular recognition or a model of cell membrane. Second, it is a potential energy-saving concentration technique. In the studies from the former standpoint, it is reasonable to employ a carrier compound of relatively

---

*Adapted from T. Araki, Y. Kubo, M. Takata, S. Gohbara, and T. Yamamoto, *Chem. Lett.*, 1011–1012 (1987), by the courtesy of the Chemical Society of Japan.

complicated but properly designed structure. When, however, LMT is studied on the latter viewpoint, it is worthwhile to find effective and costless carriers suitable for a given substrate mixture.

Despite recent developments, it seemed still rather difficult to attain effective selective transports for nonionic organic molecules [9]. Our recent results in separator- or carrier-aided CPC (see Section II) indirectly suggest that the chromatography is related to LMT, because the retention volumes (V_R) for the samples were almost linearly correlated with the equilibrium constants obtained by extraction from the closely related two-phase systems.

This section discusses the selective LMT of p-nitrophenol (NP) and 2,4,6-trinitrophenol (TNP, also called picric acid) by an extremely simple separator compound, butan-1-ol.

Prior to the LMT experiment, we studied the preparative-scale separation of the two nitrophenols by CPC (Sanki Engineering Co. Ltd., Model L-90 apparatus was used). In this CPC experiment, a mixed solvent ($CHCl_3$/butan-1-ol) was used as the stationary phase and H_2O as the mobile phase (1 ml min^{-1}), under centrifugation with a rotation speed of 700 rpm at room temperature. When the amount of butan-1-ol was 0.5% (v/v) of $CHCl_3$ (45 ml), the mixture of equal amounts of the two nitrophenols was fractionated into nuclear magnetic resonance (NMR) spectroscopically pure compounds (ratio of V_R(PN) to V_R(TNP): 5.4) with recovery efficiency of 70% (13.0 mg) and 76% (14.9 mg) (total of 5 runs) for PN and TNP, respectively, with only 150 separating microcells (50 microcells × 3 cartridges, which corresponds to an almost identical number of theoretical plates—see Section II.

LMT experiments were performed by using an armed double-cylinder glass apparatus at room temperature. The source phase and receiving phase were H_2O, and the liquid membrane was constructed by $CHCl_3$ containing 5% butan-1-ol (v/v), as in the case of CPC separation. The amounts of nitrophenols transported were determined by absorption spectra using precalibrated graphs.

No transport was observed through the liquid membrane of $CHCl_3$ alone. When, however, the membrane containing 5%-butan-1-ol was used, TNP was transported very quickly and PN slowly, as expected from the results of CPC. The rate constants of transport per unit concentration, k_{tr}(TNP) and k_{tr}(PN) (corrected for the concentration: $1.0 \times 10^{-3} M$ and $1.0 \times 10^{-2} M$ for TNP and PN, respectively), were 2.06×10^{-1} hour^{-1} and 2.46×10^{-2} hour^{-1}, respectively. The relative rate can be estimated as 8.4.

Our results indicate that these data from CPC are useful to find an effective carrier of a selective LMT system. The fundamental mechanism of CPC is the same [2] as the classical countercurrent distribution method [10] and its improvements, including droplet countercurrent chromatography (DCCC) [11] or planet-coil centrifugal countercurrent chromatography [12], and other partition data have been accumulated using a variety of two-phase systems. Hence, researchers in LMT seeking energy-saving concentration of weakly ionic or nonionic compounds can obtain guidance also from these countercurrent or partition data. At the same time, preparative-scale separations by the separator-aided CPC can be developed by using the data of LMT systems. Note that the detailed phase transfer mechanisms in CPC and LMT are not fully identical; it is necessary to consider the difference. The studies on this problem are under way in our laboratory.

IV. SEPARATION OF LIGHTER RARE EARTH METAL IONS BY CPC WITH DI(2-ETHYLHEXYL) PHOSPHORIC ACID*

A. Introduction

Recently, application of centrifugal force [2,13–17] in the continuous extraction method has opened a new quick technique for separation, isolation, purification, and concentration of a variety of organic [2,16–21] and inorganic ions (see Section II). The results of centrifugal countercurrent chromatography were found capable of being successfully applied for costless selective liquid-membrane transport systems (see Section III). The work described in this section was the first report showing the usefulness of this type of technique for lighter rare earth (RE) metal ions.

Centrifugal continuous extraction methods are now developing under two names: (1) centrifugal countercurrent chromatography (CCC) and (2) centrifugal partition chromatography (CPC). These two chromatographies differ mainly in their machinery for applying centrifugal force in the continuous extraction processes. Namely, the former employs Ito's seal-free planet-coil centrifuge [16,17] and the latter Murayama's microcell cartridges together with rotary seal joints [2]. The fundamental chemical aspects involved in both of these chromatographies are, however, almost identical.

For the separation of rare earth metal ions, the single-stage liquid liquid extraction using di(2-ethylhexyl)phosphate (D2EHPA) or related phosphates [22–30] has been studied extensively, and its applications to a column chromatography [31] and a 20-stage mixer–settler method [32] have already been reported. Recently, extractions by using liquid surfactant membranes have also been investigated [33–35]. The results of separation using a centrifugal countercurrent method reported here can be used in advancing the studies on purification and/or concentration of rare earth metal ions by this type of quick and effective centrifugal countercurrent chromatographic technique, with a view toward industrial-scale applications.

---

*Adapted from T. Araki, T. Okazawa, Y. Kubo, H. Ando, and H. Asai, *J. Liq. Chromatogr.*, *11*: 267–281 (1988), by the courtesy of Marcel Dekker, Inc.

B. Experimental Procedure

1. Materials

Lighter rare earth metal ions were obtained from commercial suppliers and used without further purification: $LaCl_3 \cdot 7H_2O$ (Wako, first grade), $CeCl_3 \cdot 7H_2O$ (Wako, >98%), $PrCl_3 \cdot 7H_2O$ (Kanto, first grade), $NdCl_3 \cdot 6H_2O$ (Wako, >97%), $SmCl_3 \cdot 6H_2O$ (Nacalai, first grade), and $EuCl_3 \cdot 6H_2O$ (Nacalai, extra pure grade). Di(2-ethylhexyl)phosphoric acid (D2EHPA) (Nacalai, 99%), n-heptane (Nacalai, extra pure grade), xylenol orange (Wako, analytical grade), and hexamethylenetetramine (Wako, extra pure grade) were used without further purification. Deionized water was used throughout the experiments.

2. Apparatus

Centrifugal partition chromatography was performed by using a CPC L.L. apparatus Model NMF (Sanki Engineering Ltd.) with Model 240W cartridges. The pumping system was Model LBP-V (Sanki Engineering Ltd.), and the injection valve system was a Model CPC-FCU II (Sanki Engineering Ltd.). A Hitachi spectrophotometer Model 200-20 was used for quantitative analysis of RE ions in the fractions eluted.

3. Chromatographic Procedures

Three high-resolution-type cartridges (Model 240W) containing 400 microcells per cartridge were equipped in series (total 1200 microcells). A solution of 0.1 M D2EHPA in n-heptane (stationary phase) was filled in the microcells by the following procedures: The solution of stationary phase was pumped at a maximum speed under centrifugation (400 rpm) while the mode valve was turned to "ascending mode," and then the stationary phase solution was kept recycling for 10–15 minutes to remove gaseous materials completely. Next, the valve was turned to "descending mode," and a given concentration of aqueous HCl solution (mobile phase) was pumped in a descending fashion through the microcells under centrifugal force (800 rpm), adjusting to a given elution speed (1.12 ± 0.02 ml min^{-1}). The n-heptane solution (23 ml) in the dead spaces was eluted during this procedure, and about 40 ml (containing 1.3 g of D2EHPA) of stationary phase was retained in the microcells. When "descending elution" is employed, it is necessary to carefully avoid introducing gaseous materials in the microcells; otherwise, the gaseous materials stay in the microcells and cause trouble, for example, leaking of the stationary phase out of the microcells during chromatography, together with instability in the rate of elution. After the elution was stabilized for about 1 hour, sample solution (0.54 ml) was charged through the sample injecting valve.

4. Analysis of Fractions

The eluted aqueous mobile phase was collected at the end of the chromatographic line at 5-minute intervals. From each fraction, 2-ml aliquots were taken, and aqueous solutions of 5.60×10^{-4} M xylenol orange (1 ml) and 0.5 M hexamethylenetetramine hydrochloride buffer (2 ml) were added, resulting in 5 ml of aqueous solutions of pH 5.6 containing 1.2×10^{-4} M of xylenol orange and 0.2 M hexamine buffer. Concentrations of the RE ions thus colored were determined from the intensities of absorption spectra at 420 nm [36] by comparison with the precalibrated values.

C. Results and Discussion

The samples of lighter RE ions used were aqueous solutions of $LaCl_3$, $CeCl_3$, $PrCl_3$, $NdCl_3$, $SmCl_3$, and $EuCl_3$. Under centrifugation the mobile phase (aqueous HCl solution) was pumped through the microcell cartridges (400 microcells per cartridge; total 1200 ($= 3 \times 400$) microcells were used) containing 0.1 M D2EHPA in n-heptane solution as the stationary phase. The fractions collected were analyzed by absorption spectroscopy at 420 nm after treating with xylenol orange.

Figure 4 shows the results of chromatographic complete separation of one-to-one mixture of La^{3+} and Ce^{3+} ions, an adjacent pair of the lightest RE ions, and also shows the excellent reproducibility of the experiments (curve a). In this case, total 0.02 M $RECl_3$ (0.54 ml) was charged and eluted with $0.050 N \pm 0.001 N$ aq.-HCl by "descending" method.

When Pr^{3+} ion was charged under the same conditions, it was eluted considerably later than the Ce^{3+} ion peak (curve b), and it is clear that the separation between La^{3+} and Pr^{3+} ions is also complete. The separation between Ce^{3+} and Pr^{3+} ions was very good, but the tail of the peak of Ce^{3+} ion was slightly overlapped with the peak-front of Pr^{3+} ion. Under the conditions used, it appears possible to obtain about 75% yields of pure Ce^{3+} and Pr^{3+} ions from a 1:1 mixture of these ions.

We found that the retention time of the RE ions depends strongly on the concentration of HCl in the mobile phase, consistent with the conclusion of extraction experiments that the extraction with D2EHPA involves a cation-exchanging processes based on acid–base interactions [28]:

$$3(HG)_2 + RE^{3+} \rightleftarrows RE(HG_2)_3 + 3H^+ \qquad (2)$$

where G denotes the phosphate anion. A typical example of HCl-concentration dependence is shown in Fig. 4 (curve c), in which the separation between La^{3+} and Ce^{3+} ions was further improved with a mobile phase of aqueous $0.040 N$ HCl.

The separation between La^{3+} and Nd^{3+} ions was found to be also complete

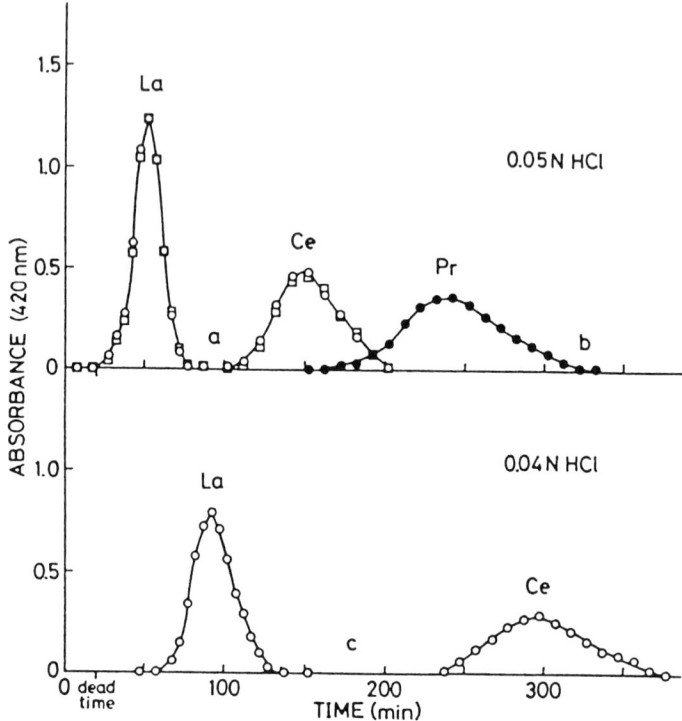

Figure 4 Centrifugal partition chromatograms of $LaCl_3$, $CeCl_3$, and $PrCl_3$ at room temperature with D2EHPA/n-heptane stationary phase. Curve (a) shows reproducibility of chromatography by using a binary mixture (0.54 ml) of $LaCl_3$ (0.01 M) and $CeCl_3$ (0.01 M); mobile phase: 0.050 N HCl. First run ○, and second run □. Curve (b) is the chromatogram of 0.54 ml of $PrCl_3$ (0.01 M) alone under the same conditions as curve (a). Curve (c), the same sample as the case of curve (a); mobile phase, 0.040 N HCl.

as shown in Fig. 5 (curve a), where concentration of HCl was 0.070 N to shorten the total retention times. Under this condition, the complete separation between La^{3+} and Pr^{3+} ions was still retained. Pr^{3+} and Nd^{3+} ions showed considerably different retention times (curves a and b), but because of the broadness of the peaks, resolution of these two curves was not so good under the conditions used. When the concentration of HCl was decreased to 0.060 N, the separation of Pr^{3+} versus Nd^{3+} ions was somewhat improved, allowing concentration of these ions to a considerable extent (curves c and d). For the separation of the last pair of ions, peak broadening became significant, and it seems better to test by another stationary phase or by using a microcell cartridge with higher theoretical plate numbers.

Separation between Nd^{3+} and Sm^{3+} ions was almost complete with a 0.140

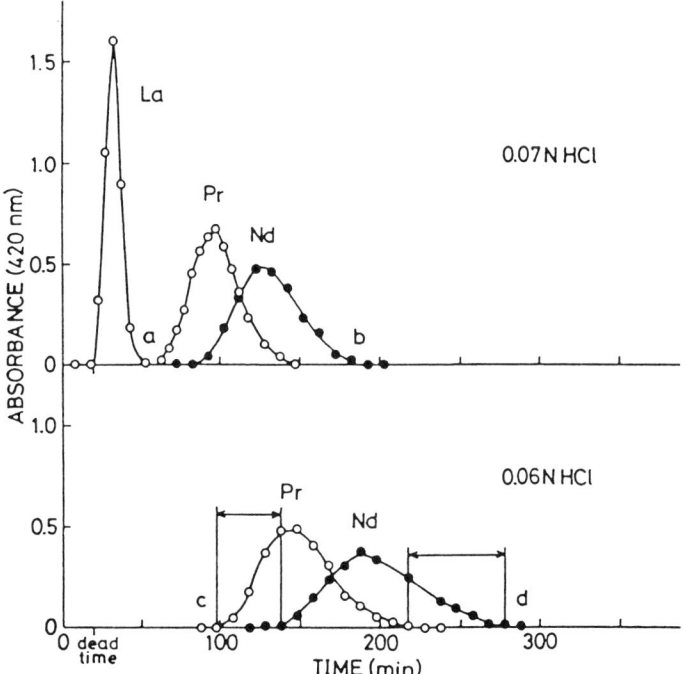

Figure 5 Centrifugal partition chromatograms of $LaCl_3$, $PrCl_3$, and $NdCl_3$ at room temperature with D2EHPA/n-heptane stationary phase. Curve (a): sample: a binary mixture of 0.54 ml of $LaCl_3$ (0.01 M) and $PrCl_3$ (0.01 M); mobile phase: 0.070 N HCl. Curve (b): sample: 0.54 ml of $NdCl_3$ (0.01 M) alone under the same conditions as curve (a). Curve (c): sample: 0.54 ml of $PrCl_3$ (0.01 M) alone, and curve (d): sample: 0.54 ml of $NdCl_3$ (0.01 M) alone; mobile phase, 0.060 N HCl. The ranges indicated on the chromatograms are those of the fractions of high-purity RE ions.

N HCl mobile phase as shown in Fig. 6 (curve a). It turned to be complete when a 0.120 N HCl mobile phase was used (curve b), although peak broadening for the Sm^{3+} ion was considerable.

Chromatography of an equimolar binary mixture of Sm^{3+} and Eu^{3+} ions by using the D2EHPA/n-heptane stationary phase resulted in extremely significant peak broadening and was not suitable for the purpose of purification. The use of D2EHPA/toluene stationary phase, however, gave quite good resolution of the mixture, as shown in Fig. 7.

Table 3 compares separation factors α obtained by our countercurrent-type chromatography with those of literature values obtained by the corresponding extraction methods using D2EHPA [31]:

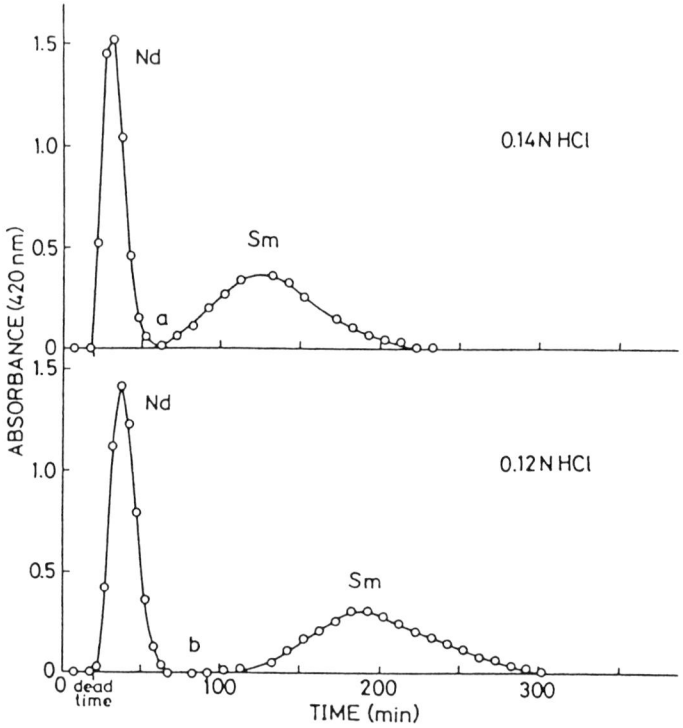

Figure 6 Centrifugal partition chromatograms of a binary mixture of NdCl$_3$ and SmCl$_3$ at room temperature with D2EHPA/n-heptane stationary phase. Sample: 0.54 ml of NdCl$_3$ (0.01 M) and SmCl$_3$ (0.01 M). Curve (a): mobile phase, 0.140 N HCl. Curve (b): mobile phase, 0.120 N HCl.

$$\alpha = \frac{t_R(RE_2)}{t_R(RE_1)} \tag{3}$$

where t_R is the apparent retention time, and RE_2 and RE_1 are RE ions of higher and lower atomic numbers, respectively.

Although the experimental conditions of these two systems are different and no detailed comparison is possible, our countercurrent chromatographic results were found to be substantially parallel to the single-stage extraction data. (Interestingly, our results from using chlorides of RE ions agreed relatively well with the extraction values obtained by perchlorates rather than chlorides; the reason for this is unknown at present.) This parallelism indicates that our countercurrent chromatography is almost just a multistage process of the extraction method; and as long

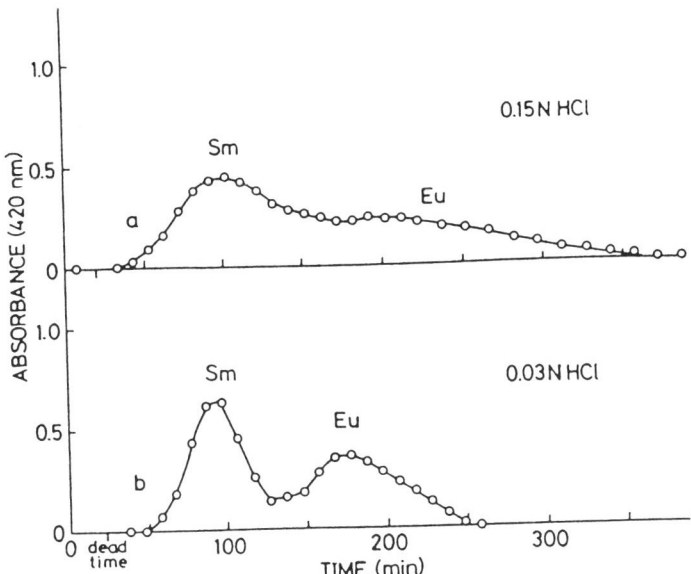

Figure 7 Centrifugal partition chromatograms of a binary mixture of SmCl$_3$ and EuCl$_3$ at room temperature. Sample: 0.54 ml of SmCl$_3$ (0.01 M) and EuCl$_3$ (0.01 M). Curve (a): stationary phase, 0.1 M D2EHPA in n-heptane; mobile phase, 0.150 N HCl. Curve (b): stationary phase, 0.1 M D2EHPA in toluene; mobile phase, 0.030 N HCl.

Table 3 Comparison of Separation Factors Between Centrifugal Partition Chromatography and Extraction Method

		Separation factor (α)		
		CPC[a]	Extraction[b]	
RE$_1$	RE$_2$	(RECl$_3$)	(RE(OCl$_4$)$_3$)	(RECl$_3$)
La	Ce	3.9–4.1	3.0	2.4
Ce	Pr	1.7	2.1	2.8
Pr	Nd	1.4	1.4	1.7
Nd	Sm	9.2–9.3	6.8[c]	5.0[c]
Sm	Eu	2.3 (2.2[d])	1.9	2.2

[a]This work: 0.1 M D2EHPA in n-heptane.
[b]0.54 M D2EHPA in toluene was used for extraction of trace amounts of Re ions [31].
[c]Calculated by the following equation: $\alpha_{Nd,Sm} = (\alpha_{Nd,Pm})(\alpha_{Pm,Sm})$.
[d]Stationary phase: 0.1 M D2EHPA in toluene.

as the data of extraction are obtained carefully, the approximate chromatographic separations can be readily predicted as in the cases of many organic samples.

As can be seen from Table 3, the separation of higher members of RE ions than Ce^{3+} from La^{3+} can be readily realized by our chromatographic technique. On comparing the behaviors of Ce^{3+}, Pr^{3+}, and Nd^{3+} ions with each other, it is clear that Ce^{3+} and Nd^{3+} ions can be separated if ca. 0.070 N HCl is used as mobile phase (see Table 4). Separation of Sm^{3+} and Eu^{3+} ions from Ce^{3+}, Pr^{3+}, or Nd^{3+} ion seems to be quite easy.

From our centrifugal countercurrent experiments using D2EHPA as stationary phase, values concerning other chromatographic behaviors of lighter RE ions can be summarized as shown in Table 4. Tabulated in Table 4 are the number of effective theoretical plates (n) and the observed peak resolution (R), which are calculated from

$$n = 5.55 \left(\frac{t_R}{w_h} \right)^2 \quad (4)$$

$$R = 2 \frac{t_R(RE_2) - t_R(RE_1)}{w(RE_2) + w(RE_1)} \quad (5)$$

where w_h is the half width of the peak and w ($= 2w_h$) is the peak width.

The limit of the use of D2EHPA is found in its significant peak broadening for all of the RE ions used, especially for the higher homologs. The values of the separation factor were high in this D2EHPA system (see Table 4), but the values of n were much lower than 100, though the number of microcells used was 1200.

Table 4 Effective Number of Theoretical Plates (n) and Peak Resolution (R) in Centrifugal Partition Chromatography[a]

RE_1	RE_2	Concn. of HCl in mobile phase (M)	$n(RE_1)$	$n(RE_2)$	$R(RE_1$ vs. $RE_2)$
La	Ce	0.04	34	77	2.0
La	Ce	0.05	14	41	1.5
La	Pr	0.07	5	31	1.4
Ce	Pr	0.05	41	54	0.77
Pr	Nd	0.06	33	41	0.44
Pr	Nd	0.07	31	28	0.37
Nd	Sm	0.12	5	21	1.4
Nd	Sm	0.14	3	12	1.1
Sm	Eu	0.15	5	18	0.56
Sm	Eu[b]	0.03	18	27	0.75

[a]Stationary phase: 0.1 M D2EHPA in n-heptane.
[b]Stationary phase: 0.1 M D2EHPA in toluene.

When the sample ions behave ideally in the stationary phase, the values of n should be quite similar to the number of microcells. (For instance, in the previous combination of alkali metal picrates and dibenzo-crown ether, we observed in Section II that the values of n almost correspond to the number of microcells (150).)

Although our separation experiments with the D2EHPA system were successful in laboratory scale, considerable loss in the n values resulted in quite low values of R, which should be improved before this chromatographic technique is applied for industrial-scale purification of RE ions. A key cause of the peak broadening is the mechanism of acid–base interaction (Eq. 2) between the sample and D2EHPA in the stationary phase. Similar considerable peak broadenings have been observed for the behavior of optically active mandelic acids with separator-aided CPC by the use of a stationary phase containing brucine, a base [37].

Our results reported here will stimulate studies to find new reagents for the stationary phase showing higher n values via interaction mechanisms of other than acid–base type. As exemplified by the separation of Sm^{3+} and Eu^{3+} ions, the nature of solvent in the stationary phase greatly alters the resolution (and number of theoretical plates). Certainly, the solvent controls the equilibria involved in the acid–base interaction, and finding an appropriate solvent is also necessary.

V. FURTHER RESULTS ON BEHAVIOR OF RARE EARTH METAL IONS IN CPC WITH DI(2-ETHYLHEXYL) PHOSPHORIC ACID*

A. Introduction

We have previously reported (Section IV; see also [38]) that adjacent pairs of lighter rare earth (RE) metal ions from $LaCl_3$ to $SmCl_3$ can be effectively separated by a centrifugal countercurrent–type chromatography (centrifugal partition chromatography, CPC) by using di(2-ethylhexyl)phosphate (D2EHPA), the most familiar extractant, as "separator" in the stationary phase. We have emphasized the importance of the adjustment of concentration of HCl ([HCl]) in the aqueous mobile phase, which is one of the key factors for obtaining effective separations. This is basically reasonable because the separation is governed by acid–base interactions of D2EHPA with the RE ions.

In addition, as exemplified in the separation of $SmCl_3$ from $EuCl_3$, the use of toluene, which is more polar than n-heptane, was effective for RE ions of higher atomic number. Independent of our work, Akiba and coworkers [39] also succeeded effective separations of RE ions of La, Pr, Nd, Sm, Eu, Gd, Tb, and Dy by the use of a related phosphonate of PC-88A (2-ethylhexyl phosphonic acid mono-2-ethylhexyl ester) as separator in kerosene solvent [39]. Although their

---

*Adapted from T. Araki, T. Okazawa, Y. Kubo, H. Asai, and H. Ando, *J. Liq. Chromatogr.*, *11*: 2473–2485 (1988), by the courtesy of Marcel Dekker, Inc.

mobile phase was aqueous $(H,Na)Cl_2CHCOO$ solutions containing 20% ethylene glycol, they also used mobile phase of higher pH value, similarly to our case, for the separation of RE ions of higher atomic number.

Since D2EHPA has long been mainly used for large-scale separating extraction of RE ions, it seemed to be important to clarify the behaviors of RE ions in CPC using this phosphate. This kind of information will be useful in replacing the familiar extraction techniques with industrial-scale CPC. Hence, some key factors other than [HCl] were examined with a view to improving separations by the use of D2EHPA separator. The target of the present study was the effect of number of microcells on the separation of $CeCl_3$, $PrCl_3$, and $NdCl_3$, and the effect of solvent polarity in the stationary phase for the separations of RE ions heavier than $EuCl_3$.

$YbCl_3$ was chosen as a typical heavier RE ion because it is the second heaviest element of the RE series that is readily available commercially, and in addition, the CPC behaviors of $LuCl_3$ will be easily extrapolated from those of $YbCl_3$. Similarly, $ErCl_3$ was chosen as a typical sample for predicting the behaviors of $TmCl_3$, $HoCl_3$ and other $RECl_3$ ions somewhat lighter than $ErCl_3$.

Finally, for comparison of separating abilities of acidic and neutral phosphates as separator, tri-n-butyl phosphate (TBP), again one of the classical extractants, was tested.

B. Experimental Procedures

1. Materials

Rare earth metal ions were obtained from commercial suppliers and used without further purification: $CeCl_3 \cdot 7H_2O$ (Wako, >98%), $PrCl_3 \cdot 7H_2O$ (Kanto, first grade), $NdCl_3 \cdot 6H_2O$ (Wako, >97%), $ErCl_3 \cdot 6H_2O$ (Wako, first grade), $YbCl_3 \cdot 6H_2O$ (Wako, 99.9%), $Nd(NO_3)_3 \cdot nH_2O$ (Wako, 99.9%), $Sm(NO_3)_3 \cdot nH_2O$ (Wako, 99.9%), $Eu(NO_3)_3 \cdot nH_2O$ (Nacalai tesque, first grade), and tri-n-butyl phosphate (TBP) (Wako, extra pure grade). Other reagents and solvents were the same as used in the study in Section IV.

2. Apparatus

Centrifugal partition chromatography was performed by using a L. L. apparatus Model NMF (Sanki Engineering Ltd.) with cartridges of Model 240W, and other apparatuses were used as in the work in Section IV.

3. Chromatographic Procedures

Effect of the Number of Cartridges for Separation of $CeCl_3$, $PrCl_3$, and $NdCl_3$ with D2EHPA Separator. Three of six high-resolution-type cartridges (Model 240W) containing 400 microcells per cartridge were equipped in series (total 1200 or 2400 microcells). Dead volume: 25 ml. Mode: descending mode.

Stationary phase: 0.1 M D2EHPA in n-heptane. Mobile phase: aqueous HCl solution (1.12 ± 0.02 ml min^{-1}); 0.06 M aq.-HCl for separation of CeCl$_3$ and PrCl$_3$; 0.07 M aq.-HCl for separation of PrCl$_3$ and NdCl$_3$. Rotation: 800 rpm. Temperature: room temperature. Sample size: 0.54 ml of 0.02 M RECl$_3$ (containing equimolar amounts of RE$_1$ and RE$_2$ ions). Detection: xylenol orange method as described in Section IV.

Behaviors of YbCl$_3$ and ErCl$_3$ with D2EHPA Separator. Number of cartridges: 3. Dead volume: 23 ml for toluene, and 24 ml for CHCl$_3$ solvents. Mode: descending mode for toluene, and ascending mode for CHCl$_3$ solvents. Stationary phase: 0.1 M D2EHPA in toluene or CHCl$_3$. Mobile phase: 1.12 ± 0.02 ml min^{-1}; 0.20 M aq.-HCl for YbCl$_3$ (toluene), 0.15 M aq.-HCl for YbCl$_3$ (CHCl$_3$), and 0.10 M aq.-HCl for separation of YbCl$_3$ and ErCl$_3$. Rotation: 1200 rpm for YbCl$_3$ (toluene), and 700 rpm for CHCl$_3$ systems. Temperature: room temperature. Sample size: 0.54 ml of 0.02 M for a binary RECl$_3$ sample (containing equimolar amounts of RE$_1$ and RE$_2$ ions), and 0.01 M YbCl$_3$. Detection: the same as in the previous case.

TBP Separator System for RE(NO$_3$)$_3$. Number of cartridges: 3. Dead volume: 25 ml. Mode: descending mode. Stationary phase: neat TBP. Mobile phase: 1.12 ±/0.02 ml min^{-1}; 1.00 M and 1.50 M Li(NO$_3$)$_3$. Rotation: 1200 rpm. Temperature: room temperature. Sample size: 0.54 ml of 0.01 M RE(NO$_3$)$_3$ (RE = Nd, Sm, Eu, and Yb), and 0.02 M for binary RE(NO$_3$)$_3$ sample (containing equimolar amounts of Nd(NO$_3$)$_3$ and Sm(NO$_3$)$_3$. Detection: the same as in the previous cases.

C. Results and Discussion

1. Effect of Number of Microcells for Separation of CeCl$_3$, PrCl$_3$, and NdCl$_3$ with D2EHPA Separator

The effect of the number of cartridges on the chromatographic behaviors of the binary mixtures of PrCl$_3$/NdCl$_3$ and CeCl$_3$/PrCl$_3$ were examined with 0.1 M n-heptane solution of D2EHPA as stationary phase. With 0.07 M aqueous HCl solution as mobile phase, peak resolution (R) of the former mixture was improved (0.62) in the six-cartridge (containing 2400 microcells) system relative to the three-cartridge (containing 1200 microcells) system (0.37) (Table 5). This is mainly due to the increases in the number of effective theoretical plates (ca. 1.5 times for PrCl$_3$, and ca. 3 times for NdCl$_3$). Since the separation factors (α) were the same for the both cases (1.4), a two-peak chromatogram was obtained with the six-cartridge system, as shown in Fig. 8.

For the case of a binary CeCl$_3$/PrCl$_3$ mixture, the effect of increasing the number of cartridges was not so important when the value of α was identical (1.7). Namely, the result of using six cartridges with 0.06 M [HCl] was only slightly better (n_{Ce} and n_{Pr} were only 1.17 and 1.13 times higher, respectively, resulting in

Table 5 Effect of Number of Microcells on Separation Parameters for Equimolar $CeCl_3/PrCl_3$ and $PrCl_3/NdCl_3$ Binary Mixtures[a]

RE_1	RE_2	Number of microcells	[HCl] (M)	$n(RE_1)$	$n(RE_2)$	R	α
Ce	Pr	1200	0.05	41	54	0.77	1.7
Ce	Pr	2400	0.06	48	61	0.85	1.7
Pr	Nd	1200	0.07	31	28	0.37	1.4
Pr	Nd	2400	0.07	47	82	0.62	1.4

[a]Stationary phase: 0.1 M D2EHPA in *n*-heptane. 400 microcells are contained in one cartridge. *n*: effective theoretical number of plates; *R*: peak resolution; α: separation factor.

only 1.10 times higher *R* value) than that of using three cartridges with 0.05 *M* [HCl] (Fig. 9). This is because chromatographic behaviors of lighter RE systems can be well controlled by the concentration of HCl in the mobile phase, and this type of [HCl]-based controlling is more convenient for practical applications than increasing the number of cartridges.

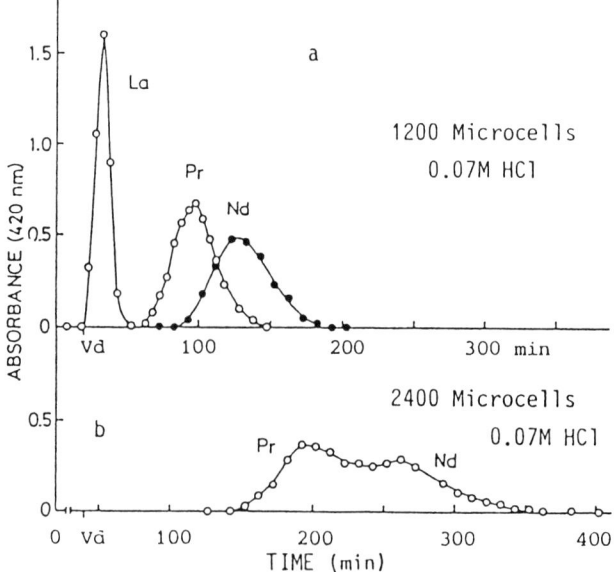

Figure 8 The effect of number of microcells on separation of an equimolar binary mixture of $PrCl_3$ and $NdCl_3$ by CPC with D2EHPA/*n*-heptane stationary phase. Curve (a): 1200 microcells; [HCl] = 0.07 *M* (see Section IV). Curve (b): 2400 microcells; [HCl] = 0.07 *M*.

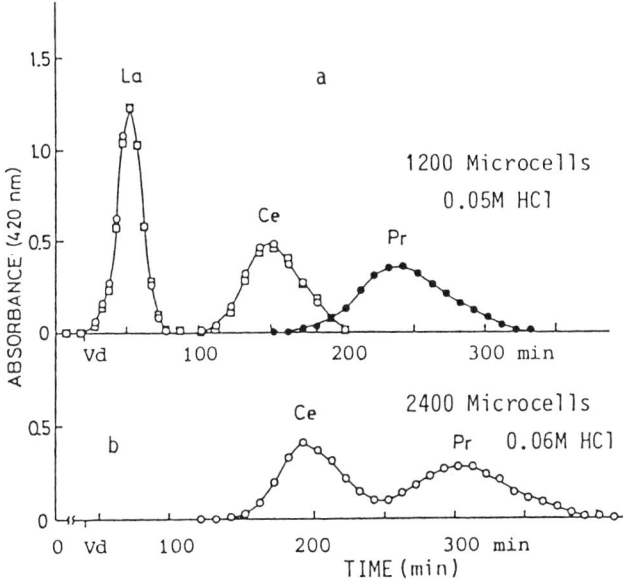

Figure 9 The effect of number of microcells on separation of an equimolar binary mixture of $CeCl_3$ and $PrCl_3$ by CPC with D2EHPA/n-heptane stationary phase. Curve (a): 1200 microcells; [HCl] = 0.05 M (see Section IV). Curve (b): 2400 microcells; [HCl] = 0.06 M.

Our conclusion from this experiment is that increasing the number of microcells is effective when the peak broadening is significant, that is, for RE ions of higher atomic numbers.

2. *Behavior of $YbCl_3$ and $ErCl_3$ with D2EHPA Separator (Effect of Solvent)*

With 0.1 M toluene solution of D2EHPA as stationary phase and with 0.20 M aqueous HCl (almost the maximum [HCl] for our CPC experiment) as mobile phase, an extremely broad peak of $YbCl_3$ was observed, and its retention value was also high, as shown in Fig. 10a. But the chromatographic behavior was extensively improved by the use of $CHCl_3$ as stationary solvent (Fig. 10b). Combining our present results with the ones of Section IV, we can conclude that more polar solvent is effective for both shortening the retention times and narrowing the peaks of RE^{3+} ions having higher atomic numbers, that is, n-heptane for La^{3+}, Ce^{3+}, Pr^{3+}, and Nd^{3+}; toluene for Sm^{3+}, Eu^{3+}, and RE^{3+} ions slightly heavier; and $CHCl_3$ for much heavier ions around Yb^{3+}.

In fact, separation of Er^{3+} from Yb^{3+} with 0.1 M D2EHPA solution in $CHCl_3$ was readily realized at [HCl] = 0.10 M (Fig. 10c). This result suggests that

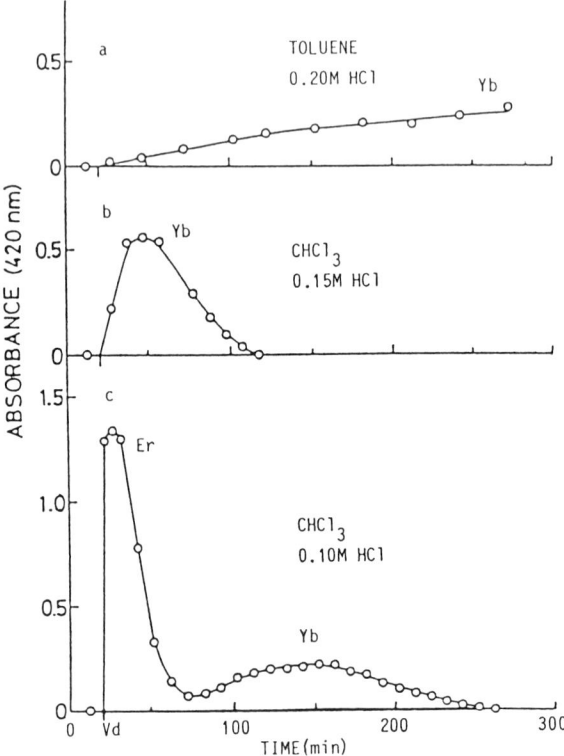

Figure 10 Behaviors of Yb^{3+} ion in CPC using toluene and chloroform as stationary solvent of D2EHPA. 1200 microcells. Curve (a): Solvent: toluene; [HCl] = 0.20 M. Curve (b): Solvent: $CHCl_3$; [HCl] = 0.15 M. Curve (c): separation of an equimolar binary mixture of $ErCl_3$ and $YbCl_3$ by the use of $CHCl_3$ solvent, where [HCl] = 0.10 M.

the whole range of RE ions will be separated with D2EHPA by adjusting the nature of solvent in the stationary phase, along with adjusting [HCl] in the mobile phase.

3. TBP Stationary Phase for $RE(NO_3)_3$

D2EHPA is an acidic separator whose key interaction in CPC has been presumed to be relatively strong acid–base type interaction, as shown in Eq. 6 (see Section IV; also see [28]; in contrast, tri-n-butyl phosphate (TBP) is a neutral phosphate also widely used in extraction of RE ions, and its extraction mechanism is described in Eq. 7 [40]:

$$RE^{3+} + 3(HG)_2 \rightleftarrows RE(HG_2)_3 + 3H^+ \quad (6)$$
$$(HG = D2EHPA)$$
$$RE^{3+} + 3NO_3^- + 3TBP \rightleftarrows RE(NO_3)_3 TBP \quad (7)$$

The TBP extraction results already published differ considerably from those of D2EHPA; for example, the order of extractability is not always parallel to the atomic number of RE ions [40,41]. In addition, there was a possibility that milder interaction of neutral TBP might cause narrowing of the chromatographic peaks of RE ions in CPC. In the present study, neat TBP was used as the stationary phase because we intended to test the maximum efficiency of TBP.

In the extraction procedure with TBP, a very high concentration (1–18 N) of aqueous HNO_3 has been employed. The use of such a highly concentrated HNO_3, however, is difficult in our CPC machine. Alternatively, aqueous solutions containing salting-out agents, that is, nitrates of Li, NH_4, Fe, Al, and Ca, or isothiocyanates of Li, Na, NH_4, have been used for the TBP extraction [42–44]. Among them, $LiNO_3$ was found to be the most effective, and moreover, the detection ability of RE ions by xylenol orange method in the presence of $LiNO_3$ was also satisfactory. Hence, we chose to use its aqueous solution as our mobile phase.

CPC results obtained with neat TBP as stationary phase and aqueous $LiNO_3$ as mobile phase, and RE^{3+} (Nd^{3+}, Sm^{3+}, Eu^{3+}, and Yb^{3+}) nitrates as sample were considerably poorer than the cases of D2EHPA separator:

1. Separation of Nd^{3+} versus Sm^{3+} gave $\alpha = 1.2$, and $n = 21$ for Nd^{3+} and 20 for Sm^{3+}, giving rise to extremely low R value ($= 0.15$).
2. Sm^{3+} and Eu^{3+} gave almost the same retention times [$22(V_d)$ + ca. 28 = ca. 50 minutes], and the shape of the rather broad peaks was almost identical.
3. The α value of Yb^{3+} versus Eu^{3+} was high ($= 1.9$), but significant broadness of these peaks ($n = 14$ for Yb^{3+} and 13 for Eu^{3+}) prevents effective separation ($R = 0.50$).
4. Retention time of heavier Yb^{3+} [$22(V_d)$ + ca. 4 = ca. 26 minutes] was much smaller than Eu^{3+} [$22(V_d)$ + ca. 28 = ca. 50 minutes]; that is, the retention times of RE ions are not parallel to their atomic numbers.

Even considering that the nitrate samples have lower affinities for the highly organic phosphate relative to the corresponding chloride samples, the neutral phosphate TBP should be considered a significantly less effective separator for CPC applications. In contrast, Akiba and coworkers [39] obtained effective separations by using PC-88A, which shows essentially similar behaviors to our D2EHPA cases (see Section IV, Ref. 38). Combining the CPC results obtained by us using D2EHPA (($RO)_2P(=O)OH$) in the Section IV and in this section with those obtained by Akiba [39] using PC-88A ($R(RO)P(=O)OH$) and with those obtained by us using TBP (($RO)_3P=O$), the acidic nature (P–OH) of the phosphorus compounds seems to be a key property for effective separations of RE^{3+} ions in the present stage.

D. Concluding Remarks

Since D2EHPA has been widely used in practical extraction of rare earth metal ions, testing of its applicability to CPC seems reasonable. On the basis of the results of Section IV (see also [38]), the behaviors of RE ions in CPC with D2EHPA as separator were further studied. (1) For the binary mixture of Pr^{3+} and Nd^{3+}, whose separation is important for RE technologies, increasing the number of microcells to obtain higher numbers of theoretical plates is shown to be effective. (2) The use of the more polar $CHCl_3$ as the stationary solvent showed promise for separation of heavier series of RE ions including Yb^{3+} and probably Lu^{3+}. (3) The acidic nature of the phosphorus-containing separator is thought to be the key property for effective separation, until a new and effective neutral separator is developed in future. Hence, [HCl]-gradient techniques, addition of an organic material in the mobile phase [39], or other experimental improvements for narrowing the peaks will be of value at the present.

VI. SEPARATING BEHAVIORS OF HEAVIER RARE EARTH METAL IONS IN CPC WITH DI(2-ETHYLHEXYL) PHOSPHORIC ACID*

A. Introduction

The importance of CPC for the separation of inorganic ions was shown in 1985 (see Section II). As an extension, Sections IV and V show CPC is useful also for the separation of lighter rare earth (RE) metal ions from $LaCl_3$ to $EuCl_3$ and preliminary tests for separation of heavier RE ions by choosing $YbCl_3$ and $ErCl_3$ (Section IV). When di(2-ethylhexyl)phosphoric acid (D2EHPA) was used as separator in the stationary phase, the separation behaviors of these lighter RE ions were well correlated with the results of extraction experiments using D2EHPA.

Namely, the key chemical equation both in CPC and extraction processes should be Eq. 8, which was developed from the earlier extraction study [28]:

$$RE^{3+} + 3(HG)_2 \rightleftarrows RE(HG_2)_3 + 3H^+ \quad (8)$$

where HG denotes D2EHPA.

So that the chromatographic behaviors of selected heavier RE ions (Yb^{3+} and Er^{3+}) would be in line with those for the lighter RE ions, we suggested the separation of heavier RE ions would be attained by suitably setting the experimen-

---

*Adapted from T. Araki, H. Asai, H. Ando, N. Tanaka, K. Kimata, K. Hosoya, and H. Narita, *J. Liq. Chromatogr.*, 13: 3673–3687 (1990), by the courtesy of Marcel Dekker, Inc.

tal conditions such as increased polarity in the stationary solvent (see Section V). In fact, CPC separation of heavier RE ions was attained with the related separator of 2-ethyl phosphoric acid mono-2-ethyl-hexyl ester (EHPA or PC-88A) [45] by assuming that the equilibrium in CPC was

$$M_{aq}^{3+} + 3(HA)_{2,org} \rightleftarrows M(HA_2)_{3,org} + 3H_{aq}^{+} \quad (9)$$

where HA denotes EHPA.

This section discusses studies on CPC behavior of the heavier RE ions of Sm^{3+}, Eu^{3+}, Gd^{3+}, Tb^{3+}, Dy^{3+}, Ho^{3+}, Er^{3+}, Tm^{3+}, Yb^{3+}, and Lu^{3+} with the D2EHPA separator, and CPC separation of this series of ions was attained in an expected way.

B. Experimental Procedures

1. Materials

The following rare earth metal ions were obtained from commercial suppliers and used without further purification: $SmCl_3 \cdot 6H_2O$ (Nacalai tesque, 99.9%), $EuCl_3 \cdot 6H_2O$ (Nacalai tesque, 99%), $GdCl_3 \cdot 6H_2O$ (Nacalai tesque, 99.9%), $TbCl_3 \cdot 6H_2O$ (Nacalai tesque, 99.9%), $DyCl_3 \cdot 6H_2O$ (Wako, >97%), $HoCl_3 \cdot 6H_2O$ (Wako, 99.9%), $ErCl_3 \cdot 6H_2O$ (Wako, >97%), $YbCl_3 \cdot 6H_2O$ (Wako, 99.9%). Two kinds of $RECl_3 \cdot 6H_2O$ (RE: Tm and Lu) were prepared from high purity oxides (Tm_2O_3 and Lu_2O_3, >99.99%, respectively; a gift from Professor H. Hashitani, Shimane University) by treating with aq.-HCl solutions at elevated temperature according to general methods for rare earth elements. The purity of the chlorides was checked by absorption spectra. Ethylene glycol (Waco, extra pure grade) was used. Other reagents and solvents were the same as used in the work of Sections IV and V.

2. Apparatus

CPC was performed with Sanki's apparatus as described in Section V. A Shimadzu double-beam spectrophotometer Model UV-180 and a Hitachi-Horiba pH meter Model F-7 II were used for determination of rare earth concentrations in the eluted fractions.

3. Chromatographic Procedures and Conditions

General chromatographic procedures have been already given in Sections IV and V. The experimental conditions used in this work are as follows: Number of cartridges: 3 (Model 240W, total microcells of $400 \times 3 = 1200$); elution mode: ascending; mobile phase: 0.02–0.14 N aq.-HCl solution with flow rate 1.10 ml min^{-1}; stationary phase: $CHCl_3$ solution containing 0.1 M D2EHPA; rotation: 700 rpm; temperature: 25 and 45°C; sample: 8 mM aq.-$RECl_3$ solution (0.54 ml) for RE = Sm, Eu(0.02 N HCl), Gd(0.02 N), Tb(0.04 N), Er(0.1 N), Tm, Yb, and

Lu(0.10–0.14 N); 4mM aq.-RECl$_3$ solution (0.54 ml) for RE = Eu(0.03 N), Gd(0.03, 0.04 N), Tb(0.03 N), Dy, Ho, and Er(0.06, 0.08 N); dead volume: 21 ml (25°C) and 20 ml (45°C). Detection: the batch type analysis of RE ions in the eluted fractions (every 5 minutes) was used as described in Sections IV and V. A fraction was diluted (2.5 times), and the analytical sample (pH 6.0) contains xylenol orange (1.12 × 10^{-4} M) and hexamine buffer (0.64 M). The RE concentrations were determined spectroscopically with the 580 nm band intensities precalibrated. This band was more sensitive for analysis of low-concentration samples than the 420 nm band used previously.

C. Results and Discussion

The previous observation suggests that stability of the RE salt formed with D2EHPA in the CPC stationary phase increases with the atomic number of the RE ion. In fact, it was effective to increase the polarity of the stationary solvent from toluene to chloroform for separation of Er^{3+}/Yb^{3+} (see Section V). Accordingly, chloroform was used in this work throughout.

Figure 11 shows the CPC diagram of Sm^{3+}, Eu^{3+}, and Gd^{3+} obtained with aqueous mobile phase of 0.02 N HCl at 25°C. Almost complete CPC separation of Sm^{3+} and Gd^{3+} was observed. The separation between Sm^{3+} and Eu^{3+} was attained in a shorter time with lower HCl concentration than in the corresponding case where toluene stationary solvent and 0.03 N HCl mobile phase were used (see Section IV).

The separation of Eu^{3+} and Gd^{3+} was partial, but one can obtain high-purity

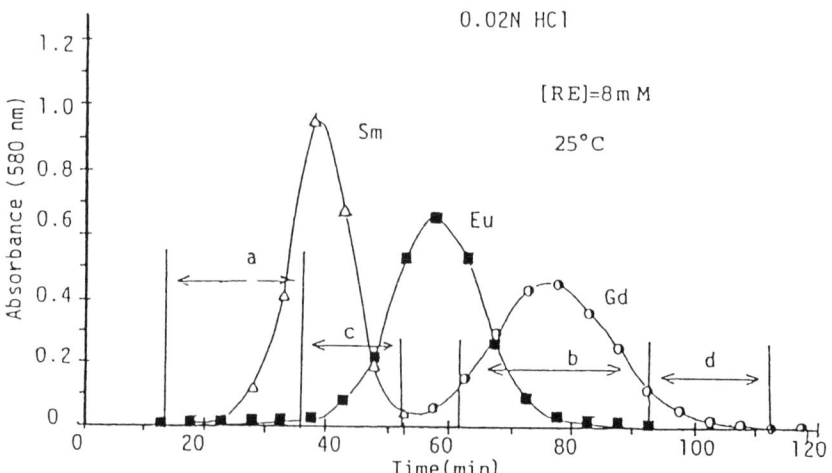

Figure 11 Chromatograms of Sm^{3+}, Eu^{3+}, and Gd^{3+} at 25°C. Stationary phase: 0.1 M D2EHPA in CHCl$_3$; mobile phase: 0.02 N aq.-HCl.

ions in the fraction ranges of c and d for Eu^{3+} and Gd^{3+}, respectively, from the 1:1 binary mixture.

Figure 12 shows the separation of Gd^{3+}, Tb^{3+}, and Dy^{3+} ions with 0.04 N HCl mobile phase at 25°C. In this case, effective separations were observed for both pairs Gd^{3+}/Tb^{3+} and Tb^{3+}/Dy^{3+}. The figure shows that even when the concentration of Tb^{3+} is twice that of Gd^{3+} and Dy^{3+}, mutual separations of these adjacent ions are attained.

As Fig. 13 shows, the use of mobile phase of slightly lower HCl concentration (0.03 N) attains almost complete mutual separation of the ternary mixture at the same time. Separation of the $Gd^{3+}/Tb^{3+}/Dy^{3+}$ set observed with the D2EHPA separator are quite similar to those observed by Akiba et al. [45] with the EHPA separator. This indicates that the basic roles of D2EHPA and EHPA are similar in view of their phosphate salt formation with RE^{3+} ions and their identical trends of change in stability for the ions. By comparing the chromatograms in Figs. 12 and 13, one can find how HCl concentration affects the separation of Eu^{3+} and Gd^{3+} in the chloroform solvent system.

Separating behaviors of Dy^{3+} and Ho^{3+} ions in CPC have not been reported. These two adjacent ions were found to be rather difficult to separate under our experimental conditions at 25°C. By elevating temperature to 45°C, however, these two peaks were partially separated with 0.06 N HCl mobile phase as shown in Fig. 14. Under this condition, the separation of Ho^{3+} and Er^{3+} was almost satisfactory.

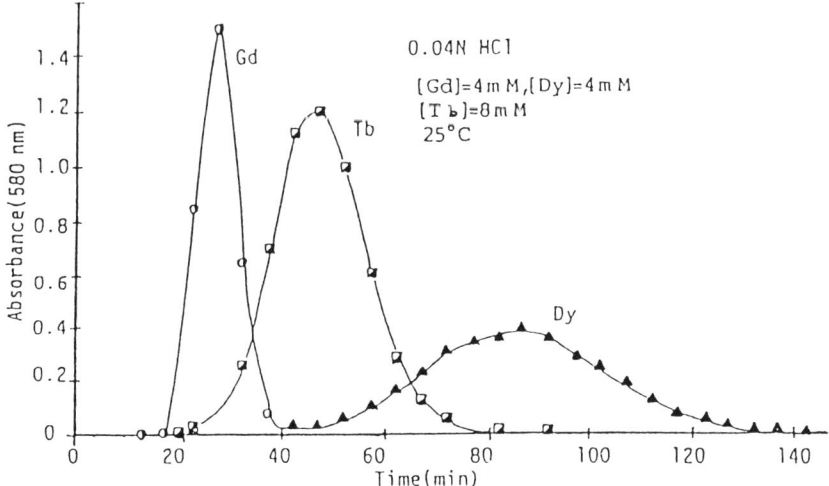

Figure 12 Chromatograms of Gd^{3+}, Tb^{3+}, and Dy^{3+} at 25°C. Stationary phase: 0.1 M D2EHPA in $CHCl_3$; mobile phase: 0.04 N aq.-HCl.

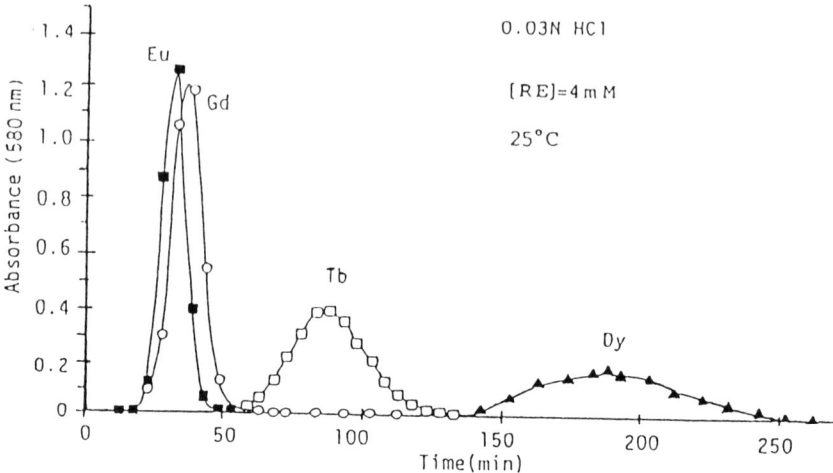

Figure 13 Chromatograms of Eu^{3+}, Gd^{3+}, Tb^{3+}, and Dy^{3+} at 25°C. Stationary phase: 0.1 M D2EHPA in $CHCl_3$; mobile phase: 0.03 N aq.-HCl.

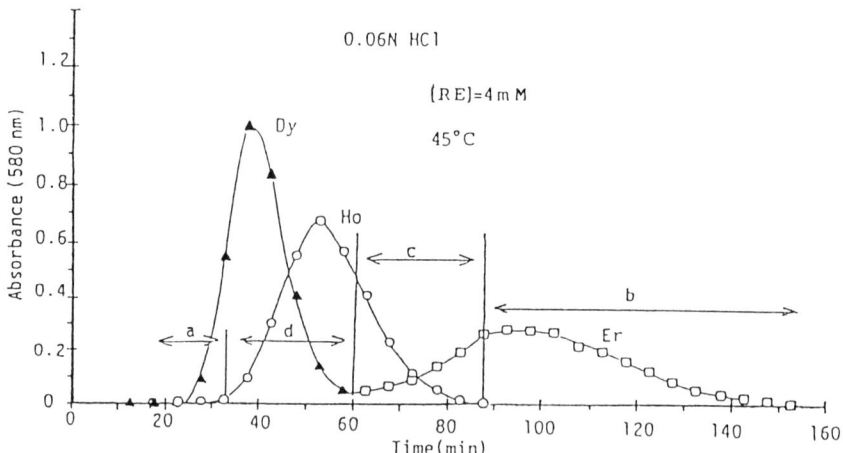

Figure 14 Chromatograms of Dy^{3+}, Ho^{3+}, and Er^{3+} at 45°C. Stationary phase: 0.1 M D2EHPA in $CHCl_3$; mobile phase: 0.06 N aq.-HCl.

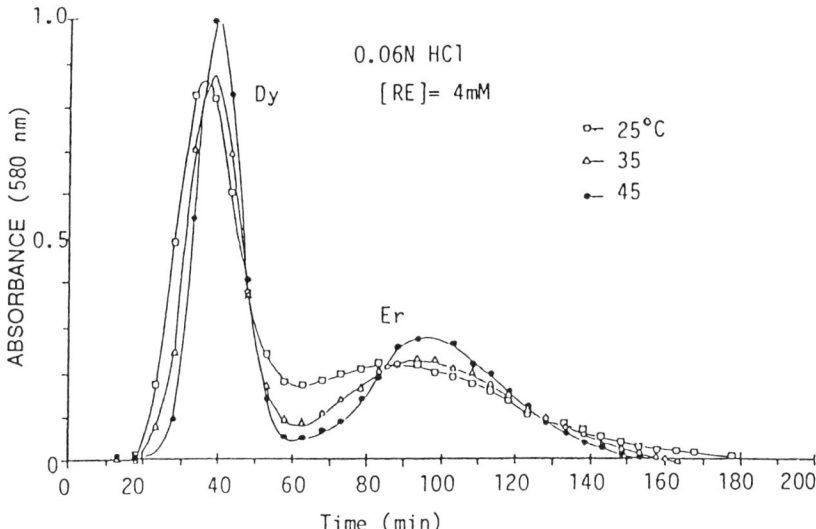

Figure 15 Effect of temperature on the behaviors of Dy^{3+} and Er^{3+} binary sample. Stationary phase: 0.1 M D2EHPA in $CHCl_3$; mobile phase: 0.06 N aq.-HCl.

The effect of temperature was examined by using a binary mixture of Dy^{3+}/Er^{3+} pair (Fig. 15) at 25°C, 35°C, and 45°C. Interestingly, the retention times (t_R) were not significantly changed with temperature, but the number of theoretical plates (n), especially for the component having higher t_R, was considerably increased with temperature.

Chromatographic peaks of the RE ions having atomic numbers higher than that of Er^{3+} were highly broad at 25°C, and at this temperature no effective separations were observed as long as aqueous HCl solution was used as mobile phase. When, however, 20% ethylene glycol [45] was added to the 0.08 N aqueous HCl mobile phase at 45°C, rather good chromatograms (Fig. 16) were obtained for the separation of Er^{3+} and Tm^{3+} ions. Almost complete separation of Tm^{3+} and Yb^{3+} ions was attained with 0.10 N HCl as mobile phase containing 20% ethylene glycol, though the separation of Er^{3+} and Tm^{3+} became unsatisfactory (Fig. 17). For comparison, chromatograms of the same components obtained without ethylene glycol are given in Fig. 18.

Separation of Lu^{3+}, the heaviest RE ion, from Yb^{3+} was only partially attained with a higher HCl concentration in mobile phase at 25°C (Fig. 19). The main reason is the broadness of these peaks, as in the case of EHPA system [45]. An attempt to obtain a chromatogram upon addition of ethylene glycol at 45°C failed because the presence of ethylene glycol prevented the color development of

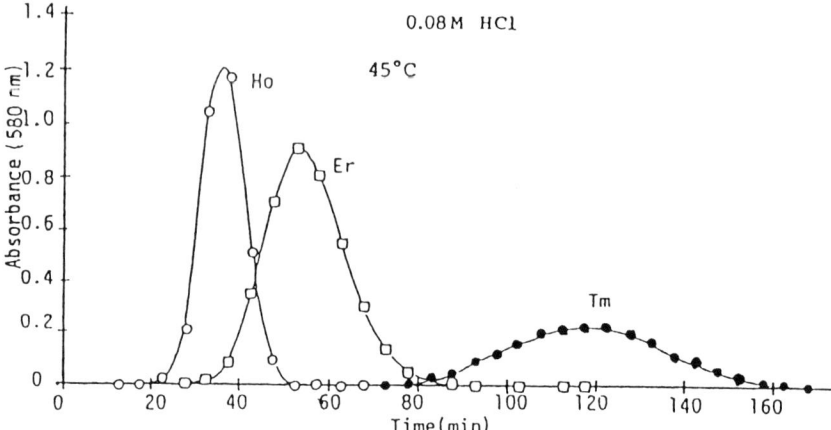

Figure 16 Chromatograms of Ho^{3+}, Er^{3+}, and Tm^{3+} at 45°C. Stationary phase: 0.1 M D2EHPA in $CHCl_3$; mobile phase: 0.08 N aq.-HCl containing 20% ethylene glycol.

Lu^{3+} ion by xylenol orange. When, however, the systematic chromatographic behaviors of the series of intermediate or heavy RE^{3+} ions described above are considered, one hypothesizes the separation of Yb^{3+} and Lu^{3+} should be attained by the use of experimental conditions of around 45°C with mobile phase of 0.12–0.14 N HCl containing 20% ethylene glycol, and by employing a suitable detection method other than color development with xylenol orange.

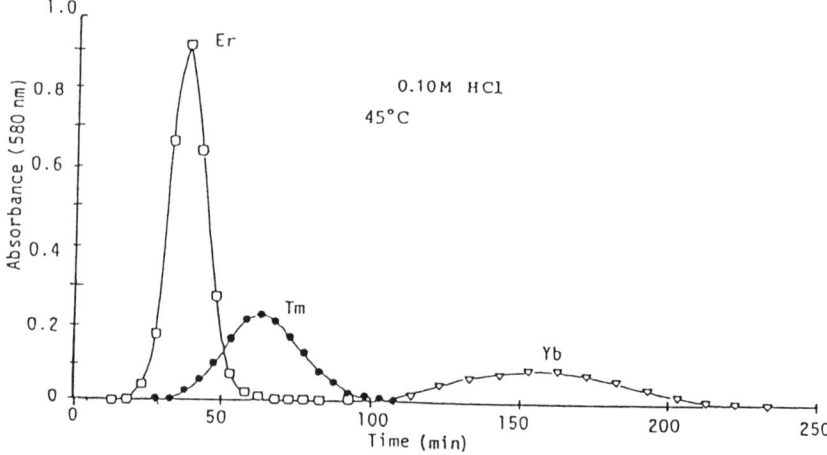

Figure 17 Chromatograms of Er^{3+}, Tm^{3+}, and Yb^{3+} at 45°C. Stationary phase: 0.1 M D2EHPA in $CHCl_3$; mobile phase: 0.10 N aq.-HCl containing 20% ethylene glycol.

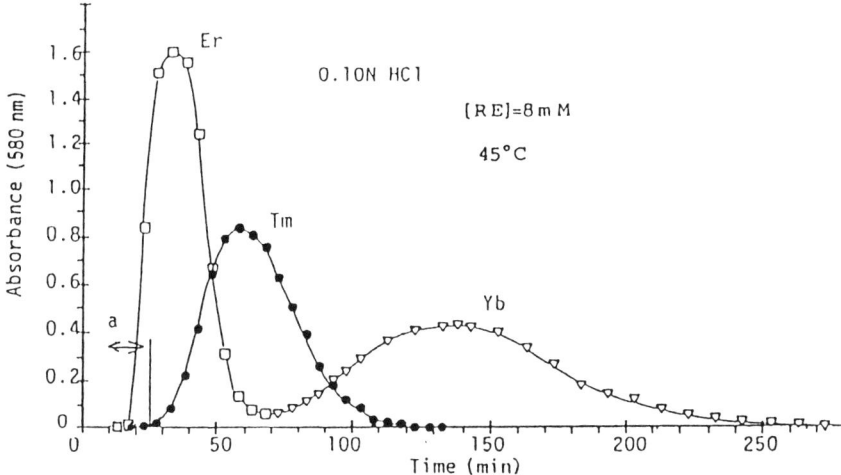

Figure 18 Chromatograms of Er^{3+}, Tm^{3+}, and Yb^{3+} at 45°C. Stationary phase: 0.1 M D2EHPA in $CHCl_3$; mobile phase: 0.10 N aq.-HCl alone.

The chromatographic data for separation of heavier RE^{3+} ions obtained in this work are summarized in Table 6. Separation parameters are defined as follows:

$$\alpha = \frac{t_{R,2} - t_0}{t_{R,1} - t_0} \tag{10}$$

$$Rs = \frac{2(t_{R,2} - t_{R,1})}{w_1 + w_2} \tag{11}$$

$$n = 5.54 \left(\frac{t_R}{w_h}\right)^2 \tag{12}$$

where α, Rs, and n are the separation factor, resolution, and number of theoretical plates, respectively. Retention time and width of the jth peak are denoted by $t_{R,j}$ and w_j, respectively. Half width of a peak and dead time are shown by w_h and t_0, respectively.

As mentioned above, the Rs values for Sm^{3+}, Eu^{3+}, Gd^{3+}, Tb^{3+}, and Dy^{3+} can be improved by elevating the temperature or by addition of ethylene glycol in the mobile phase. It is notable that the trends and values of separation factor (α) observed for D2EHPA are almost comparable with the corresponding cases for EHPA [45], though details of the chromatographic conditions differ from each other. This fact suggests again that the essential roles of D2EHPA and EHPA are similar with respect to the phosphate salt formation as observed in extraction studies. Hence, D2EHPA and EHPA are almost equally useful for the CPC technique when suitable modifications of the operating conditions are made.

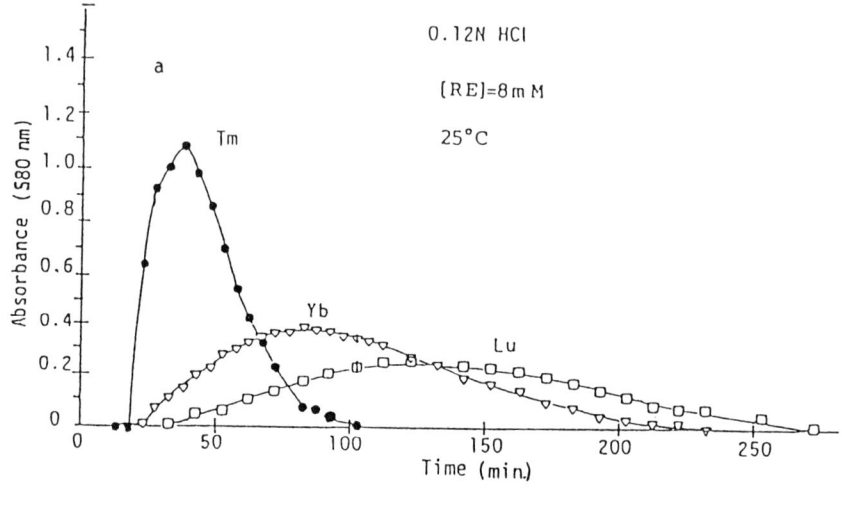
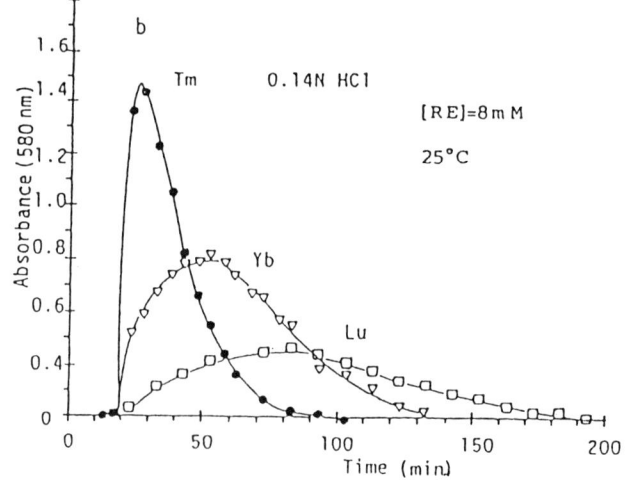

Figure 19 Chromatograms of Tm^{3+}, Yb^{3+}, and Lu^{3+} at 25°C. Stationary phase: 0.1 M D2EHPA in $CHCl_3$; mobile phase: (a) 0.12 N and (b) 0.14 N aq.-HCl alone.

Table 6 Separation Parameters

RE$_1$	RE$_2$	[HCl] (N)	Temperature (°C)	n RE$_1$	n RE$_2$	Rs (RE$_1$/RE$_2$)	α
Sm	Eu	0.02	25	15	30	0.72	2.08
Eu	Gd	0.02	25	30	33	0.42	1.43
Eu	Gd	0.03	25	6	8	0.19	1.40
Gd	Tb	0.03	25	8	21	1.08	4.3
Gd	Tb	0.04	25	3	8	0.6	3.61
Tb	Dy	0.03	25	21	42	1.10	2.6
Tb	Dy	0.04	25	8	13	0.61	2.46
Dy	Ho	0.06	45	6	11	0.40	1.94
Ho	Er	0.06	45	11	15	0.65	2.43
Ho^a	Er^a	0.08	45	8	13	0.51	2.10
Er	Tm	0.10	25	1	3	0.35	4.06
Er^a	Tm^a	0.08	45	13	26	1.0	3.06
Er	Tm	0.10	45	3	7	0.4	2.4
Er^a	Tm^a	0.10	45	8	8	0.5	2.37
Tm	Yb	0.10	25	3	7	0.58	3.8
Tm	Yb	0.10	45	7	12	0.65	2.89
Tm^a	Yb^a	0.10	45	8	21	0.87	3.11
Tm	Yb	0.12	25	12	4	0.36	3.20
Tm	Yb	0.14	25	<1	1	0.28	5.60
Yb	Lu	0.14	25	1	3	0.20	2.02
Yb	Lu	0.12	25	4	5	0.22	1.35

^{a}20% ethylene glycol was added to the mobile phase.

VII. LIQUID-MEMBRANE TRANSPORT OF LIGHTER RARE EARTH METAL IONS FACILITATED BY DI(2-ETHYLHEXYL) PHOSPHORIC ACID: COMPARISON WITH THE RESULTS OF CORRESPONDING CPC*

A. Introduction

The selective liquid-membrane transport (LMT) of rare earth (RE) ions is a challenging problem driving the development of new techniques for a variety of analytical tools, including RE ion-selective sensors. Izatt and coworkers [46] have reported the liquid-membrane transport of Eu^{2+} ion with a crown ether carrier. The present section deals with selective LMT of lighter RE metal ions with

---

*Adapted from T. Araki, T. Okazawa, Y. Kubo, H. Ando, and H. Asai, *J. Liq. Chromatogr.*, *11*: 2487–2506 (1988), by the courtesy of Marcel Dekker, Inc.

D2EHPA carrier, which was attained based on the corresponding countercurrent-type chromatographic behaviors.

Recently, we developed a technique of "carrier- (or separator-) aided centrifugal partition chromatography (CPC)" (see Section II), which is an extension of the centrifugal countercurrent continuous extraction method [2,12]. The classical countercurrent distribution method [10] has been developed as a continuous extraction technique in which a mixture of sample solutes distributes between two immiscible liquid phases (solvents) simply by the difference in partition coefficients. When, however, a separating reagent or a carrier compound added to the stationary liquid phase of the CPC plays a crucial role for the separation problem, the chromatography comes to have a fundamental similarity to the LMT system [8,47], as both of these separation system involve the following chemical processes (see Section III and Ref. 48):

$$S + Y \rightarrow [S\text{----}Y] \rightarrow Y + S \qquad (1)$$

where S denotes a sample molecule in CPC and a substrate molecule in LMT, Y is a separator in CPC or a carrier in LMT, $[S\text{---}Y]$ designates any interactions between S and Y leading to effective separation or molecular recognition.

The carrier-aided CPC can effectively be used when the sample components have identical or very similar partition coefficients for the solvent pair used. An example is that alkali metal picrates were able to separate (see Section II) with a separator of benzo-crown ether, which had been known as an effective carrier in selective LMT of the alkali metal picrates [8,47].

The research field of LMT has been extensively developed in the last decade, but one frequently encounters difficulty in finding an effective carrier for selective transport of neutral molecules or ions with larger size, and researchers have been forced to design complicated carrier molecules. Recently, on the basis of the result of CPC data, we achieved the selective transport of picric acid and p-nitrophenol by using a simple and costless compound of butan-1-ol (see Section III). At the same time, we also proved (Section III) a direct correlation between CPC an LMT in the two liquid-phase system used.

More recently, we found that lighter rare earth metal ions can be effectively separately by CPC containing di(2-ethylhexyl)phosphoric acid (D2EHPA) as separator (see Section IV). Hence, in the present section we applied the same liquid system to LMT, aiming to test the selective transport of RE^{3+} ions with this costless carrier molecule of D2EHPA.

B. Fundamental Aspects of the Relation Between CPC and LMT

Actually, there is a common chemical behavior of Eq. 1 between LMT and CPC. When, however, the mechanistic difference between LMT and CPC (Fig. 20) is considered, the relationship between LMT and CPC is expected to be more

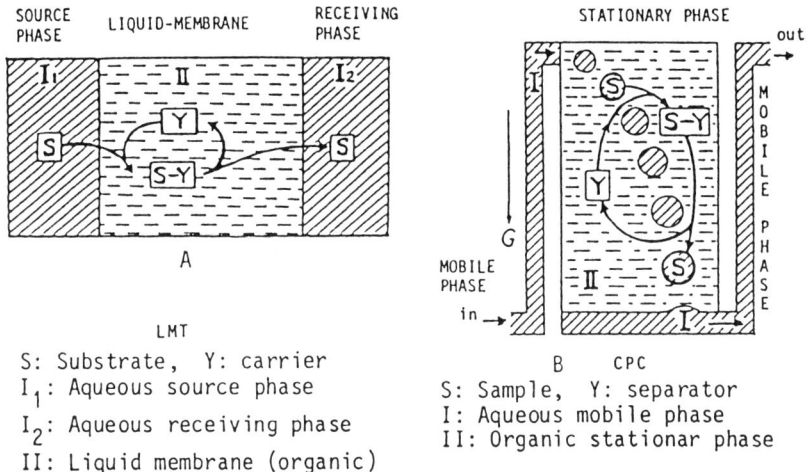

Figure 20 Fundamental illustrations of LMT and CPC. For simplicity, shown are only the systems of organic phases as liquid-membrane and stationary phase. A: LMT system with aqueous source and receiving phases. B: CPC system with aqueous mobile phase in a descending mode.

complex, as summarized in Table 7 [48]. In Table 7 it is assumed, for simplification, that partition effects of the solvent are completely negligible: no transport through a liquid membrane occurs in the absence of the carrier molecule, and retention volume of the sample agrees with the value of the dead volume if a stationary phase containing the solvent alone is used in the corresponding CPC. Hence, we also assume that the results of LMT (as rate of transport) and CPC (as retention time) depend solely on the interaction equilibria in the formation and dissociation of the [S---Y] state, and that the phase transfer processes involved in both systems depend only on these equilibria. (For practical comparison of the data of LMT with CPC, the interaction equilibria are conventionally designated as "ready" and "difficult," and "rapid" and "slow" instead of more correct descriptions of "small or large equilibrium constants.")

As can be seen from the Table 7, the results of LMT and CPC will correlate well in the cases 2 and 3. These cases are behaviors of alkali metal picrates with crown ether (see Section II) and of nitrophenols with butan-1-ol (see Section III). However, in the other cases, 1, 4, and 5, the correlation between LMT and CPC data will not be straightforward. (For more detailed discussion, see Section IX.D.) In our preliminary tests, stronger [S---Y] interactions such as acid–base interaction as for dl-mandelic acid (S) with brucine (Y) was found to be case 3 with significant peak broadening in CPC [49].

Table 7 Guidelines for Predicting the Results of CPC and LMT on the Basis of Phase Transfer Processes

	Phase transfer processes		CPC	LMT
No.	(In) aq. to org.	(Out) org. to aq.	(Retention value)	(Rate of transport)
1	No	—	Very low	No transport
2	Ready	Ready	Low	High
3	Ready	Difficult	High (tailing)	Low
4	Difficult	Ready	Low (leading)	Low
5	Difficult	Difficult	Very broad peak	Very low
6	Interface only	—	High or low (depends on the partition coefficient)	No transport

Since the interaction of the RE ions with D2EHPA is a type of acid–base interaction (Eq. 13) and significant peak broadening was observed in CPC (see Section IV), we assumed the behavior in these substrate/carrier pairs might be classified as case 3 or 4:

$$3(HG)_2 + RE^{3+} \rightleftarrows RE(HG_2)_3 + 3H^+ \quad (13)$$
$$G = (2\text{-ethylhexyl-O})_2P(=O)OH$$

If it is case 3, it seemed possible to realize selective transport of the RE ions by LMT with D2EHPA carrier; but if it is case 4, only slow transport would be observed. Further, if there are factors other than Eq. 13, correlation between LMT and CPC might be much more complicated. Then, we tested the LMT of RE ions by the use of the same two liquid-phase systems as used in Section IV.

C. Experimental Procedures

1. Materials

Lighter rare earth (RE) metal ions were obtained from commercial suppliers and used without further purification: $LaCl_3 \cdot 7H_2O$ (Wako, first grade), $CeCl_3 \cdot 7H_2O$ (Wako, >98%), $PrCl_3 \cdot 7H_2O$ (Kanto, first grade), $NdCl_3 \cdot 6H_2O$ (Wako, >97%), $SmCl_3 \cdot 6H_2O$ (Nacalai tesque, first grade), and $EuCl_3 \cdot 6H_2O$ (Nacalai, extra pure grade). Di(2-ethylhexyl)phosphoric acid (D2EHPA) (Nacalai, 99%), n-heptane (Nacalai, extra pure grade), xylenol orange (Wako, analytical grade), and hexamethylenetetramine (Wako, extrapure grade) were used without further purification. Deionized water was used throughout the experiments. (Compare the materials used in Section IV.)

2. Apparatus

For analysis of RE ions, a Hitachi spectrograph Model 200-20 was used, and pH measurements were carried out with a Horiba pH meter Model H-7LD. Centrifugal partition chromatography was performed as in Section IV by using a CPC-L.L. Model NMF (Sanki Engineering Ltd.) apparatus with Model 240W cartridges.

3. Liquid-Membrane Transport Procedures

A double-cylinder glass apparatus (Fig. 21) was designed for liquid-membrane transport experiments in which liquid membrane is the upper organic layer (15 ml, 0–0.10 M D2EHPA in n-heptane) and source (containing 3–30 mM $RECl_3$), and receiving phases are constructed in the two lower aqueous layers (each 5 ml, 0.01–0.20 M aq.–HCl). The upper layer was stirred (ca. 150 rpm) at the top of the layer at room temperature with a stainless steel stirring bar connected with a Yanako electrode-head motor (Model P10-RE), which was rotated by a variable-rotation motor Model K-1033 (TOP, Ltd.).

4. Analysis of RE Ions in Aqueous Fractions

Analysis of RE ions in the source and receiving phases were carried out at 5-minute intervals as described in Section IV. From each fraction, 2-ml aliquots were taken, and aqueous solutions of 5.60×10^{-4} M xylenol orange (XO) (1 ml) and 0.5 M hexamethylenetetramine hydrochloride buffer (2 ml) were added, resulting in 5 ml of aqueous solutions of pH 5.6 containing 1.12×10^{-4} M of XO and 0.2 M hexamine buffer. Concentrations of the RE ions thus color-developed were determined from the intensities of absorption spectra at 420 nm [36] by comparison with the precalibrated values.

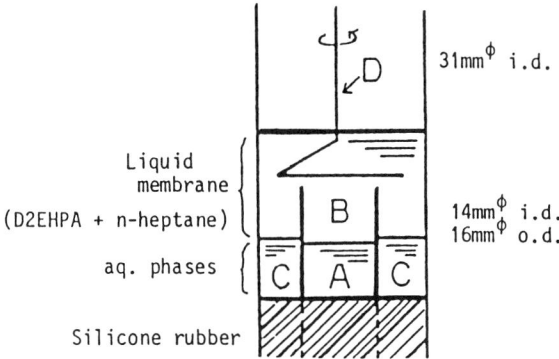

Figure 21 LMT apparatus for upper liquid-membrane system.

5. Chromatographic Procedures

Three cartridges (Model 240W: high-resolution type) containing 400 microcells per cartridge were equipped in series (total 1200 microcells). A 0.1 M D2EHPA solution in n-heptane (total ca. 40 ml n-heptane containing 1.3 g of D2EHPA) was used as stationary phase, and a given concentration of aqueous HCl solution (mobile phase, elution speed (1.12 ± 0.02 ml min^{-1})) was pumped through the microcells in a descending fashion under centrifugal force (800 rpm). The sample (0.54 ml) was charged through the sample injecting valve. Analytical procedures of RE ions were identical with those employed in the liquid-membrane transport experiments. (Compare the procedures of Section IV.)

D. Results and Discussion

1. LMT of Rare Earth Metal Ions Under Constant Initial HCl Concentration

When n-heptane solvent alone was used as the liquid membrane without D2EHPA, no transport was observed for all of the RE ions. Namely, the transport of the RE ions was confirmed to be facilitated only when the substrate interacts with this carrier compound in the liquid membrane.

In principle, the effect of carrier concentration can be dual: (1) At high concentration of D2EHPA in n-heptane, the rate of transfer from aqueous source phase to liquid membrane will be increased, resulting in a higher rate of transport. (2) When excess amounts of carrier are present, however, the rate of transfer from the organic phase to the aqueous receiving phase will be decreased, leading to a lower transport rate. By using 0.01 M NdCl$_3$ in 0.05 M aq.-HCl solution as a typical substrate, the effect of carrier concentration on the transport rate was tested (Fig. 22). The apparent rates (V) of transport were estimated from slopes of the time profile of Fig. 22, and they are plotted against the carrier concentrations (Fig. 23). From these results, the optimum carrier concentration was established as 0.02 M.

Rate of transport of a substrate through a liquid membrane also depends on the initial concentration of substrate in the source phase. In general, a higher rate is attained with a higher substrate concentration. For the case of NdCl$_3$ in 0.05 M aq.-HCl solution, the relation between the rate of transport and the initial substrate concentration found is illustrated in Fig. 24. The results in Fig. 24 suggest that a substrate concentration higher than 0.01 M causes a rapid saturation of the substrate in the liquid membrane and exceeds the ability of the system to give a smooth diffusion from the organic phase to the receiving aqueous phase, resulting in unexpectedly large fluctuations of the rate data. Hence, we chose 0.01 M substrate concentration (the turning point in Fig. 24c) as an optimum.

2. Dependence of Initial HCl Concentration in the Source Phase on the Transport Rate of RE Metal Ions

In view of the chemical mechanism of interaction between RE^{3+} ions with D2EHPA (Eq. 13), transport behaviors in our system were expected to depend on

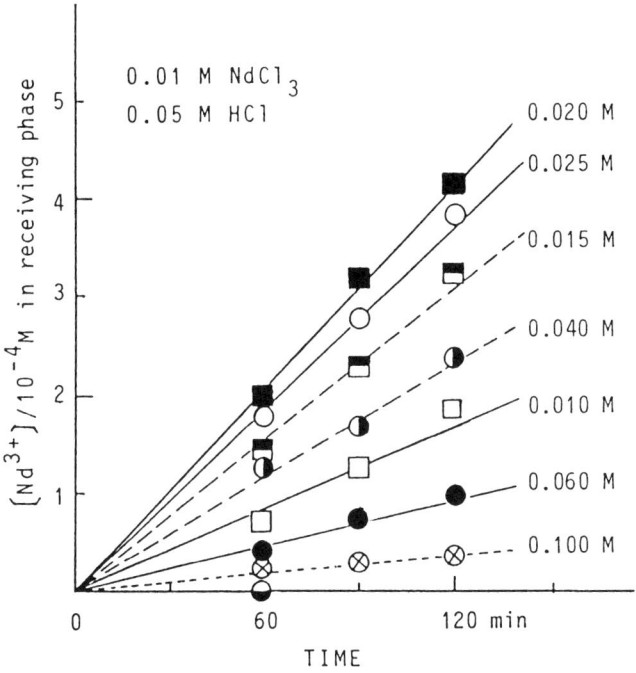

Figure 22 Effect of carrier concentration on the time profile of transport of $NdCl_3$. $NdCl_3$: 0.01 M in 0.05 M aq.-HCl. Liquid membrane: D2EHPA in n-heptane.

the initial concentration of HCl in the aqueous source phase. Concentration of HCl in aqueous phase was, in fact, also the key factor in the CPC of RE^{3+} ions employing the corresponding two-phase systems (see Section IV).

As expected, our LMT systems were strongly affected by the concentration of HCl in the source phase (Fig. 25). With 0.01 M HCl, the transport rates of all RE^{3+} ions were very low and a slight selectivity was observed only for La^{3+} (Fig. 25a,b). With 0.05 M HCl, however, the rates increased significantly for Nd^{3+}, Pr^{3+}, and Ce^{3+} ions, and to a lesser extent for Sm^{3+}, La^{3+}, and Eu^{3+} ions (Fig. 25c,d). With 0.10 M HCl, the transport rates of La^{3+}, Ce^{3+}, Pr^{3+}, and Nd^{3+} ions slowed down, but those of Sm^{3+} and Eu^{3+} ions increased (Fig. 25e,f). With 0.15 M HCl, the behaviors of ions were somewhat modified from those shown in Fig. 25e,f: the transport rate of Eu^{3+} ion was considerably higher than that of Sm^{3+} ions (Fig. 25g,h). With 0.20 M HCl transport rates of all of the RE^{3+} ions decreased, but the selectivity of Sm^{3+} or Eu^{3+} ion from other RE^{3+} ions remained effective (Fig. 25i,j). Figure 26 summarizes the dependencies of transport rate of the HCl concentration for all of the RE^{3+} ions tested. Under these concentrations,

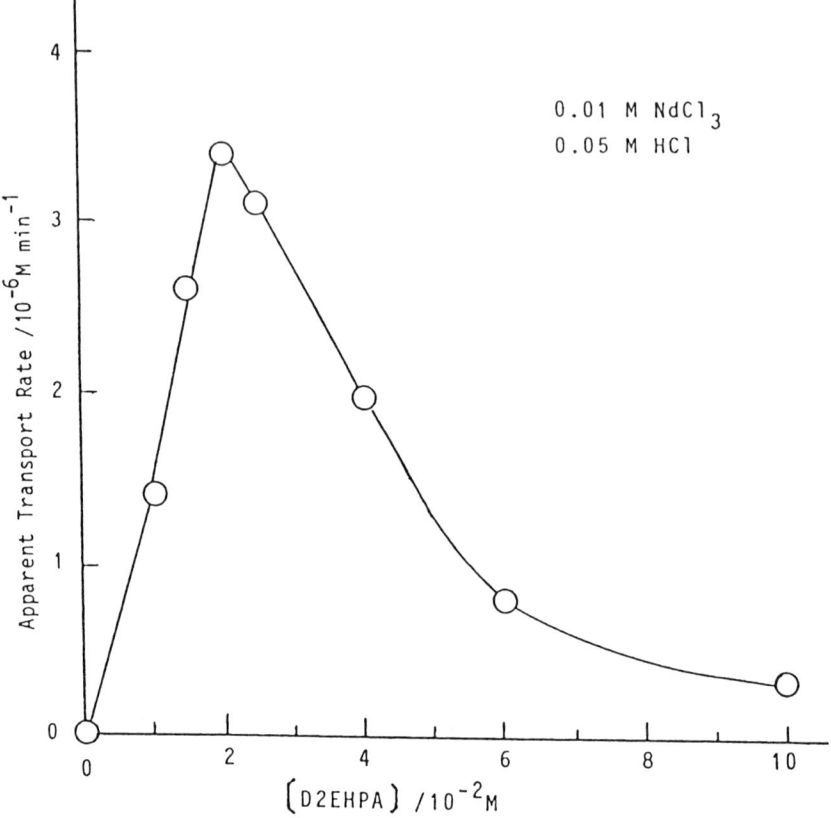

Figure 23 Plot of apparent rates (V) versus carrier concentrations for $NdCl_3$ (cf. Fig. 22).

it is possible to realize selective transport for various pairs of RE^{3+} ions, as summarized in Table 8.

3. Comparison of LMT with CPC for RE Metal Ions

Our LMT results of RE^{3+} ions by using D2EHPA carrier in heptane indicate that La^{3+} is most effectively transported with 0.03 M HCl; Nd^{3+}, Pr^{3+}, and Ce^{3+} with 0.05 M HCl; Sm^{3+} with 0.10 M HCl; and Eu^{3+} with 0.15 M HCl; that is, the optimum HCl concentrations for individual RE^{3+} ions increase roughly with the increasing order of atomic number of RE^{3+} metal ions (Fig. 26).

In Section IV, we observed that results of the corresponding CPC of the RE

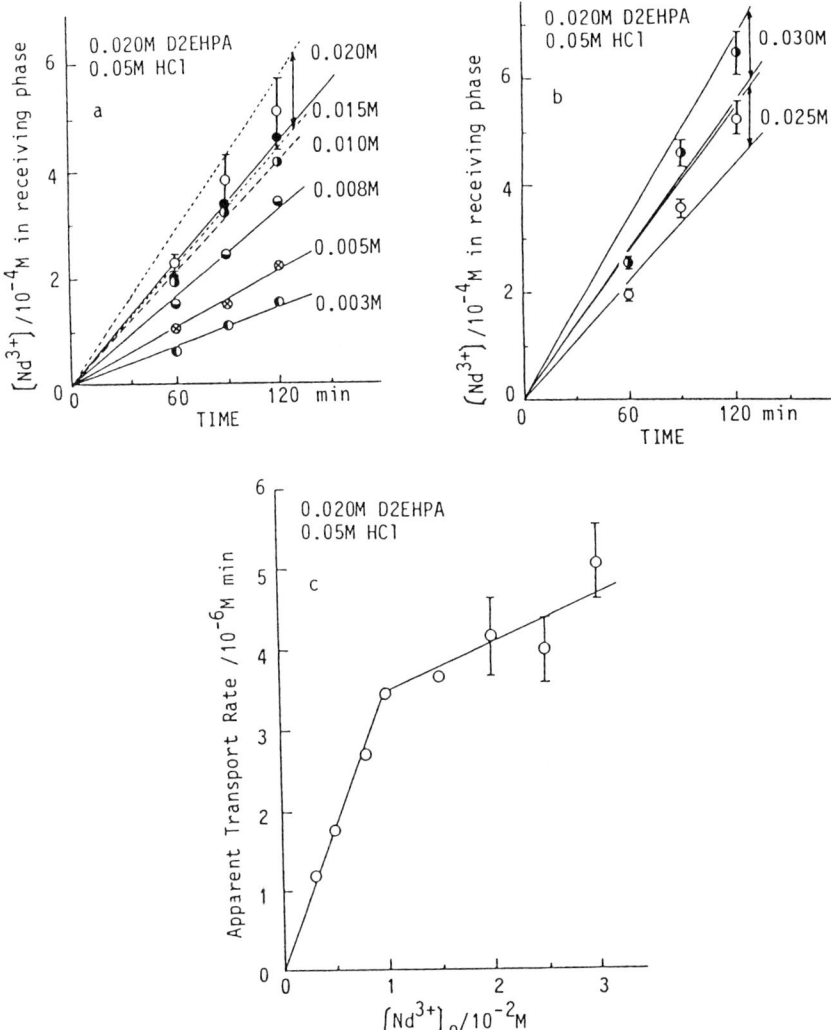

Figure 24 Effect of initial substrate concentration on the transport rate of $NdCl_3$. HCl concentration: $0.05\ M$; carrier concentration in the liquid membrane: $0.02\ M$ in heptane. (a) Time profile for 0.003–$0.020\ M$. (b) Time profile for 0.025–$0.030\ M$. (c) Plot of apparent rates (V) versus initial substrate concentrations.

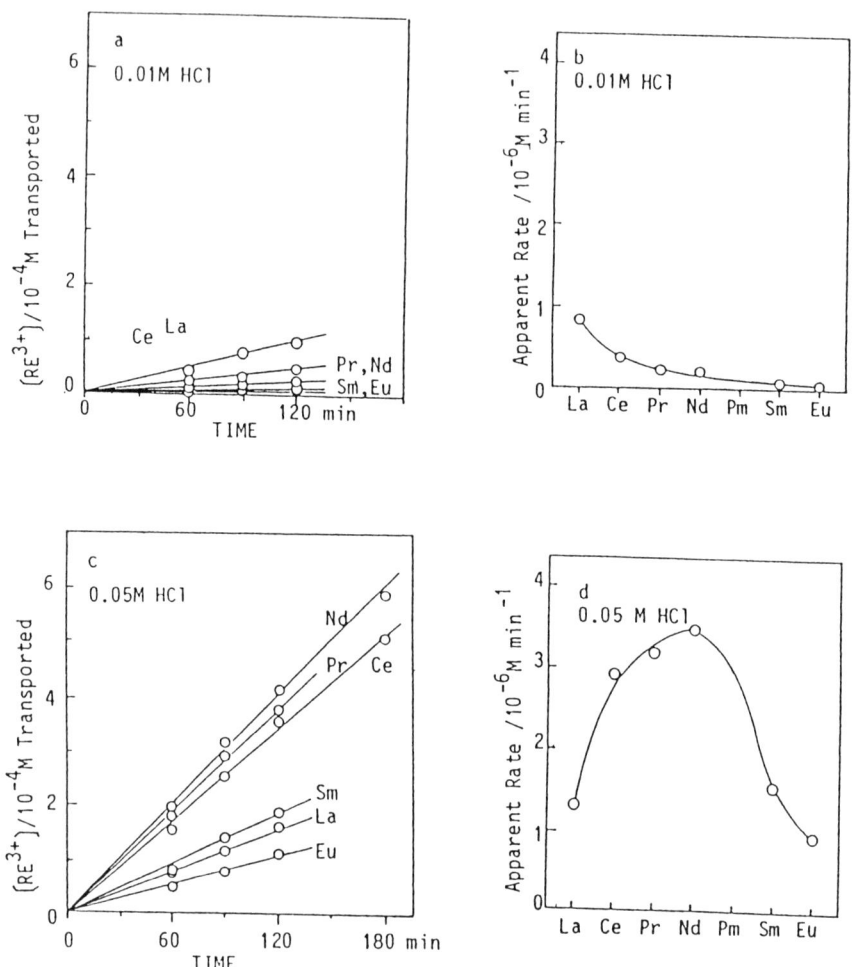

Figure 25 Effects of HCl concentration on the transport rates of a series of RE^{3+} ions. Carrier concentration in the liquid membrane: $0.02\ M$ in heptane. Substrate concentration: $0.01\ M$ in aq.-HCl solution. (a), (c), (e), (g), and (i) show time profiles, and (b), (d), (f), (h), and (j) show the corresponding apparent rates (V) vs. HCl concentration. [HCl]: (a) $0.01\ M$; (b) $0.01\ M$; (c) $0.05\ M$; (d) $0.05\ M$; (e) $0.10\ M$; (f) $0.10\ M$; (g) $0.15\ M$; (h) $0.15\ M$; (i) $0.20\ M$; (j) $0.20\ M$.

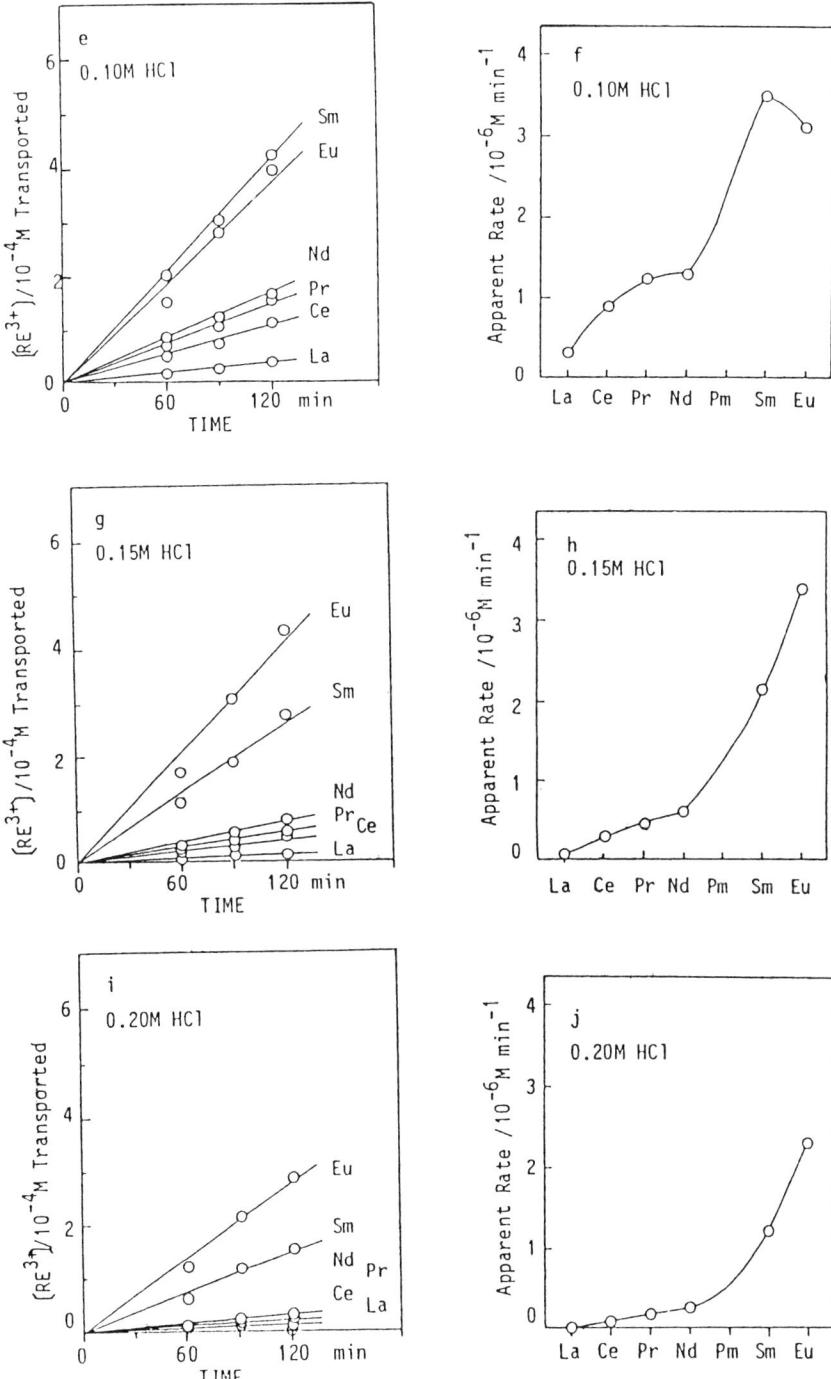

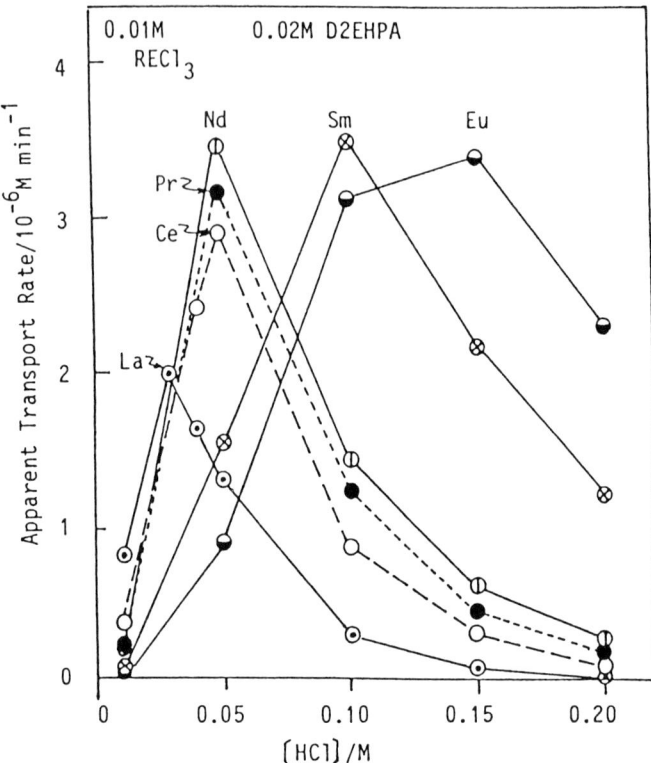

Figure 26 Summary of effects of HCl concentration on the transport rates of a series of lighter RE^{3+} ions.

Table 8 Selectivities in Liquid-Membrane Transport for Pairs of RE^{3+} Ions

	La	Ce	Pr	Nd	Sm	Eu
La	—	yes[a]	yes[a]	yes[a]	yes[b]	yes[c]
Ce		—	slight	slight	yes[b]	yes[c]
Pr			—	slight	yes[b]	yes[c]
Nd				—	yes[b]	yes[c]
Sm					—	yes[d]

[a]With aq.-HCl solution of 0.05 M.
[b]With aq.-HCl solution of 0.10–0.15 M.
[c]With aq.-HCl solution of 0.15 M.
[d]With aq.-HCl solution of 0.20 M.

metal ions used here were well correlated with the corresponding extraction results [31]. And in many cases, LMT behaviors have been shown to correlate with those of the corresponding extraction experiments. Our CPC results also showed that the optimum HCl concentrations increased with the atomic number of RE^{3+} ions, as shown in Fig. 27. This behavior is nicely consistent with our present observations in the corresponding LMT systems.

That no transport in LMT was observed without D2EHPA and no effective retention times were recorded in CPC in the absence of D2EHPA (Section IV) indicates that our LMT and the corresponding CPC systems satisfy the assumption of complete negligibility of the partition effect due to the mother solvent. Namely, both the LMT and CPC systems examined are almost solely the results of interaction of $[S\text{---}Y]$ in the organic phases.

Hence, we can conclude that the behaviors of RE^{3+} ions in the presence of D2EHPA belong to case 3 in Table 7 as expected, and no other significant factor is involved in the major parts of the phase transfer processes between the aqueous and organic phases. It is important that, although significant peak broadening is observed in CPC (case 3), its direct application to the LMT systems leads to achievement of promising selective ion transport. In Table 9, relative transport rates obtained in the LMT system using n-heptane solution of D2EHPA and aqueous solution are compared with the separation factors (α) obtained in the corresponding CPC. Both of the results show qualitatively parallel trends in selectivity of adjacent couples of RE^{3+} ions.

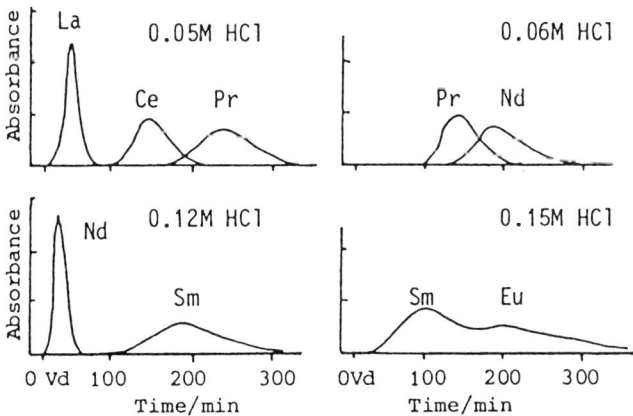

Figure 27 D2EHPA-aided CPC separations of lighter $RECl_3$ used in this Section. Cited from Section IV. Stationary phase: 0.1 M D2EHPA in n-heptane; mobile phase: aq.-HCl solution. Centrifugation: 800 rpm at room temperature. Elution: descending mode. Microcells: $400 \times 3 = 1200$.

Table 9 Comparison of Relative Transport Rates ($V(RE_2)/V(RE_1)$) in LMT with Separation Factors (α) in CPC[a]

		LMT[b]		CPC[c]	
RE_1	RE_2	[HCl] (M)	$V(RE_2)/V(RE_1)$	[HCl] (M)	α
La	Ce	0.05	2.2	0.05	4.0 ± 0.1
Ce	Pr	0.05–0.10	1.2 ± 0.2	0.05	1.7
Pr	Nd	0.05–0.10	1.1 ± 0.1	0.06	1.4
Nd	Sm	0.10	2.4	0.12	9.2 ± 0.1
Sm	Eu	0.15	1.6	0.15	2.3

[a]Organic phase: *n*-heptane solution of D2EHPA; aqueous phase: aq.-HCl solution.
[b]0.02 M D2EHPA.
[c]0.10 M D2EHPA (see Section IV).

There can be derived the further important prediction that many types of carrier compounds other than D2EHPA, which have been used in the related extraction procedures for refining RE ions, will be effectively applied to selective LMT techniques, leading to some useful sensors for RE ions in the future. Since several types of structurally simple and costless compounds have been used in the extraction techniques, it is likely not necessary to always design the guest-recognizing host molecules with complicated structures, where practical application of an LMT system is sought.

More recently, we have observed that the separation of Er^{3+} and Yb^{3+} (the second heaviest RE metal) can be readily attained by CPC with D2EHPA in $CHCl_3$ (see Section V). This means that the selective LMT of a series of heavier RE metal ions will also be realized readily by the use of a liquid membrane containing D2EHPA in $CHCl_3$.

E. Conclusion

According to our idea that LMT and separator-aided CPC are basically correlated with each other, selective LMT of lighter RE^{3+} ions was found to be attained with a simple and costless carrier of D2EHPA. In this case, rather good correlation between CPC and LMT was observed. Our present observations might be extended to realize the selective LMT of heavier RE metal ions including Er^{3+}/Yb^{3+} pair. (These problems will be described in Section VIII.)

Since LMT and CPC techniques are based on extraction behavior, a variety of costless, simple, and effective carriers can be used for sensing applications of LMT in future. This will cover the selective transport of versatile molecular and ionic substrates, which has been highly difficult with macrocyclic host carriers.

VIII. LIQUID-MEMBRANE TRANSPORT OF HEAVIER RARE EARTH METAL IONS BASED ON CPC RESULTS WITH DI(2-ETHYLHEXYL) PHOSPHORIC ACID*

A. Introduction

In previous sections we have shown that separator-aided centrifugal partition chromatography (CPC) can be regarded as a type of multistage liquid membrane transport (LMT) system. This is because the key processes in these techniques can be represented by a common equilibrium,

$$S + Y \rightarrow [S\text{----}Y] \rightarrow Y + S \qquad (1)$$

where S denotes a substrate and a sample species in LMT and CPC, respectively, and Y a carrier in LMT and a separator in CPC. The symbol $[S\text{---}Y]$ represents the interacting states in the liquid membrane (LMT) or in the stationary phase (CPC) (see Sections II, III, and VII and Ref. 48).

This concept was tested for lighter series of rare earth (RE) metal ions: La^{3+}, Ce^{3+}, Pr^{3+}, Nd^{3+}, Sm^{3+}, and Eu^{3+} (see Section VII). In the last section, the LMT system was constructed by the two-phase liquid system, which was almost the same as used in CPC. Namely, the aqueous phase (the source and receiving phases in LMT, and the mobile phase in CPC) was HCl solution and the organic phase (the liquid membrane in LMT and the stationary phase in CPC) was n-heptane containing di(2-ethylhexyl) phosphoric acid (D2EHPA). The results showed that the LMT and CPC were well correlated, and selective LMT was attained for several combinations of the lighter RE ions.

In Section VI, we examined the CPC behaviors of heavier RE ions (Gd^{3+}, Tb^{3+}, Dy^{3+}, Ho^{3+}, Er^{3+}, Tm^{3+}, Yb^{3+}, and Lu^{3+}) by using D2EHPA as separator in $CHCl_3$ stationary solvent. The CPC separation results, summarized in Table 6, indicated there were rather high values of separation factor (α) for these RE ions.

This section discusses the investigation of LMT behaviors of these heavier RE ions by the use of the LMT system corresponding to the CPC experiments of Section VI. The results indicate that there are several RE combinations that can be used for selective LMT procedures such as ion-sensing devices.

B. Experimental Procedures

1. Materials

The reagents used in this study were identical to those described in Section VI.

---

*Adapted from T. Araki, H. Asai, N. Tanaka, K. Kimata, K. Hosoya, and H. Narita, *J. Liq. Chromatogr.*, *13*: 3689–3704 (1990), by the courtesy of Marcel Dekker, Inc.

Table 10 Comparison of Liquid-Membrane Transport Data at 30° with Centrifugal Partition Chromatographic Data at 25°C

RE_1	RE_2	Concentration of HCl (N)	LMT (V_1/V_2)	CPC (α)
Gd	Tb	0.03	2.98	4.21
Gd	Tb	0.04	2.61	3.66
Dy[a]	Ho[a]	0.06	1.76	0.94
Ho[a]	Er[a]	0.06	2.06	1.43
Er	Tm	0.10	2.53	4.05
Tm	Yb	0.10	1.84	3.79
Tm	Yb	0.12	2.05	1.20
Yb	Lu	0.12	1.33	0.61

[a]At 45°C

2. Apparatus

Detection of RE metal ions in the aqueous phase was carried out with a Shimadzu double-beam spectrophotometer Model UV-180 and a Hitachi-Horiba pH meter Model F-7 II.

3. Liquid-Membrane Transport Procedures

An H-type glass apparatus (Fig. 28) was immersed in a thermostated bath, where the phases A (5 ml) and C (5 ml) are aqueous source and receiving phases containing HCl (0.01–0.20 N), respectively, and the phase B (15 ml) is the liquid membrane composed of $CHCl_3$ solution containing D2EHPA (2–20 mM) as carrier. The concentration of RE ions in the source phase was 1–10 mM. The liquid membrane phase was gently stirred (150 rpm) with the aid of a Yanaco rotating electrode head Model P10-RE equipped with a small magnet.

The aliquots of the receiving aqueous phase were taken up after 60, 90, and 120 minutes, diluted, and analyzed spectroscopically at 580 nm as described for the corresponding CPC (see Section VI).

C. Results and Discussion

The experiments of LMT were carried out by employing almost identical procedures as in Section VII, except for the use of an H-type apparatus and chloroform organic phase as liquid membrane. The constitution of the two liquid-phase system was almost comparable with that of the corresponding CPC. Namely, the source and receiving phases were aqueous HCl solutions (corresponds to the

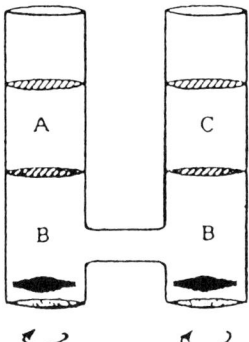

Figure 28 An H-type apparatus used in the LMT experiments. *A*: source phase; *B*: liquid membrane; *C*: receiving phase.

mobile phase in CPC) and the liquid membrane was a chloroform solution of D2EHPA (corresponds to the stationary phase in CPC).

1. Effect of Concentration of D2EHPA in Chloroform

It was confirmed that no transport was observed when a liquid membrane without D2EHPA was used. The typical results of the effect of D2EHPA concentrations on the rate of transport for Ho^{3+} ion are shown in Fig. 29. The rate increased steeply with increase in the D2EHPA concentration from 8 to 16 mM, and the rate-enhancing effect became smaller at concentrations higher than 18 mM D2EHPA.

2. Effect of Initial Substrate Concentration on the Transport Rate

The effect of initial substrate concentration on the rate of transport is exemplified by the case of Ho^{3+} ion at a D2EHPA concentration of 15 mM and a HCl concentration of 0.03 N (Fig. 30). The apparent rates of transport (V) were calculated from the slopes of time-conversion plots. The general trend in the effect of initial substrate concentration on V was similar to the cases of lighter RE ions in Section VII.

3. Effect of Concentration of HCl in the Aqueous Source Phase on the Transport Rates of RE Ions

As in the case of LMT of lighter RE ions (Section VII), the concentration of HCl in the aqueous source phase was of prime importance for transport of heavier RE ions. In Figs. 31–35, the variation of transport rates of RE ions from Gd^{3+} to Lu^{3+} with change in the HCl concentration are illustrated. The relations of V versus

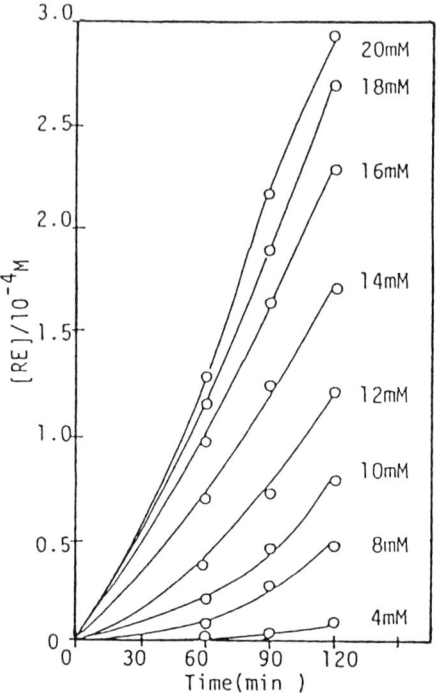

Figure 29 Effect of concentration of D2EHPA on the rate of transport of Ho^{3+} ion at 30°C. $[Ho^{3+}]$: 5 mM, [HCl] in aqueous phases: 0.03 N.

atomic number of RE ions are shown for some selected HCl concentrations (Figs. 31b, 33b, and 34b).

4. Correlation Between LMT and CPC

Figure 36 summarizes the behavior of the heavier RE ions examined by plotting V versus HCl concentration. Underneath are the corresponding CPC chromatograms (see Section VI) obtained with an identical two-phase system as LMT. The HCl concentrations that give rise to good separation of the RE ions in CPC produce rather large differences in V values in LMT. The optimum V values in LMT were shifted toward higher HCl concentrations with increasing atomic number of RE ions, as observed in the cases of lighter series of RE ions (Section VII). In CPC, the optimum peak separation with reasonable peak shapes was also obtained at higher HCl concentrations as the atomic number of RE ions increased. In these respects, the behaviors of heavier RE ions in LMT and CPC are well correlated.

The observation that the maximum V values in LMT are present with respect

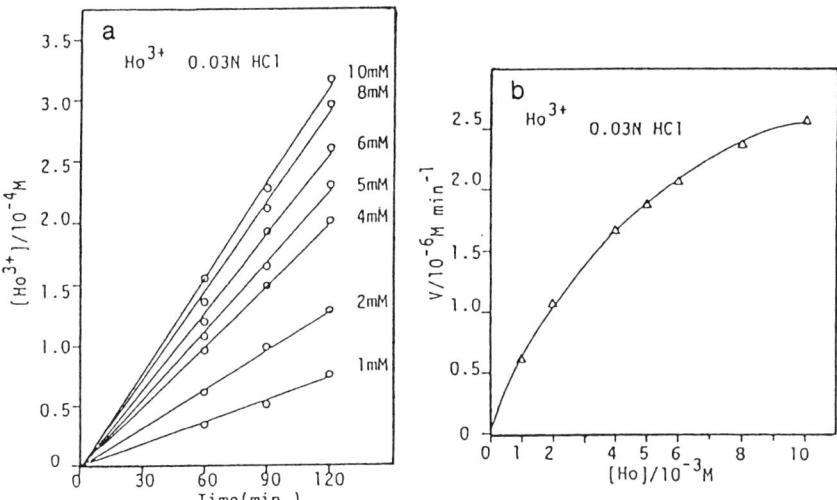

Figure 30 Effect of initial concentration of Ho^{3+} ion on the rate of transport. (a) $[Ho^{3+}]$ in the receiving phase vs. time. (b) Apparent transport rate constant vs. initial $[Ho^{3+}]$. [D2EHPA]: 15 mM, [HCl] in the aqueous phases: 0.03 N.

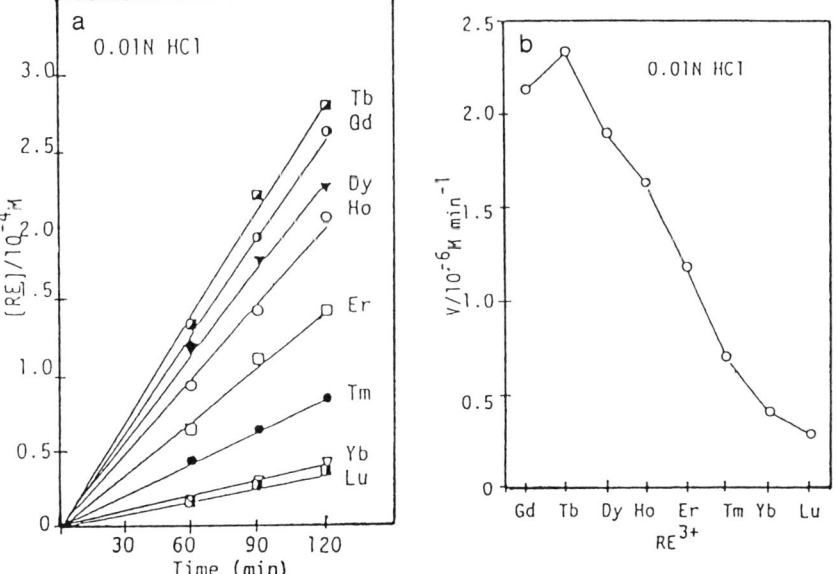

Figure 31 LMT of RE ions with D2EHPA separator and 0.01 N aqueous HCl phases at 30°C. (a) Transport rate data. (b) Transport rate constants vs. atomic number or RE.

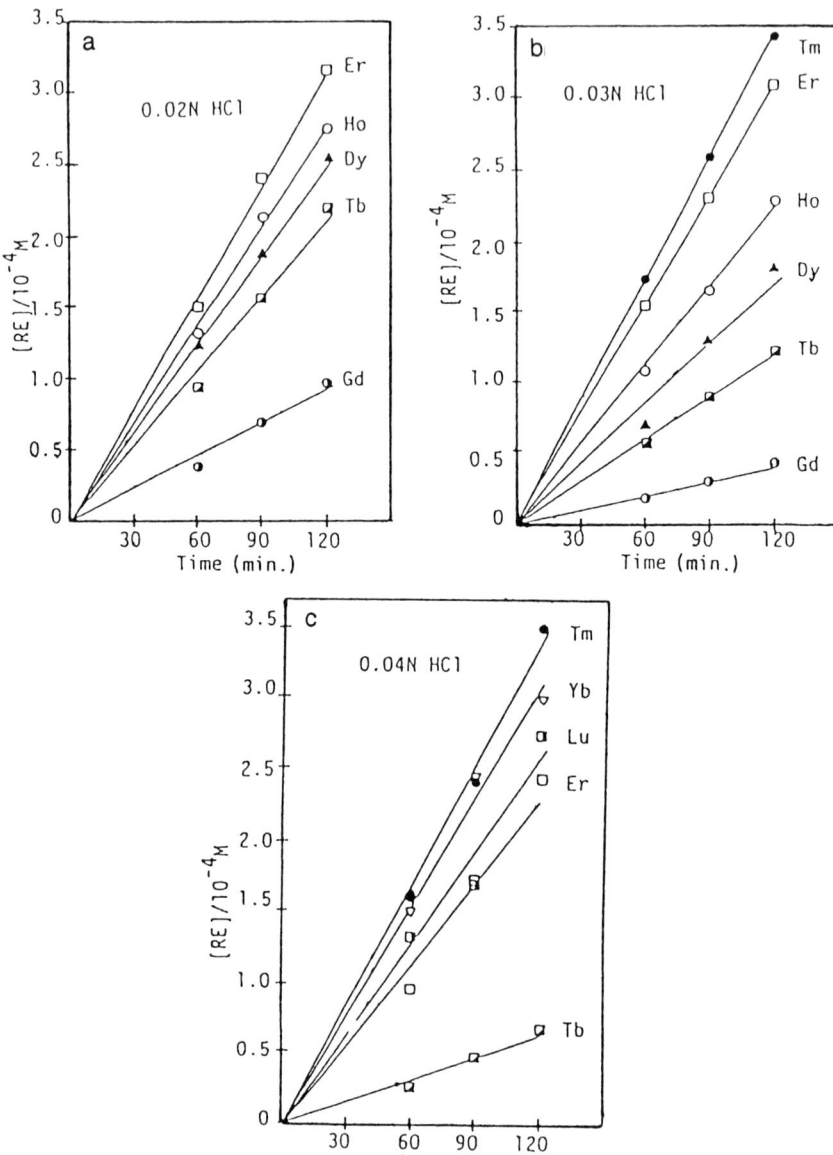

Figure 32 LMT of RE ions with D2EHPA separator at 30°C. Concentration of HCl in the aqueous phases: (a) 0.02 N; (b) 0.03 N; (c) 0.04 N.

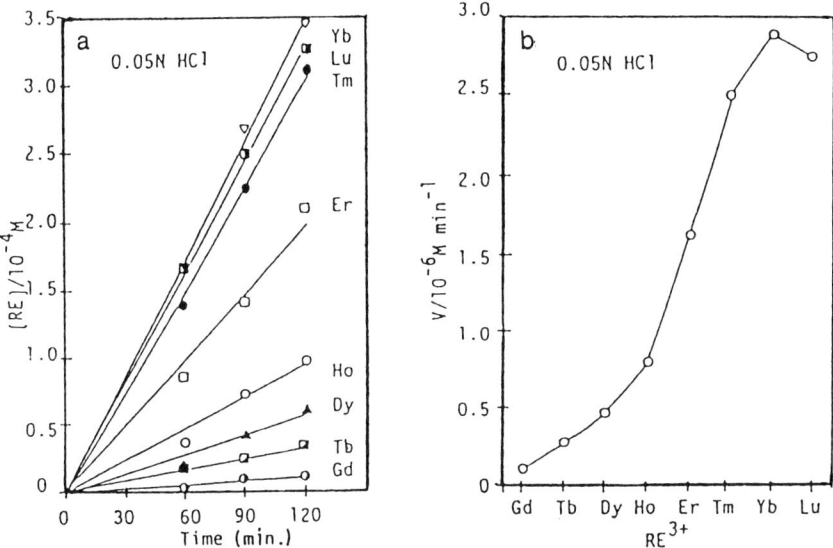

Figure 33 LMT of RE ions with D2EHPA separator and 0.05 N aqueous HCl phases at 30°C. (a) Transport rate data. (b) Transport rate constants vs. atomic number of RE.

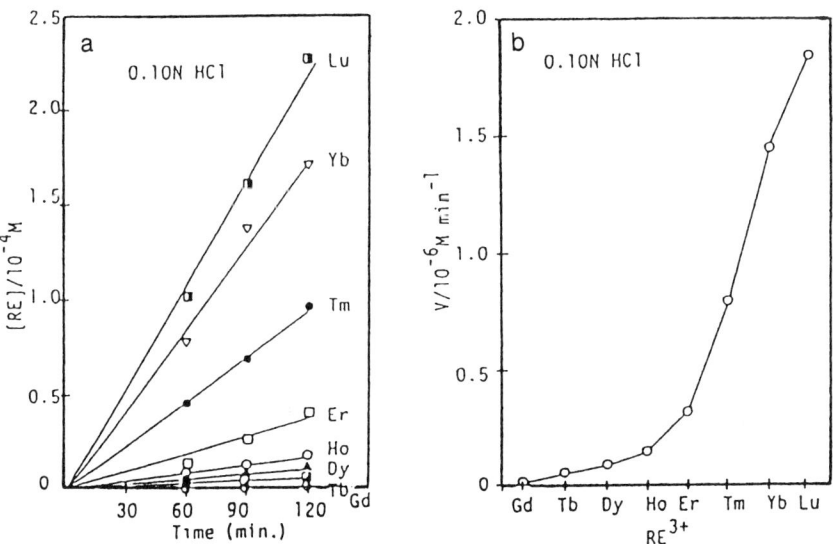

Figure 34 LMT of RE ions with D2EHPA separator and 0.10 N aqueous HCl phases at 30°C. (a) Transport rate data. (b) Transport rate constants vs. atomic number of RE.

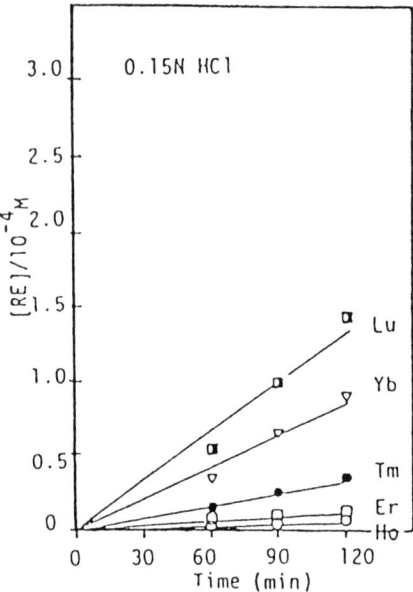

Figure 35 LMT of RE ions with D2EHPA separator and 0.15 N aqueous HCl phases at 30°C.

to HCl concentration seems to be associated with the broad peaks in CPC (see Section VI). At a HCl concentration higher than its maximum, the equilibrium

$$3(HG)_2 + RE^{3+} \rightleftarrows RE(HG_2)_3 + 3H^+ \qquad (13)$$

will be shifted left. In this situation, the substrate ion cannot transferred into the liquid membrane so easily, resulting in a low transport rate. On the other hand, this situation corresponds to fast elution of a sample in CPC, according to the expectation described in Table 7. In fact, the leading edge of a very broad peak in CPC came to start immediately after the dead time as the HCl concentration increased (see Section VI). At the same time, the elution of tailing edge of the peak became faster with increased HCl concentration.

Table 10 is a comparison of CPC and LMT data for adjacent pairs of RE ions by the use of almost comparable two liquid-phase conditions. The data are also illustrated in Fig. 37. Except for the Tm^{3+}/Yb^{3+} pair (0.10 N HCl), a very good correlation between the ratio of rate constants (V_1/V_2) in LMT and the separation factor (α) in CPC was obtained. This supports our assumption that the key equilibria in LMT and CPC are common, that is, Eq. 13.

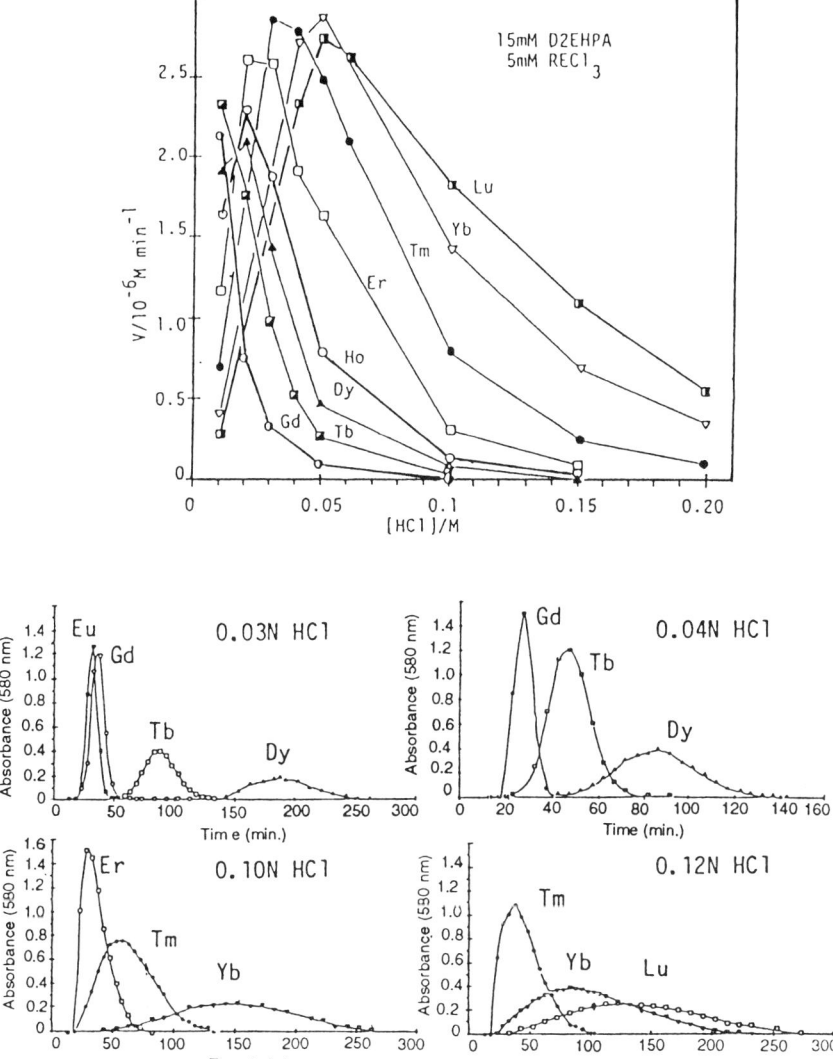

Figure 36 Relation between transport rate constant and HCl concentration for a series of heavier RE ions (top diagram). The corresponding typical CPC chromatograms (see Section VI) are shown underneath.

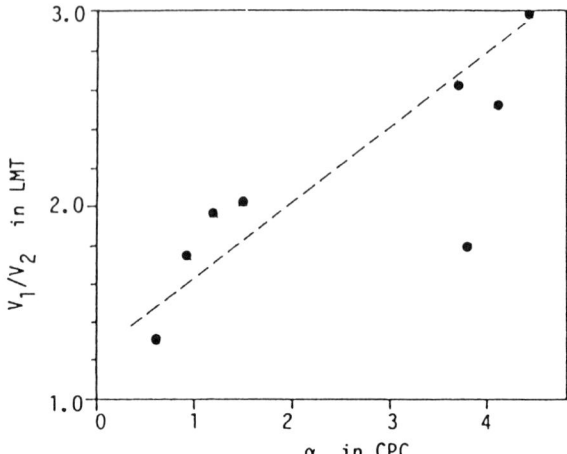

Figure 37 Correlation between V_1/V_2 in LMT and α in CPC.

5. Selective LMT for Heavier RE Ions

From Fig. 36 one can find several combinations of RE ions that are promising for selective LMT processes such as ion sensing by the use of D2EHPA. Figure 36 shows that in some cases it is possible to invert the order of transport of RE ion sets. For example, in the case of the Yb^{3+}/Gd^{3+} pair, Yb^{3+} is selectively transported with 0.04–0.07 N HCl aqueous phases, but Gd^{3+} can be selectively transported with 0.01 N HCl. It is similar for the Yb^{3+}/Tb^{3+}, Yb^{3+}/Ho^{3+}, Lu^{3+}/Gd^{3+}, Lu^{3+}/Tb^{3+}, Lu^{3+}/Ho^{3+}, and Lu^{3+}/Er^{3+} pairs. The LMT selectivities are also tabulated in Table 11 for convenience.

D. Conclusion

Liquid-membrane transport of heavier rare earth metal ions from Gd^{3+} to Lu^{3+} was facilitated by di(2-ethylhexyl) phosphoric acid as well as the lighter series (La^{3+}-Eu^{3+}) was (see Section VII). Thus, the behavior of a whole series of rare earth metal ions in the bulk liquid-membrane transport system was unveiled throughout by completion of this work. All the rare earth metal ions were found to be in accord with a common equilibrium representation, Eq. 13. The chemical processes in CPC for the whole series of rare earth metal ions with the corresponding two liquid-phase system containing D2EHPA also follow the same equation (see Section VI). Hence, the correlation between LMT and CPC was almost satisfactory. This supports our original concept that CPC can be regarded as a kind of multistage LMT system (see Section III and Ref. 48).

Table 11 Selectivities in Liquid-Membrane Transport for Pairs of Heavier RE^{3+} Ions

	Tb	Dy	Ho	Er	Tm	Yb	Lu
Gd	yes[a]	yes[b]	yes[b]	yes[e]	yes[g]	yes[i]	yes[m]
Tb		slight[c]	yes[d]	yes[f]	yes[g]	yes[i]	yes[n]
Dy			slight[c]	yes[c]	yes[h]	yes[j]	yes[o]
Ho				yes[c]	yes[h]	yes[k]	yes[p]
Er					slight[c]	yes[l]	yes[q]
Tm						yes[l]	yes[r]
Yb							slight[s]

[a–s]With aqueous solutions of 0.02 N (a), 0.02–0.03 N (b), 0.05 N (c), 0.03 N (d), 0.02–0.05 N (e), 0.03–0.05 N (f), 0.03–0.06 N (g), 0.05–0.06 N (h), 0.01 N and 0.04–0.07 N (i), 0.04–0.1 N (j), 0.02 N and 0.05–0.1 N (k), 0.1 N (l), 0.01 N and 0.03–0.1 N (m), 0.01 N and 0.04–0.1 N (n), 0.05–0.1 N (o), 0.02 N and 0.05–0.1 N (p), 0.02 N and 0.1 N (q), 0.03 N and 0.1 N (r), 0.1–0.15 N (s).

CPC behaviors of these metal ions have been reported [45] by using EHPA, and the key process has been presumed to be an equilibrium essentially the same as the Eq. 13, and the CPC behaviors were found to closely resemble those employing D2EHPA (see Section VI). So, we can venture to propose that LMT of rare earth metal ions using EHPA carrier will give a result similar to that of this work, using D2EHPA.

A highly familiar and costless phosphate carrier, D2EHPA (and probably also EHPA), was shown to be capable of realizing selective LMT of several combinations of rare earth metal ions. This will encourage the development of more sophisticated ion-sensing molecular systems in rare earth metal applications. Combinations of macrocyclic structures [46] and a phosphate group may be candidates for an effective carrier.

IX. CPC AS MULTISTAGE LIQUID-MEMBRANE TRANSPORT SYSTEMS*

A. Introduction

In basic research laboratories, the selectivity efficiencies of carriers are usually tested by employing single-stage batch-type liquid-membrane transport (LMT) systems. As descried in the other chapters of *Liquid Membranes: Chemical*

---

*Adapted from the Chapter 7-5 written by T. Araki, in *Liquid Membranes: Chemical Applications* (T. Araki and H. Tsukube, eds.), CRC Press Inc., Boca Raton, Florida, pp. 199–213 (1990), by the courtesy of CRC Press Inc.

Applications, excellent carriers have been found for specific substrates. It will be quite difficult, however, to find the carriers having almost perfect selectivity for the majority of other substrates. Hence, in order to increase the separation efficiencies so that they are practical, multistage liquid membrane processes must be developed. The emulsion liquid membrane (ELM) process appears to be multistage in nature, and series arrangements of several unit modules can serve as a multistage system. Several multistage modules of supported liquid membrane (SLM) are already quite familiar. This section will concentrate on a third possible chromatographic multistage process, a new candidate for multistage operation of batch-type liquid-membrane transport systems.

B. Types of Multistage Liquid-Membrane Transport Systems

1. Why Liquid Membrane?

After lengthy fundamental investigations on material separation and concentration using the processes of mass transfer through liquid membranes, industrial application is now a close goal. The most attractive feature of liquid membrane compared with solid polymer membrane is its extremely high mobility as a liquid, which leads to very rapid interfacial and in-phase diffusion. Pore-free permeation thus occurs quickly. Ultra-thin liquid films can be quite readily prepared, and very small amounts of highly selective carrier compounds can be impregnated in the liquid membranes.

2. Emulsion Liquid Membrane or Liquid Surfactant Membrane

In the ELM processes, there are cases where a substrate species penetrates successively into many microemulsion globules while it is being transferred from the supply to the receiving vessel. This process thus seems to be multistage in nature. In 1966 Li of the Esso group patented [50,51] a desalination process using ELM; in 1971 Cussler [52] achieved active transport of Na ion; and in 1986, Marr and his collaborators [53,54] constructed the first commercial-scale plant for recovery and concentration of Zn ion from viscose-waste solution. Comprehensive reviews on the basic concept of the latter [53,54] and basic data for the plant have been reported [55,56]. One reason for the good efficiency of these processes may be the intrinsic multistage nature of a flux through microglobules of the emulsion. The dynamic molecular motions in a surfactant layer of an emulsion globule cause instability of the emulsion. Extensive efforts to overcome drawbacks in the ELM processes are now being made by several research groups.

3. Supported Liquid Membrane

Another promising multistage liquid membrane process is the SLM system, in which liquid membranes are impregnated in the microscopes and/or thin portions of a polymer hollow fiber and/or polymer sheet. Use of porous cellulose acetate as the supporting material for an immobilized film of water was tested in the 1960s to accelerate the transport of O_2 [57], for desalination [58], and then for CO_2–O_2 separation [59]. The SLM process permits the use of physically stable ultra-thin liquid films and the ready construction of multistage modules, as in the solid membrane process, and is also still under investigation.

4. Chromatography Related to Liquid-Membrane Transport

Although mechanical treatment for the facilitated mass transport in liquid membrane systems using planar sheet models can basically and quantitatively predict their operating results as shown by Ward [60] and Cussler [52], more work with sphere and cylinder models is required for the ELM and SLM processes, respectively [62]. The two types of processes also involve severe inherent drawbacks for practical operation, mainly due to problems of operational stability preventing their smooth development for industrial use [61–65].

The ELM processes are often operated using mixer–settler type continuous extraction devices [32,66,67]. Other apparatuses for the continuous process have also been designed to develop certain chromatographic multistage effects. The SLM process can be operated in modules, thereby resembling liquid chromatography. This, however, has rarely been referred to as liquid-membrane chromatography, because the net separation efficiency for multicomponent fractionation is much less than for familiar chromatography.

Chromatography is a technique in which the effect of a unit process is amplified by thousands of repetitions; for example, recent high-performance chromatographies permit complete separation even if the separation factor of two given samples is <1.1. Liquid membranes coated on the particles of solid porous support have been used in gas–liquid partition chromatography, but this method is not covered in this chapter.

In the 1950s, Craig's [68] classical continuous multistage extraction process, the "countercurrent distribution method," an original form of liquid–liquid two-phase extraction chromatography, was in common use in biochemical laboratories, and a great quantity of separation/isolation data were compiled [10,69–73]. This classical process, however, was quite complicated and extremely time consuming.

Recently, Craig's method has been greatly modernized for rapid and easy operation by applying centrifugal force [2,13–17]. Two types of chromatographic machines are now commercially available for laboratory use, and excellent fractionation/isolation/enrichment results have been obtained in bioorganic and other laboratories [18–21,74–88]. Industrial-scale chromatographic apparatuses are also available. The fundamental theory of modern centrifugal countercurrent-type chromatographies is almost identical to that established by the classical method [10]. Only minor modifications are necessary for the effects of centrifugal force when maximum efficiency is desired [2,18,74,89,90].

Although centrifugal countercurrent-type chromatography in the standard form (only partition effect between two solvents) is not directly correlated with the facilitated LMT systems, Araki et al. (see Section II) showed that when a carrier or a separator is impregnated in the stationary phase, the separation behaviors become closely related to the facilitated LMT results. Hence, the new chromatography, especially "separator-aided" centrifugal countercurrent–type chromatography, seems to be a third promising possibility of multistage liquid membrane transport process for separation/isolation/purification. The following descriptions will focus on this subject.

C. Types of Centrifugal Countercurrent Extraction Methods

1. Centrifugal Partition Chromatography

Centrifugal partition chromatography (CPC) was introduced by Murayama et al. [2] of the Sanki Engineering group in 1982. The machinery characteristic of CPC are the simple and compact construction of microcell cartridges and the use of rotatory seal joints. In the handy laboratory-scale equipment, 50 or 400 microcells are arranged in series in a Teflon cartridge block (ca. 50 × 40 × 200 mm), and a maximum of 12 cartridges (containing a maximum of 4800 microcells) can be set in series on a rotatory disk. The basic features of the unit microcell are depicted in Fig. 38a. Upon rotation of the disk, one of the two immiscible liquid phases acts as the stationary phase as a result of the centrifugal force applied. The researchers can select an optimum from four types of elution modes: organic eluent/aqueous stationary phase (ascending or descending) or aqueous eluent/organic stationary phase (ascending or descending). A schematic sketch of the cartridge construction of the rotary disk is given in Fig. 38b.

The machinery construction of CPC is closely related to that of Craig's continuous extraction method, that is, a series arrangement of extraction compartments. Hence, Craig's distribution theory can be almost directly applied, permitting estimation of the ideal number of theoretical plates quite readily.

2. Centrifugal Countercurrent Chromatography

The present style of high-speed centrifugal countercurrent chromatographic apparatus (CCC) was developed by Ito [15] of NHLBI somewhat earlier than CPC. This was an epochal change from the slow countercurrent chromatography, including droplet countercurrent chromatography (DCCC), to today's high-speed techniques. Ito's apparatus is called a coil-planet centrifuge because a coiled tube(s) is rotated outside a rotating diaphragm (Fig. 39b) to attain the unilateral hydrodynamic equilibrium in a tube (Fig. 39a). This apparatus is operated without either rotary seal joints or microcell compartments. Again, researchers can choose the optimum operational mode as in the case of CPC. Estimation of the ideal number of theoretical plates, however, is rather difficult in this noncompartment system. Length and combination of the tube(s) seem to be the user's empirical choice. Several mechanical modifications have been advanced by Ito's group.

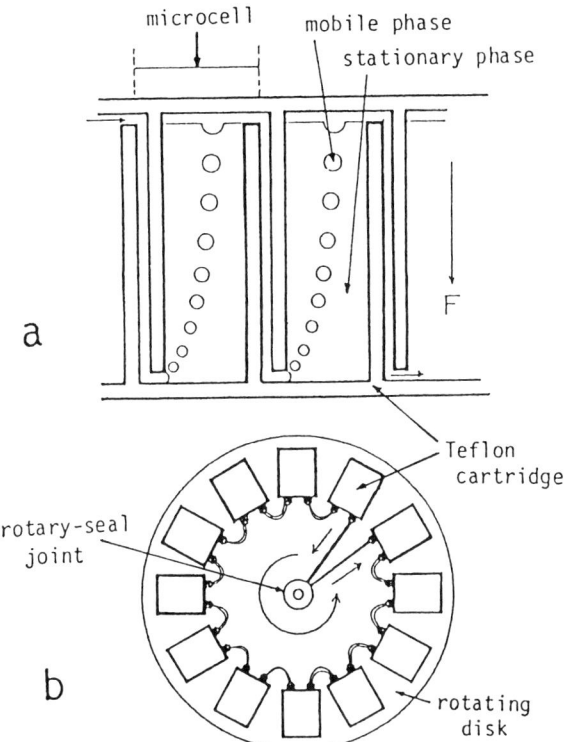

Figure 38 The basic processes of centrifugal countercurrent-type chromatographies. (a) A unit microcell in CPC. (b) Cartridges containing microcells are arranged on a rotary disk in CPC. (From Ref. 2.)

D. Correlation Between Centrifugal Partition Chromatography and Liquid Membrane Transport System

1. Basic Concept

The behavior of a single unit in separator-aided CPC (CCC has basically similar unit behavior) is illustrated in Fig. 40a, in comparison with that of a single-stage facilitated LMT system (Fig. 40b). The basic chemical mechanism for the both techniques can be described as (see Section III and Ref. 48)

$$S + Y \rightarrow [S\text{----}Y] \rightarrow Y + S \tag{1}$$

where S denotes a sample and a substrate for CPC and LMT, respectively, and Y denotes a separator and a carrier for CPC and LMT systems, respectively. $[S\text{---}Y]$ represents interactions between S and Y in the stationary phase in CPC and in the

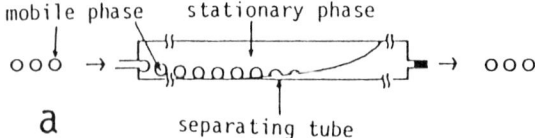
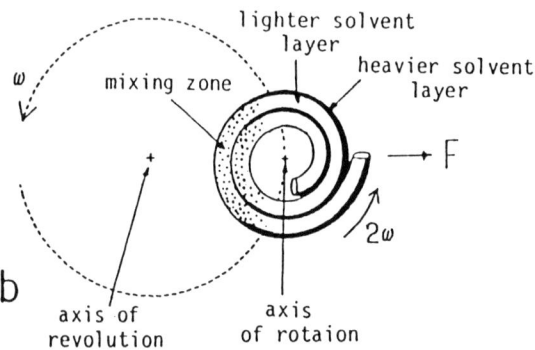

Figure 39 Schematic sketches of the mechanical constructions of centrifugal countercurrent chromatographies (CCC). (a) A separating tube in CCC. (b) Coil-planet centrifuge apparatus in which the coiled tube(s) is rotated outside a rotating diaphragm. (From Ref. 18.)

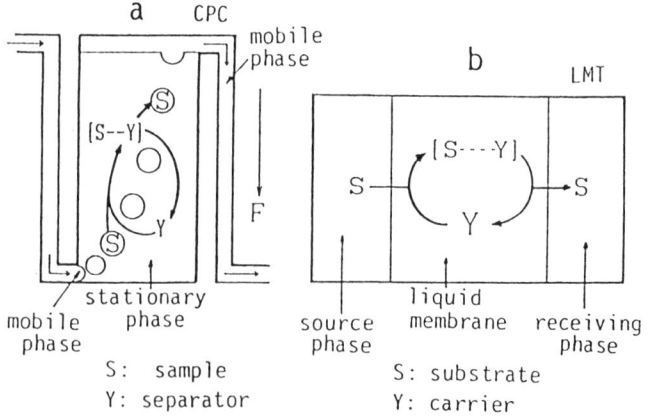

Figure 40 Unit behaviors in separator-aided CPC (a) and the single-stage facilitated LMT systems (b).

liquid membrane. The [S---Y] includes hydrogen bonding, dipolar, coordinating, chelating, and any other interactions effective for discriminating the components S in the CPC and LMT systems. From Eq. 1, it is clear that effective CPC and LMT results will be obtained when Y is effectively circulated in the stationary phase or in the liquid membrane via complexation (at the first interface) and dissociation (at the second interface). This feature is well documented in the literature of LMT systems.

There is an important difference, however, between the LMT and CPC systems: S transferred into the stationary phase in CPC from the mobile phase forms [S---Y] and then the dissociated S goes back to the mobile phase (Fig. 40a). In contrast, S transferred into the liquid membrane in LMT to form [S---Y] with Y goes into the source and receiving phases after dissociation. Effective transport is attained when the dissociation at the receiving phase/liquid membrane interface exceeds that at the source phase/liquid membrane interface (Fig. 40b).

In the facilitated LMT system, in general, Y is necessary for transporting S from the source phase to the liquid membrane. The separator-aided CPC is effective when a set of samples S_1 and S_2 has the same retention volume (V_R) in the normal partition-type CPC without Y. In the separator-aided CPC, $V_R = 0$ is not a necessary condition, but Y in the stationary phase is solely responsible for production of the difference in V_R values between S_1 and S_2. In a typical case where the [S---Y] is a very weak interaction, S will have a very small V_R value in CPC. In the corresponding LMT system, however, the transport will require a rather long time.

Thus, in spite of their chemical similarity (Eq. 1), the observed behaviors in LMT and CPC will not always be parallel, due to their physical differences. To conventionally compare separator-aided CPC with the corresponding facilitated LMT system, it is assumed that S_1 and S_2 cannot permeate into either the stationary phase or the liquid membrane in the absence of Y, and that the [S---Y] interaction is the sole factor causing the difference in the transfer behaviors of S_1 and S_2. Under these assumptions, one can obtain a rough guideline (Table 12) (see Section VII and Ref. 48) of the results of CPC and the corresponding LMT by considering the ease of phase transfers involved.

For simplification, Table 12 shows only the case where Y is impregnated in the organic phase of a w/o/w two liquid-phase system in which S is transferred from w to o to w. In principle, the same pattern is applied for an o/w/o type system. (If Table 12 is treated more precisely, the term *ready* or *difficult* should be expressed by equilibrium constants (and also diffusion parameters, if necessary).) Although Table 12 is only a rough guide, it is very helpful in predicting the behaviors in CPC from results obtained by the corresponding LMT, and vice versa.

2. Examples

The crown ether–facilitated selective LMT of alkali metal picrates [8,47] is a key finding in the modern use of liquid membranes as host–guest molecular recogni-

Table 12 Relation Between CPC (Aqueous Mobile and Organic Stationary Phases) and LMT (Aqueous Source and Receiving Phases with Organic Liquid Membrane) in View of Phase Transfer Processes

	Phase transfer processes		Expected results	
	Aqueous to organic phase	Organic to aqueous phase	Retention time in CPC	Transport rate in LMT
A	No	—	Very short	0
B	Ready	Ready	Short	High
C	Ready	Difficult	Long[a]	Low
D	Difficult	Ready	Short[b]	Low
E	Difficult	Difficult	Very broad	Very low
F	Boundary[c]	—	—[c]	0

[a]Tailing peak.
[b]Leading peak.
[c]Permeation of S occurs at the boundary interface layer only; i.e., no permeation into the inside of organic phases.
[d]The retention time varies depending on the relative partition coefficients between the two immiscible phases.
Sources: Ref. 48 and Section VII.

tion processes. Originally, the transport selectivity of alkali metal ions showed direct correlation with extraction results obtained in fundamental organic host–guest chemistry.

Recently, the system containing dibenzo-18-crown-6(Y)/alkali picrates (S) in $CHCl_3$ was applied to CPC containing 50 × 3 microcells (ideal number of theoretical plates = 150) (see Section II). Although 10% butan-1-ol was added to the $CHCl_3$ phase in CPC, the retention values obtained by CPC were shown to be well correlated with $\log K$ (Fig. 41), where K is the equilibrium constant obtained by corresponding simple extraction experiments [91,92]. In addition, the separation factor ($V_R(K^+)/V_R(Na^+)$) observed in CPC was 1.99 ($\log K(K^+/Na^+)$ = 2.0), resulting in effective separation/purification of these ions from the equimolar mixture (Fig. 42). Other binary mixtures having $V_R(M_1)/V_R(M_2)$ = 1.25–1.55 (Table 13) will be effectively purified if a larger number of microcells is used in the CPC. No sets of alkali picrates were able to separate in the CPC without the crown ether. Thus, CPC of the crown ether/alkali picrate system was found to be indirectly correlated with the corresponding facilitated LMT system.

Direct comparison of separator-aided CPC with the corresponding LMT system was first made for separation/isolation of picric acid (TNP) and p-nitrophenol (NP) (see Section III). In this case, no separation in CPC and no transport in LMT were observed using $CHCl_3$ organic phases alone. Addition of 5%–

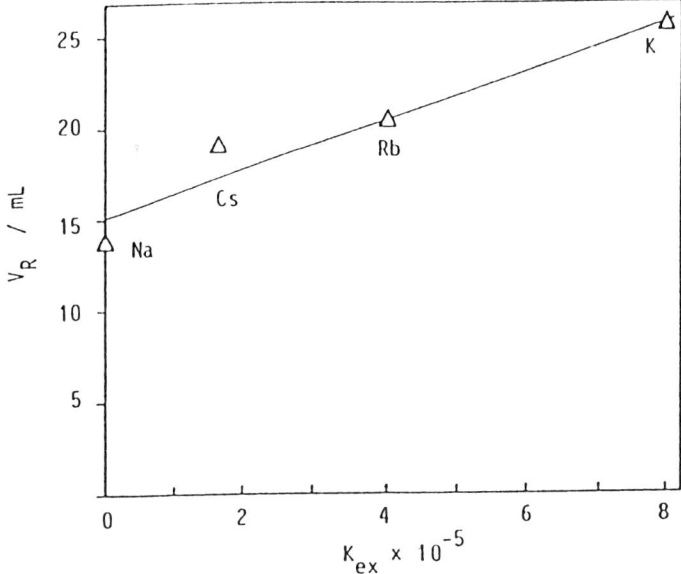

Figure 41 Plot of the V_R values obtained by CPC vs. $\log K$ obtained [91,92] by the corresponding simple extraction experiments. Organic phase: dibenzo-18-crown-6 (Y) in $CHCl_3$/n-BuOH (9 + 1, v/v); aqueous phase: M^+-picrates in H_2O (see Section II).

butan-1-ol (Y) to the organic phase, however, gave rise to complete separation of the two samples in CPC (t_R(NP)/t_R(TNP) = 5.4, where t_R is the retention time), allowing ready isolation of NMR spectroscopically pure components. Use of the same $CHCl_3$/5% butan-1-ol liquid membrane resulted in selective transport of the pair of substrates (k_{TNP}/k_{NP} = 8.4, where k is the apparent rate constant of transport). In both techniques, NP stayed for a longer period in the organic phases than TNP.

Separation and purification of lanthanide ions are important in separation chemistry. Many approaches [22–30] using di(2-ethylhexyl)phosphoric acid (D2EHPA), tri-n-butyl phosphate (TBP), 2-ethylhexyl phosphoric acid mono-2-ethylhexyl ester (PC-88A), and related phosphorus compounds as extractant have been tested by techniques such as liquid–liquid extraction, column chromatography [31], 20-stage mixer–settler method [32], and extraction by ELM [33–35]. Industrial purification of lanthanides is now being carried out on the basis of the fundamental extraction data obtained.

CPC of lighter rare earth metal ions ($RECl_3$ in aq.-HCl) was performed using D2EHPA as separator in heptane (see Section IV and Ref. 38). Although the number of theoretical plates observed was low, good separations were obtained for

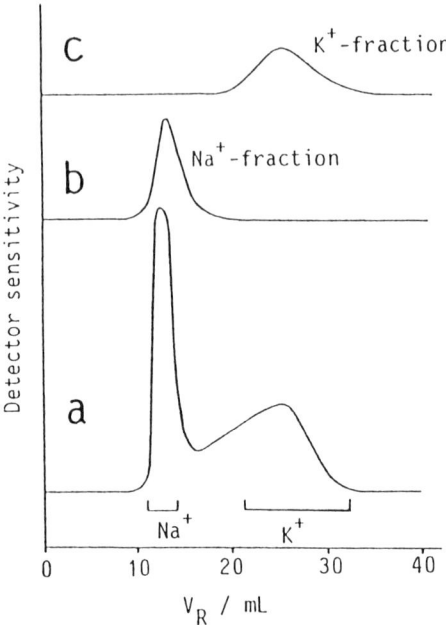

Figure 42 Chromatograms of CPC for separation of Na$^+$ and K$^+$ ions using dibenzo-18-crown-6 as Y in the 50 × 3 (= 150) microcells (see Section II). (a) Equimolar binary mixture. (b) Chromatogram of the Na$^+$ fraction obtained by the first CPC (a). (c) Chromatogram of the K$^+$ fraction obtained by the first CPC (a). $t_R = t_{R(\text{apparent})} - t_{\text{dead}}$.

Table 13 Comparison of CPC Separations with the Corresponding Extraction Equilibrium Constants for the System of Dibenzo-18-Crown-6 (Y) in CHCl$_3$/Alkali Metal Picrates (S)

M^+-picrates		$V_R(M_1{}^+)/V_R(M_2{}^+)$		
$M_1{}^+$	$M_2{}^+$	Binary sample	Single sample	$\log(K_1/K_2)$
Li$^+$	Na$^+$	1.00	1.01	—
K$^+$	Na$^+$	1.99	1.86	2.0
Rb$^+$	Na$^+$	1.52	1.49	1.7
Cs$^+$	Na$^+$	1.50	1.38	1.3
Cs$^+$	Li$^+$	1.46	1.36	—
K$^+$	Rb$^+$	1.26	1.25	0.3

Sources: Section II, and Refs. 91, 92 for $\log(K_1/K_2)$ values.

the adjacent series of lighter RE ions from La to Sm by adjusting the HCl concentration in the mobile phase (Fig. 43). Parallelism between separation factors obtained by CPC and those by extraction was reasonable. Toluene, which is more polar than heptane, was an effective stationary solvent for the separation of Sm/Eu ions. Further study indicated that $CHCl_3$ (more polar than toluene) was effective for separations of the heavier RE ions such as an Er/Yb pair (Fig. 43) (see Section V and Ref. 93).

The reason for peak broadenings is that the key reaction of RE ions with D2EHPA is a rather strong acid–base interaction (i.e., organic-to-aqueous transfer is difficult) as shown in Eq. 13, following [31]. PC-88A was found to be an effective separator, especially for heavier RE ions, when eluted with the aqueous mobile phase containing 20 vol% of ethylene glycol, which is buffered at a desired pH value with 0.1 M (H,Na)Cl$_2$CHCOO (Fig. 44) [45,94]. Considerable peak broadening was also observed because the key reaction is the same type as [45]

$$3(HG)_2 + RE^{3+} \rightleftarrows RE(GH_2)_3 + 3H^+ \tag{13}$$

where G and RE denote phosphate anion and rare earth ions, respectively.

TBP was not an effective separator for $RE(NO_3)_3$ because its key reaction is different from D2EHPA and is not adjustable by H^+ concentration (see Section V and Ref. 93):

$$RE^{3+} + 3NO_3^- + 3TBP \rightleftarrows RE(NO_3)_3TBP \tag{14}$$

For RE ions, the direct comparison of CPC with the corresponding single-stage LMT system was studied using a D2EHPA/heptane liquid membrane and aq.-HCl phase (see Sections VII and VIII and Ref. 93). The optimum HCl concentration for the transport becomes higher with an increase in the atomic number of the RE (Fig. 45), in parallel to CPC. This parallelism was kept for the heavier RE members [95,96]. Hence, this LMT system is found to be correlated with the corresponding CPC, and selective LMT can be carried out by adjusting the HCl concentrations. Namely, in applications to purification/isolation/enrichment, the countercurrent-type chromatographies will be of significant value as one type of multistage LMT system.

Along the same lines, many sets of heavier rare earth metal ions can be selectively transported through the liquid membrane systems containing D2EHPA. Again, the transport behaviors of the metal ions have been shown to correlate with those of CPC. These were discussed in Sections VI and VIII; the relevant experiments of CPC and LMT for heavier series ions had been undertaken while the original form of this Section was in print.

E. Outlook

Recent development of high-speed CPC and CCC is a modernization of the continuous multistage liquid–liquid extraction method. When a suitable separator

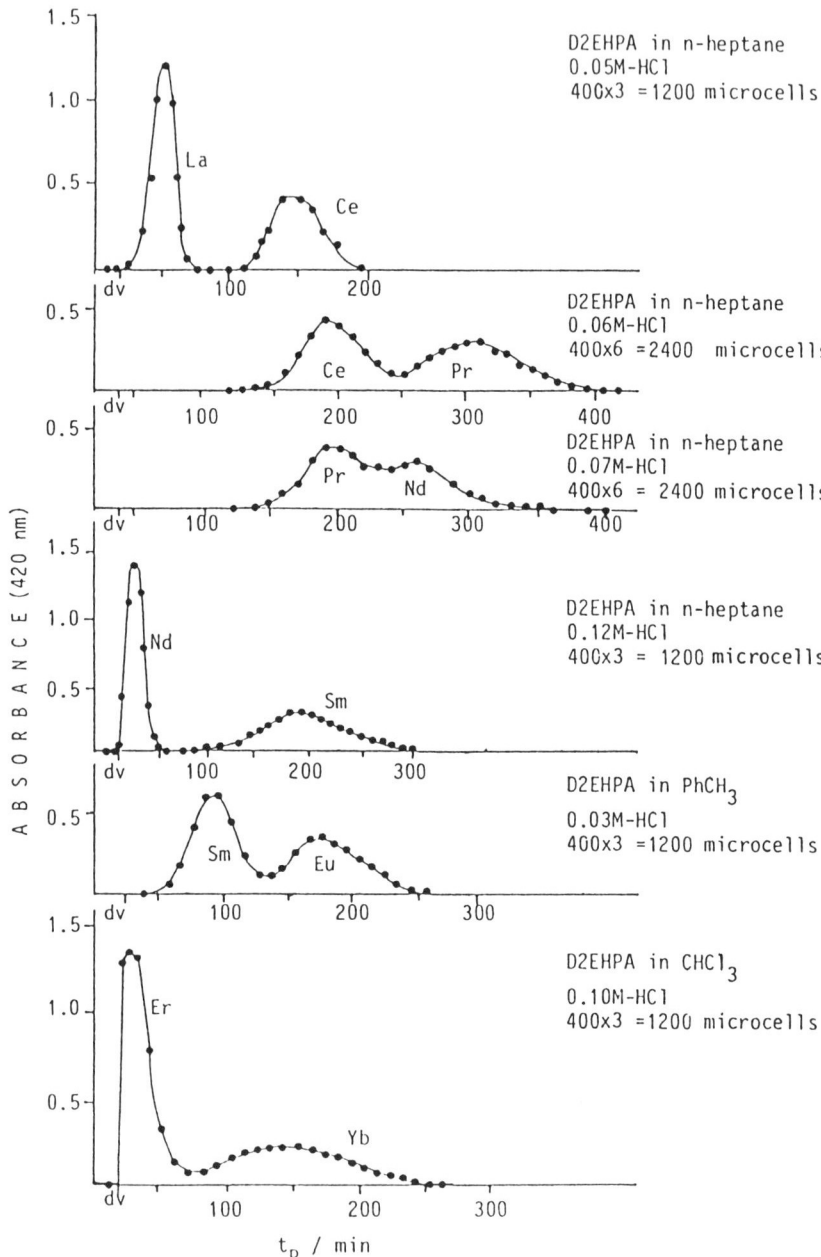

Figure 43 CPC separations of RE ions by using D2EHPA as Y and eluted by aqueous HCl solutions (see Sections IV and V and Refs. 38 and 93). The t_R values include dead time (dv: dead volume).

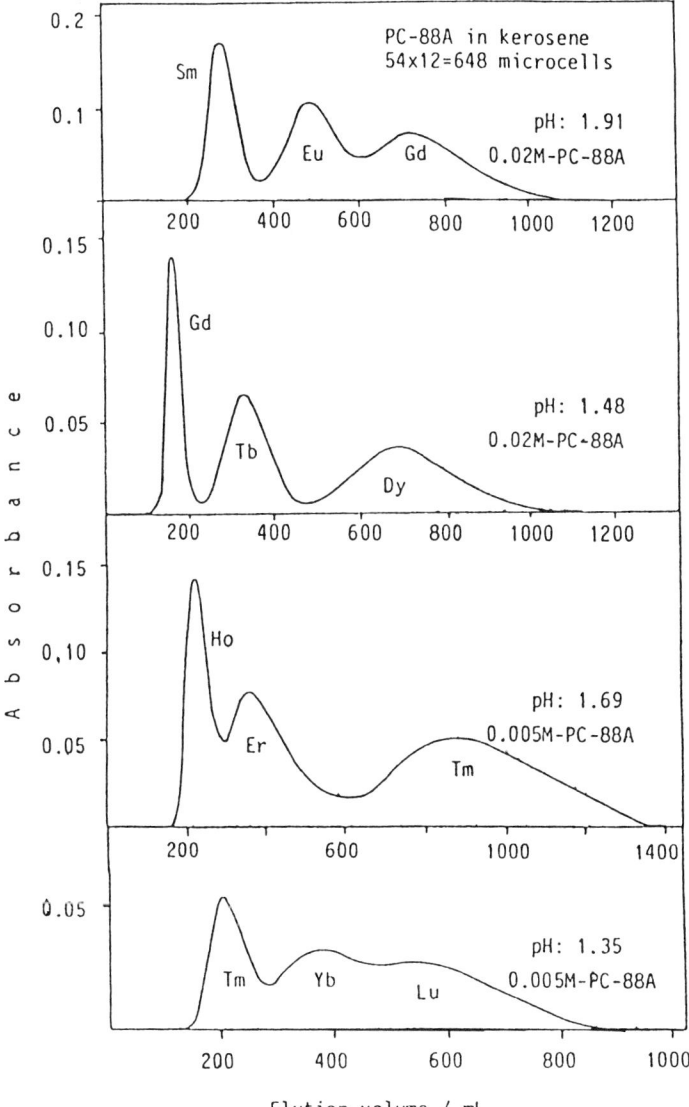

Figure 44 CPC separations of RE ions by using PC-88A as Y and eluted by aqueous solution containing 20 vol% of ethylene glycol buffered at a desired pH with 0.1 M (H,Na)Cl$_2$CHCO$_2$ [45]. The values t_R include dead time.

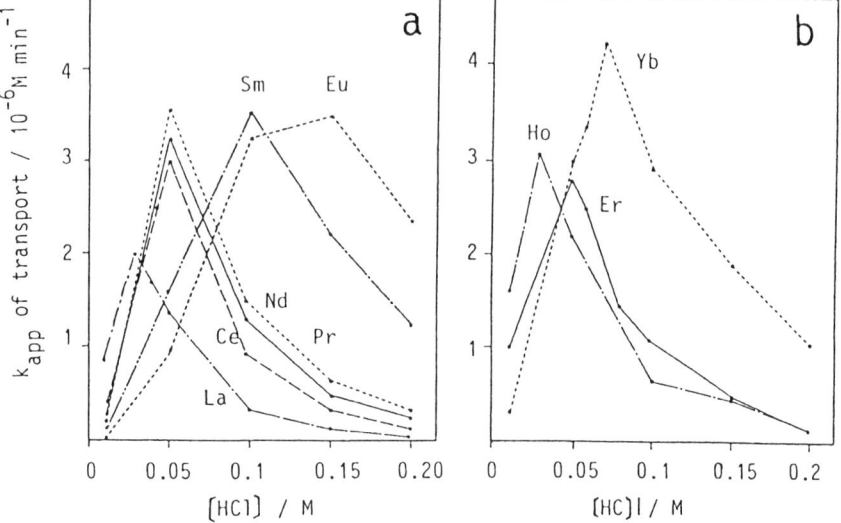

Figure 45 Single-stage LMT results for RE ions by applying the corresponding liquid–liquid system to the CPC shown in Fig. 43. Liquid membrane: (a) n-heptane containing D2EHPA as Y (see Section VII and Ref. 93); (b) $CHCl_3$ containing D2EHPA as Y (see Section VII and Ref. 94).

or carrier is impregnated in the stationary liquid phase, the chromatographic behaviors have in some cases been shown to be correlated with the corresponding facilitated LMT systems. CPC and CCC can be relatively easily upgraded to an industrial scale. These chromatographic techniques will be further used in the future to complement the continuous ELM or SLM modules [53,54,95–102], which are advancing at present. Some of the severe problems now encountered in the ELM or SLM processes may be solved by the excellent multistage performances of CPC and CCC. In turn, since a large quantity of data, especially for biologically important molecules, has been accumulated during the long history of countercurrent distribution methods, that of a chromatographic nature will contribute to advancing the fields of liquid membrane science and technology.

ACKNOWLEDGMENTS

The work represented by the articles edited for this chapter was supported by the Ministry of Education, Science, and Culture for Grant-in-Aid Research (Grants No. 61470026 and partly No. 60400009) and by Mitsui Interbusiness Research Institute. I thank my collaborators in all this work. We thank Sanki Engineering Co. Ltd. for their kind help. For the work in Sections VI and VIII, we thank

Professor Hiroshi Hashiyani of the Chemistry Department of Shimane University for the kind gift of Tm_2O_3 and Lu_2O_3, and Dr. Yasuo Kubo of the same department for his helpful discussions. Thanks are also given to Dr. Yoshio Hatanaka of Unocal Japan, Ltd. for the kind gift of RE samples.

REFERENCES

1. H. Tanaka, K. Hosoya, N. Tanaka, H. Narita, and T. Araki, *Proc. Chem. Soc. Jpn.* 66 Annual Meeting, Nishinomiya, Japan, Preprints, p. 140 (Sept. 27–30, 1993).
2. W. Murayama, T. Kobayashi, Y. Kosuge, H. Yano, Y. Nunogaki, and K. Nunogaki, *J. Chromatogr.*, *239*: 643 (1982).
3. W. Murayama, Y. Nunogaki, and K. Nunogaki, *Symposium on Solvent Extraction*, Shizuoka, Japan, preprints, p. 99 (1982); also presented at the 11th International Congress on Biochemistry (1979) and at the 18th International Symposium on Advances in Chromatography (1982).
4. C. J. Petersen, *J. Am. Chem. Soc.*, *89*: 7017 (1967).
5. A. Sadakane, T. Iwachido, and K. Toei, *Bull. Chem. Soc. Jpn.*, *49*: 645 (1978).
6. I. M. Kolthoff, *Anal. Chem.*, *51*: R1 (1979).
7. A. I. M. Keulemans, *Gas Chromatography* (C. G. Verver, ed.), Reinhold, New York, p. 96 (1957).
8. C. F. Reusch and E. L. Cussler, *AIChE J.*, *19*: 736 (1973).
9. H. Tsukube, K. Takagi, T. Higashiyama, T. Iwachido, and N. Hayama, *J. Inclus. Phenom.*, *2*: 103 (1985).
10. L. C. Craig and D. Craig, *Techniques of Organic Chemistry*, Vol. 3, Interscience, New York (1950).
11. T. Tanimura, J. O. Pisano, Y. Ito, and R. L. Bowman, *Science*, *109*: 540 (1970).
12. Y. Ito, M. A. Weinstein, I. Aoki, R. Harada, E. Kimura, and K. Nunogaki, *Nature*, *212*: 985 (1966).
13. Y. Ito, I. Aoki, E. Kimura, K. Nunogaki, and Y. Nunogaki, *Anal. Chem.*, *41*: 1579 (1969).
14. Y. Ito and R. L. Bowman, *Science*, *173*: 420 (1971).
15. Y. Ito, *J. Chromatogr.*, *214*: 122 (1981).
16. Y. Ito, *J. Biochem. Biophys. Meth.*, *5*: 105 (1981).
17. Y. Ito, J. Sandlin, and W. G. Bowers, *J. Chromatogr.*, *244*: 247 (1982).
18. Y. Ito, High-speed countercurrent chromatography, *CRC Crit. Rev. Anal. Chem.*, *17*(1): 65, and literature cited therein (1986).
19. I. A. Sutherland, D. Heywood-Waddington, and Y. Ito, *J. Chromatogr.*, *384*: 197 (1987).
20. R. C. Bruening, E. M. Oltz, J. Furukawa, and K. Nakanishi, *J. Am. Chem. Soc.*, *107*: 5298 (1985).
21. T. Okuda, T. Yoshida, T. Hatano, K. Yazaki, R. Kira, and Y. Ikeda, *J. Chromatogr.*, *362*: 375 (1986).
22. D. F. Peppard, P. R. Gray and G. W. Mason, *J. Phys. Chem.*, *57*: 294 (1953).
23. D. F. Peppard, G. W. Mason, J. L. Maier, and W. J. Driscoll, *J. Inorg. Nucl. Chem.*, *4*: 334 (1957).

24. D. F. Peppard, J. R. Ferraro, and G. W. Mason, *J. Inorg. Nucl. Chem.*, 7: 231 (1958).
25. D. F. Peppard, *The Rare Earths* (F. H. Spedding and A. H. Daane, eds.), John Wiley and Sons, New York, p.38 (1961).
26. N. E. Topp, *Chemistry of the Rare-Earth Elements*, Elsevier, Amsterdam, p. 31 (1965).
27. T. Goto and M. Smutz, *J. Inorg. Nucl. Chem.*, 27: 1369 (1965).
28. S. G. K. Nair and M. Smutz, *J. Inorg. Nucl. Chem.* 29 1787 (1967).
29. T. G. Lenz and M. Smutz, *J. Inorg. Nucl. Chem.*, 30: 621 (1968).
30. J. Shiokawa, A. Matsumoto, K. Takatsuji, S. Morito, and Y. Hirashima, *Kogyo Kagaku Zasshi (Japan)*, 74: 14 (1971).
31. T. B. Pierce, P. F. Peck, and R. S. Hobbs, *J. Chromatogr.*, 12: 81 (1963).
32. R. W. Rahn and M. Smutz, *Ind. Eng. Chem., Proc. Des. Develop.*, 8: 289 (1969).
33. C. Jiang, J. Yu, and Y. Zhu, *Chem. Abstr.*, 98: 111303u (1983).
34. J. Yu, S. Wang, C. Jiang, and Y. Zhu, *Chem. Abstr.*, 101: 154033d (1984).
35. M. Teramoto, T. Sakuramoto, T. Koyama, H. Matsuyama, and Y. Miyake, *Separat. Sci. Technol.*, 21: 229 (1986).
36. K. Ueno, *Kireito Tekitei-ho* [Chelated Titration Methods], Nanko-do, Tokyo and Kyoto, p. 133 (1979).
37. T. Araki, Y. Kubo, S. Gohbara, and T. Okazawa, *8th IUPAC Conf. on Phys. Org. Chem.*, Tokyo, Abstracts, p. 132 (Aug. 1986).
38. T. Araki, T. Okazawa, Y. Kubo, H. Ando, and H. Asai, *Proc. Chem. Soc. Jpn. Chugoku-Shikoku Div.*, Matsue, Japan, Preprints, p. 46 (Nov. 22–23, 1987).
39. S. Sawai, S. Nakamura, K. Akiba, and W. Murayama, *6th Symposium on Solvent Extraction*, Osaka, Japan, Preprints (Dec. 3–5, 1987).
40. E. Hesford, E. E. Jackson, and H. A. C. McKay, *J. Inorg. Nucl. Chem.*, 9: 279 (1959).
41. J. Bochinski, M. Smutz, and F. H. Spedding, *Ind. Eng. Chem.*, 50: 157 (1958).
42. M. Majdan, *Chem. Abstr.*, 96: 166204b (1982).
43. T. M. Norina, E. B. Mikhlin, V. N. Nikonov, A. M. Reznik, and T. A. Afonia, *Chem. Abstr.*, 84: 77322y (1976).
44. D. Usubaliev, V. V. Serebrennikov, Z. Ismanova, and K. Kanapiyaeva, *Chem. Abstr.*, 106871f (1973).
45. K. Akiba, S. Sawai, S. Nakamura, and W. Murayama, *J. Liq. Chromatogr.*, 11: 2517 (1988).
46. P. R. Brown, R. M. Izatt, J. J. Christensen, and J. D. Lamb, *Membr. Sci.*, 13: 85 (1983).
47. K. H. Wong, K. Yagi, and J. Smid, *J. Membr. Biol.*, 18: 379 (1974).
48. T. Araki, *Kagaku [Chemistry]*, 41: 512 (1986).
49. T. Araki and Y. Kubo, *Proc. 1st CPC Colloq. (Progress and Application of CPC)*, Kyoto, Japan, Preprints, p. 7 (Jan. 13, 1986).
50. N. N. Li, U.S. Patent 3,454,489 (1966).
51. N. N. Li and N. J. Somerset, U.S. Patent 3,410,794 (1968).
52. E. L. Cussler, *AIChE J.*, 17: 1300 (1971).
53. R. Marr and A. Kopp, *Chem. Eng. Tech.*, 52: 399 (1980).
54. R. Marr and A. Kopp, *Ind. Chem. Eng.*, 22: 44 (1982).

55. J. Draxler, W. Furst, and R. Marr, *Proc ISEC '86*, *1*: 1 (1986).
56. W. Furst, J. Draxler, and R. Marr, *Proc World Congr. III Chem. Eng.*, *3*: 331 (1986).
57. P. F. Scholander, *Science*, *131*: 585 (1960).
58. S. Loeb and S. Sourirajan, *Adv. Chem. Ser.*, *38*: 117 (1963).
59. W. J. Ward III, and W. L. Robb, *Science*, *156*: 1481 (1967).
60. W. J. Ward III, *AIChE J.*, *16*: 405 (1970).
61. R. D. Noble and J. D. Way, *ACS Symp. Ser. (Liquid Membranes)*, *347*: 1 (1987).
62. J. D. Way, R. D. Noble, T. M. Flynn, and E. D. Sloan, *J. Membr. Sci.*, *12*: 239 (1982).
63. R. M. Izatt, G. A. Clark, J. S. Bradshaw, J. D. Lamb, and J. J. Christensen, *Separat. Purif. Meth.*, *15*: 21 (1986).
64. R. M. Izatt, H. G. C. Lind, R. L. Bruening, J. S. Bradshaw, J. D. Lamb, and J. J. Christensen, *Pure Appl. Chem.*, *58*: 1453 (1986).
65. M. Teramoto, Y. Miyake, and H. Matsuyama, *Hyomen*, *25*: 715 (1987).
66. W. D. Jamrack, D. H. Logsdail, and G. D. C. Short, *Process Chemistry*, Vol. 2 (F. R. Bruce, J. M. Fletcher, and H. H. Hyman, eds.), Progress in Nuclear Energy, Ser. 3, Pergamon Press, New York, p. 320 (1958).
67. R. E. Treybal, *Liquid Extraction*, McGraw-Hill, New York (1963).
68. L. C. Craig, *J. Biol. Chem.*, *155*: 519 (1944).
69. E. Hecker, *Verbindungsverfahren im Laboratorium*, Verlag Chemie, Berlin (1955).
70. J. R. Weisiger, *Organic Analysis*, Vol. 2, Interscience, New York (1953).
71. H. M. Rauen and W. Stamm, *Gegenstromverteilung*, Springer-Verlag, Berlin (1953).
72. F. A. V. Metzsch, *Angew. Chem.*, *65*: 586 (1953).
73. A. Tsugita, Koryu-bunpai ho, *Jikken Kagaku Koza*, 2, Kisogijutsu II, Chem. Soc. Jpn., Maruzen, Tokyo, p. 303 (1956).
74. N. B. Mandava and Y. Ito, eds., *Countercurrent Chromatography: Theory and Practice*, Marcel Dekker, New York, and references cited therein (1987).
75. T. Okuda, T. Yoshida, and T. Hatano, *J. Liq. Chromatogr.*, *11*: 2447 (1988).
76. M. R. Jirousek and R. G. Salmon, *J. Liq. Chromatogr.*, *11*: 2507 (1988).
77. Y. Onji, Y. Aoki, Y. Yamazoe, Y. Dohi, and T. Moriyama, *J. Liq. Chromatogr.*, *11*: 2537 (1988).
78. A. Foucault and K. Nakanishi, *J. Liq. Chromatogr.*, *11*: 2455 (1988).
79. S. Kusumoto, N. Kusunose, and T. Shiba, *Proc. 2nd CPC Colloq. (Progress and Application of CPC)*, Kyoto, Japan, p. 61 (June 4, 1988).
80. Y.-W. Lee, C. E. Cook, and Y. Ito, *J. Liq. Chromatogr.*, *11*: 37 (1988).
81. Y.-W. Lee, Y. Ito, Q.-C. Fang, and C. E. Cook, *J. Liq. Chromatogr.*, *11*: 75 (1988).
82. M. Knight, J. D. Pineda, and T. R. Burke, Jr., *J. Liq. Chromatogr.*, *11*: 119 (1988).
83. R. H. Chen and J. E. Hochlowski, *J. Liq. Chromatogr.*, *11*: 191 (1988).
84. T. I. Mercado, Y. Ito, M. P. Strickler, and V. J. Ferrans, *J. Liq. Chromatogr.*, *11*: 203 (1988).
85. B. Diallo and M. Verhaelen, *J. Liq. Chromatogr.*, *11*: 227 (1988).
86. T.-Y. Zhang, X. Hua, R. Xiao, and S. Kong, *J. Liq. Chromatogr.*, *11*: 233 (1988).
87. H. S. Freeman, Z. Hao, S. A. McIntosh, and K. P. Mills, *J. Liq. Chromatogr.*, *11*: 251 (1988).

88. W. Murayama, Y. Kosuge, N. Nakaya, Y. Nunogaki, K. Nunogaki, J. Cazes, and H. Nunogaki, *J. Liq. Chromatogr.*, *11*: 283 (1988).
89. D. W. Armstrong, *J. Liq. Chromatogr.*, *11*: 2433 (1988).
90. Y. Kosuge, W. Murayama, and Y. Nunogaki, *Proc. 1st CPC Colloq. (Progress and Application of CPC)*, Kyoto, Japan, p. 30 (Jan. 13, 1988).
91. A. Sadakane, T. Iwachido, and K. Toei, *Bull. Chem. Soc. Jpn.*, *45*: 432 (1972).
92. A. Sadakane, T. Iwachido, and K. Toei, *Bull. Chem. Soc. Jpn.*, *51*: 629 (1978).
93. T. Araki, T. Okazawa, and Y. Kubo, *Proc. 2nd CPC Colloq. (Progress and Application of CPC)*, Kyoto, Japan, p. 67 (June 4, 1988).
94. T. Araki, T. Okazawa, H. Asai, and Y. Kubo, *Proc. Chem. Soc. Jpn., Chugoku-Shikoku-Kyushu Div.*, p. 54 (Oct. 10, 1988).
95. N. N. Li, *Membrane Process Ind. Biomed., Proc. Symp.*, p. 175 (1971).
96. W. C. Babcock, R. W. Baker, D. J. Kelly, and L. Chapella, *Proc. ISEC '80*, p. 80 (1980).
97. R. Chiarizia and P. R. Danesi, *Separat. Sci. Technol.*, *22*: 641 (1987).
98. M. Nakano, K. Takahashi, and H. Takeuchi, *J. Chem. Eng. Jpn.*, *20*: 326 (1987).
99. K. Takahashi, M. Nakano, and H. Takeuchi, *Kagaku Kogaku Ronbunshu*, *13*: 657 (1987).
100. T. Kataoka, T. Nishiki, M. Yamaguchi, and Y. Zhong, *J. Chem. Eng. Jpn.*, *20*: 410 (1987).
101. W. R. Dworzak and A. J. Naser, *Separat. Sci. Technol.*, *22*: 677 (1987).
102. M. Teramoto, H. Matsuyama, H. Takaya, and S. Asano, *Separat. Sci. Technol.*, *22*: 2175 (1987).

11
Centrifugal Partition Chromatographic Separations of Metal Ions

S. Muralidharan and H. Freiser
University of Arizona, Tucson, Arizona

I. INTRODUCTION

Solvent extraction has been a method of choice for difficult metals separations problems for almost a century, incorporating selectivity and sensitivity, ease, and convenience; it is a powerful separation technique applicable both to trace and macro levels of materials[1]. Extraction methods were used first by analytical chemists, but since the 1940s, starting with the Manhattan Project, process-scale separations of metal ions have expanded from use with uranium and fission products to industrial-scale separation of copper and other metals of commercial and technological interest. Because of their great sensitivity, extraction techniques can be used to recover metals from *extremely dilute solutions*, which makes them ideal for environmental remediation. The continuing vigor of this interesting field is attested to by the continuing high publication rate of research papers from active groups around the world appearing in such relatively new journals as *Separation Science and Technology* and *Solvent Extraction and Ion Exchange*, and a new journal, *Solvent Extraction Research and Development*, introduced in 1994, as well as in many other publications. It is certain that much more innovative and fruitful research in both fundamental and applied areas will be conducted in this field into the twenty-first century.

The separation of metal ions by single-stage methods, especially closely related ions such as the tervalent lanthanides, poses daunting challenges even to the most selective of extractants. Thus, the use of multistage methods is necessary for their separation. Extraction chromatography involving silica surfaces coated or derivatized with a suitable ligand has been used for some time for this purpose [2]. This procedure has its limitations, such as low load capacity and column bleeding and degradation. Further, metal complexation and dissociation reactions, especially with chelating ligands, tend to be slow, limiting the efficiencies of multistage separations of metal ions. The separation of kinetic contributions to the inefficiencies of separations, from mass transfer and diffusion, is difficult with the solid–liquid systems.

Centrifugal partition chromatography (CPC), a truly liquid–liquid multistage countercurrent distribution technique, avoids many of these difficulties encountered with extraction chromatography. Most important, it enables us to understand the influence of the slow chemical kinetic steps in limiting the CPC efficiencies of metal separations. The quantity channel equivalent of a theoretical plate (CETP), a measure of the CPC *inefficiency*, is the tool to such an understanding. The variation of CETP with flow rate of the mobile phase and concentrations of species in the aqueous and mobile phases can identify not only the slow chemical kinetic step leading to the observed inefficiencies, but also the loci of such as a step, that is, in the bulk aqueous, or at the aqueous–organic interface, or both. A direct correlation exists between the CETP and the half-life ($t_{1/2}$) of the slow kinetic step determined by independent kinetic measurements. It is also possible to determine the average size of the mobile phase droplets from such a correlation.

Our research at the Strategic Metals Recovery Research Facility (SMRRF) in the separation of metals using CPC has dealt with three families of metal ions, namely, the tervalent lanthanides, the platinum group metals (PGM), and the transition metals. The significant results are summarized in this chapter. The objectives of our work are twofold: (1) develop practical separation methods for various families of metal ions under relatively mild conditions, and (2) elucidate the fundamental factors such as chemical kinetics and interfacial activities of ligands that influence the efficiencies of these separations. The lanthanides and the transition metals have been separated using the chelating ligands—a phosphinic acid (bis(2,4,4,-trimethylpentyl)phosphinic acid; Cyanex 272) and an acyl pyrazolone (1-phenyl-3-methyl-4-benzoyl-5-pyrazolone; HPMBP). With Cyanex 272, complete separation of the adjacent tervalent lanthanides as well as the separation of a mixture of heavy and light lanthanides in a single run employing a pH gradient in the mobile phases were achieved. The PGM (Pt, Pd, Rh, and Ir) have been separated using the monodentate ligand trioctylphosphine oxide (TOPO) and by ion pair formation with protonated TOPO, that is, HTOPO$^+$ and tetraheptylammonium p-toluenesulfonate (QpTS). We have shown that the oxidation states and chloro species of Pt could be separated using HTOPO$^+$ and QpTS.

The experimental CETP values of the metal ions were shown to have a mass transfer and diffusion component (CETP$_{dif}$) and a chemical kinetic component (CETP$_{ck}$) that could be separated. A direct correlation was found between the CETP$_{ck}$ of Pd(II) and the half-life for the dissociation of its complex

PdCl$_2$(TOPO)$_2$, clearly indicating that kinetic steps with half-lives of milliseconds or less could limit the efficiencies of CPC separations. The dependence of the CETP$_{ck}$ values of the lanthanides on the mobile phase flow rate, pH, and the concentration of Cyanex 272 indicated the complex chemistry associated with this dimeric chelating ligand and the lanthanides. The study of the variation of CETP$_{ck}$ of Ni(II) with HPMBP as the ligand clearly demonstrated the influence of the interfacial activity of the ligand on the size of the mobile phase droplets and hence the CPC efficiencies. The optimum rotational speeds, mobile phase flow rates, and the stationary to mobile phase volume ratios necessary to minimize CETP$_{dif}$ for a variety of organic–aqueous phase pairs, with the organic phase as the stationary phase and the aqueous phase as the mobile phase and vice versa, were determined using three types of analytes, namely, an organic analyte (3-picoline), an ion pair (Q$_2$IrCl$_6$), and a complex (PdCl$_2$(TOPO)$_2$). The sections in the chapter are organized reflecting separations and the derivation of kinetic information from CETP.

All the CPC experiments were conducted with a Sanki Co. (Kyoto, Japan) assembly consisting of a Model SPL centrifuge containing six analytical/semipreparative cartridges, each having 400 channels (2400 total channels), a Model CPC FCU-V loop injector, and a Model LBP-V pump. The temperature was maintained at 25°C at a rotational speed of 800 rpm. The chromatograms were recorded using a Schoeffel Model 770 detector with a 0.1 ml cell volume and 8 mm path length. The PGM were detected directly at the appropriate wavelength maxima for their anionic chloro species [5–7], and the lanthanides and Ni(II) by postcolumn derivatization with the appropriate metallochromic indicator—Arsenazo III and 4-(2-pyridylazo)resorcinol (PAR), respectively [4,12].

II. SEPARATION OF METAL IONS

A. Separation of Lanthanides

The closeness of the properties of the tervalent lanthanides makes their separation very difficult, requiring multistage methods. Among the various families of ligands that were investigated in our laboratory by the batch solvent extraction method [3], acidic organophosphorus extractants provided the best extraction equilibrium constants and separation factors. The dimeric chelating ligand Cyanex 272 performed best among the organophosphorus extractants. This ligand was used in the heptane stationary phase in CPC to separate adjacent lanthanides at a stationary phase to mobile phase volume ratio of 0.18, and a aqueous mobile phase flow rate of 1–2 ml/min [4]. The tervalent lanthanides are extracted into heptane as their 1:3 metal:ligand dimer complexes:

$$M^{3+} + 3(HL)_{2,o} \overset{K_{ex}}{\rightleftharpoons} M(HL_2)_{3,o} + 3H^+ \qquad (1)$$

where M^{3+} represents the tervalent lanthanide, HL represents the ligand, and the subscript o the species in the organic phase.

A typical example of the separation of a mixture of Nd, Sm, and Eu (2 ppm each) is shown in Fig. 1. A mixture of light and heavy lanthanides could be separated in a single run by employing a pH gradient in the mobile phase (Fig. 2). The separation of tervalent lanthanides using acyl pyrazolones is currently in progress.

B. Separation of PGM

The separation of Pd(II), Pt(II), Pt(IV), Rh(III), Ir(III), and Ir(IV) and the chloro species of Pt(II), namely, $PtCl_2$, $PtCl_3^-$, and $PtCl_4^{2-}$, were achieved in these studies. A very important practical aspect of our work is that these separations were performed under relatively mild conditions in contrast to traditional methods of the separation of PGM, which involve harsh conditions such as high acidity. These separations were performed by complexation with TOPO [5], and by ion pairing with protonated TOPO (HTOPO$^+$) [6] and tetraheptyl p-toluenesulfonate (QpTS) [7]. The CPC chromatogram for the separation of Pd(II) and Pt(II) with TOPO in the heptane stationary phase, with a stationary to mobile phase volume ratio of 0.2, at a mobile phase pH of 3, and at 0.5 M chloride concentration is shown in Fig. 3. The separation was performed as a function of the concentrations of TOPO and Cl$^-$ and pH. The complex $PdCl_2$ (TOPO)$_2$ was exclusively formed irrespective of the nature of the Pd(II) chloro species in the aqueous phase:

$$PdCl_n^{(n-2)-} + 2\,TOPO \overset{K_{ex}}{\rightleftharpoons} PdCl_2(TOPO)_2 + (n-2)Cl^- \quad (2)$$

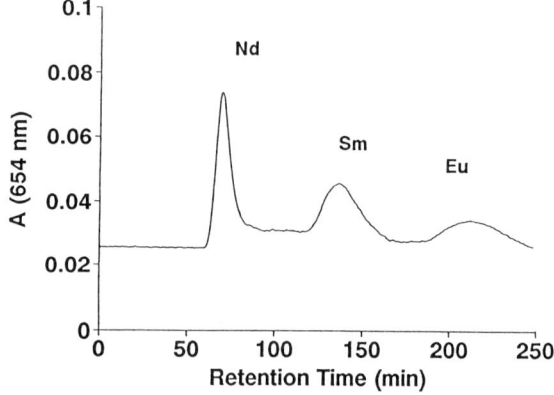

Figure 1 Separation of 2 ppm each of Nd, Sm, and Eu with 0.1 M Cyanex 272 at pH = 2.1 in heptane at $V_s/V_m = 0.18$ and flow rate 1.5 ml/min.

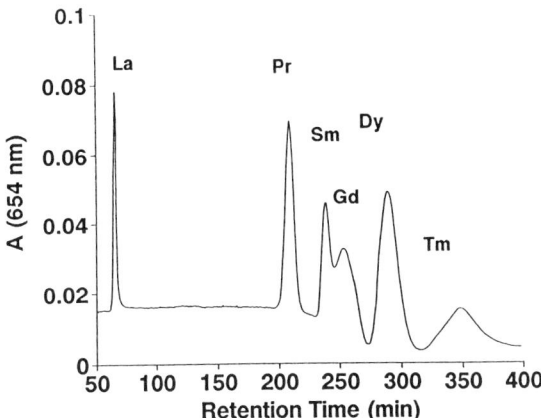

Figure 2 Separation of lanthanides by the use of pH gradient with 0.1 M Cyanex 272 at $V_s/V_m = 0.18$ and flow rate 1 ml/min. The concentrations and pH of elution are La (2 ppm; 2.5), Pr (6 ppm; 2.1), Sm (4 ppm; 1.87), Gd (4 ppm; 1.71), Dy (10 ppm; 1.58), and Tm (8 ppm; 1.4).

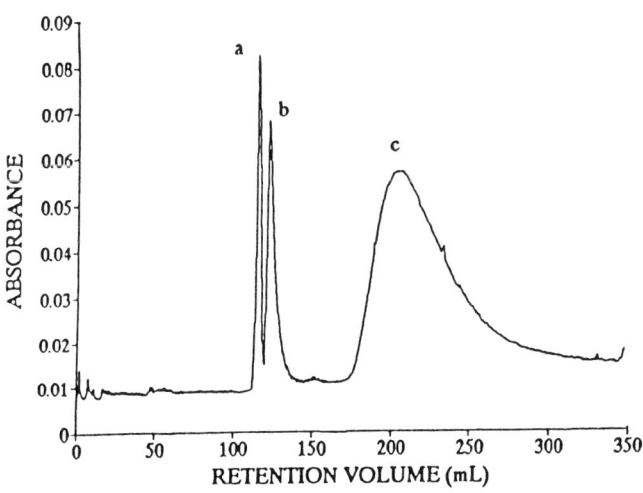

Figure 3 Separation of Rh(III), Ir(III), Pt(II), and Pd(II) using 0.5 M TOPO in heptane, 0.08 M Cl$^-$, $V_s/V_m = 0.2$, and flow rate 0.87 ml/min. (a) 5 × 10^{-4} M Rh(II) and Ir(III); (b) 10^{-4} M Pt(II); (c) 10^{-3} M Pd(II).

where $n = 2-4$. The K_{ex} values for $PdCl_2$, $PdCl_3^-$, and $PdCl_4^{2-}$ are respectively 794.3 M^{-2}, 2.75 M^{-1}, and 0.14. A single peak was observed in the CPC chromatogram of Pd(II) at any concentration of Cl^-, as its hydrolytic equilibria are rapid. The corresponding values for the Pt(II) species are 48 M^{-2}, 0.047 M^{-1}, and 0.018 respectively, clearly indicating the better extractibility of Pd(II) over Pt(II), also evident in Fig. 3.

The ligand TOPO could be protonated at HCl concentrations greater than 0.1 M. The HTOPO$^+$ extracted $PdCl_4^{2-}$ and $PtCl_4^{2-}$ as the ion pairs $(HTOPO)_2MCl_4$ (M = Pt or Pd):

$$MCl_4^{2-} + 2H^+ + 2\,TOPO_o \overset{K_{ex}}{\rightleftarrows} (HTOPO)_2MCl_{4,o} \tag{3}$$

The chromatogram of the separation of $RhCl_6^{3-}$, $PdCl_4^{2-}$, and $PtCl_4^{2-}$ by HTOPO$^+$ is shown in Fig. 4. The K_{ex} values of Pd(II) and Pt(II) are 93.3 M^{-4} and 1961 M^{-4}, respectively, indicating that Pd(II) elutes ahead of Pt(II) in the ion pair separation while the opposite is true in the separation by complexation. While

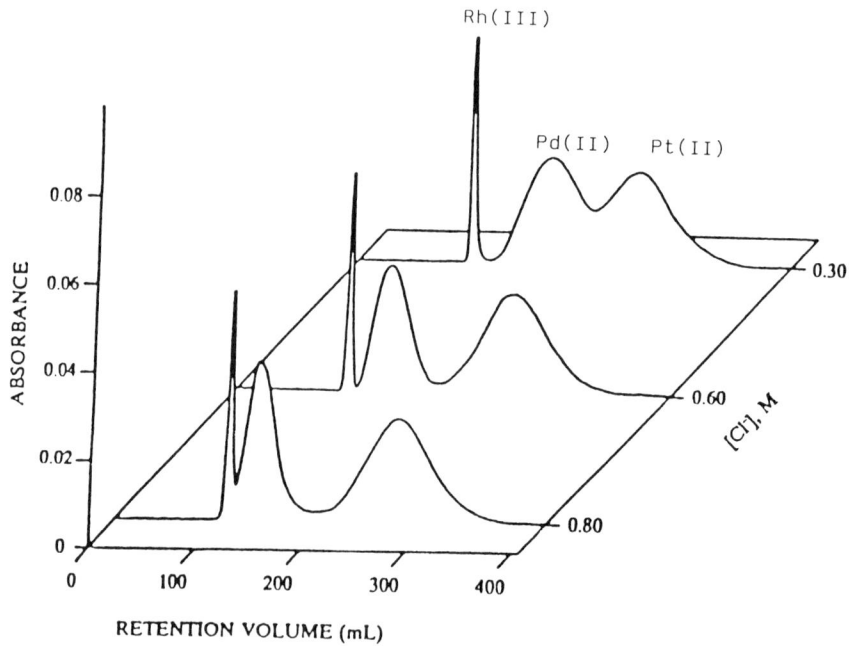

Figure 4 Separation of 10^{-4} M Rh(III), 10^{-3} M Pd(II), and 10^{-4} M Pt(II) as their chloro anions with HTOPO$^+$, as a function of Cl^-. 0.5 M TOPO in heptane, $V_s/V_m = 0.2$, 0.1 M HCl, and flow rate 4 ml/min.

the chromatogram of Pt(II) is exclusively due to the formation of $(HTOPO)_2PtCl_4$, the chromatogram of Pd(II) is due to the formation of $(HTOPO)PdCl_3$ as well. In fact, under the experimental conditions employed in these separations, this is the major Pd ion pair species extracted. The extraction equilibrium constant for $(HTOPO)PdCl_3$ is 18.25 M^{-1}. Similarly, Pt(IV) and Ir(IV) could be separated by $HTOPO^+$ by ion pair formation with their MCl_6^{2-} species. The K_{ex} values for the Pt(IV) and Ir(IV) species are 1576 M^{-4} and 8035 M^{-4}, respectively.

The ion pair extraction of the chloro anions of PGM by QpTS in 1,2-dichloroethane stationary phase also afforded the separation of these species under relatively mild conditions. The overall extraction equilibrium for the MCl_6^{2-} series is

$$MCl_6^{2-} + 2QpTS_o \overset{K_{ex}}{\rightleftharpoons} Q_2MCl_{6,o} + 2pTS^- \quad (4)$$

The K_{ex} values for Pt(IV) and Ir(IV) are 3890 M^{-2} and 7760 M^{-2}, respectively, and despite the closeness of these values, baseline separation of these can be obtained.

The CPC behavior of Pt(II) and Pd(II) are dependent on the concentration of Cl^- in the mobile phase. In the absence of any added NaCl ([HCl] = 0.0032 M) in the mobile phase, Pt(II) exhibited two peaks and Pd(II) exhibited a broad peak. When 0.1 M NaCl was added to the mobile phase, Pt(II) exhibited three peaks and Pd(II) a less broad peak. These observations could be explained on the basis of the hydrolytic equilibria of the chloro anions of Pd(II) and Pt(II), that is, the formation of MCl_2, MCl_3^-, and MCl_4^{2-} species, the fraction of which are dependent on the Cl^- concentration in the aqueous phase. The hydrolytic equilibrium, that is, the interconversion of the chloro species, is rapid for Pd(II), resulting in a broad peak at low Cl^- concentrations, where all the species tend to be present, and a narrower peak at high Cl^- concentrations, where $PdCl_4^{2-}$ is the predominant species. Such an equilibrium is slow in the case of Pt(II), resulting in the multiple peaks. The CPC chromatogram of the chloro species of Pt(II) as a function of the concentration of Cl^- at a QpTS concentration of 0.002 M in 1,2-dichloroethane and stationary to mobile phase volume ratio of 1 is shown in Fig. 5. It may be seen that increasing the Cl^- concentration results in an increase in the peak due to $PtCl_4^{2-}$. At the high concentrations of Cl^-, QCl and QpTS both compete for the extraction of the Pt(II) species. It is possible to determine the extraction equilibrium constants for the extraction of $PtCl_3^-$ and $PtCl_4^{2-}$ by QCl and QpTS from these chromatograms. The $K_{ex,3}$ and $K_{ex,4}$ values for QCl are 52 and 1.25×10^4, respectively, and the corresponding values for the extraction by QpTS are 3.9 and 76, respectively. Though the extraction equilibrium constants for QCl are much higher than those for QpTS, the latter is a better extractant for practical separations, as the back-extraction with this can be achieved under much milder conditions. These K_{ex} values yield 800 as the ratio of the equilibrium constants of formation of the QpTS an QCl (Q = tetraheptyl) ion pairs in 1,2-dichloroethane.

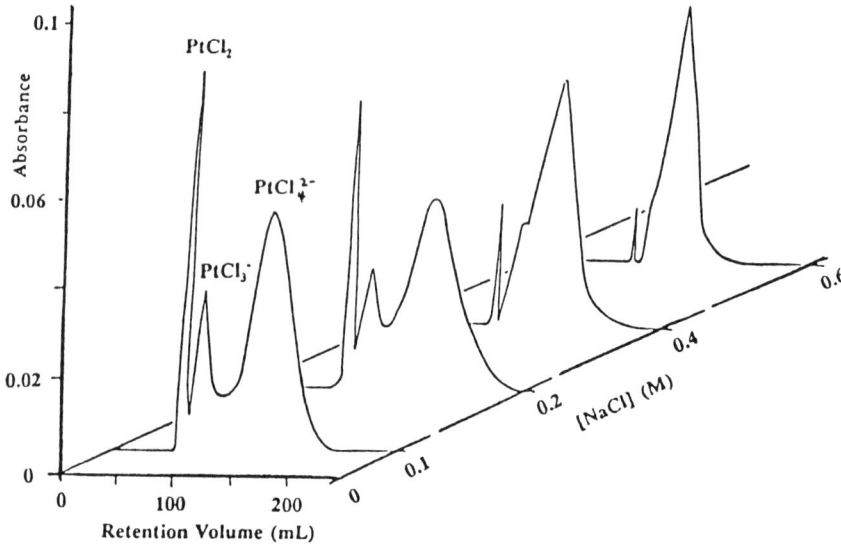

Figure 5 The separation of the different chloro species of Pt(II) as function of the concentration of Cl⁻ in the 1,2-dichloroethane–H$_2$O phase pair. $V_s/V_m = 1$, $0.002\,M$ QpTS, flow rate 2 ml/min.

This value is similar to the ratio for the tetrabutyl ion pairs in the CHCl$_3$–H$_2$O system, which is about 300 [8].

III. CHEMICAL KINETICS AND CPC EFFICIENCY

A. The Platinum Group Metals

It was evident from the separations of lanthanides and PGM that their experimental CETP values (CETP = 2400/N; N the number of theoretical plates calculated in the usual manner from retention volume and CPC band width) were much larger compared with that for an organic analyte at identical distribution ratios (D) [4,5, 9,11]. This is illustrated in Fig. 6, where the CPC chromatograms of 3-picoline and PdCl$_2$(TOPO)$_2$ at the same D values are shown. This clearly indicated that factors other than mass transfer and diffusion were responsible for the additional bandwidths in the case of the metal ions. The most likely factor is the slow kinetics of back-extraction of the metal ions, as the extraction reactions are usually rapid. In order to test this hypothesis, 3-picoline was used as the model compound for the determination of the CPC bandwidth due to mass transfer and diffusion (CETP$_{dif}$), and the CETP values due to slow chemical kinetics (CETP$_{ck}$) were derived by expressing the experimental CETP (CETP$_{obs}$) as a sum of CETP$_{dif}$ and CETP$_{ck}$:

$$\text{CETP}_{\text{obs}} = \text{CETP}_{\text{dif}} + \text{CETP}_{\text{ck}} \tag{5}$$

The CETP_{ck} values determined by varying the concentrations of the species in the aqueous and organic phases clearly showed that the slow back-extraction kinetics of the metal complexes was indeed responsible for the broad CPC chromatograms. The dependence of CETP_{ck} on the concentrations of species in the aqueous and organic phases could be used to derive a mechanism for the dissociation of the metal complexes. For example, the mechanism of the dissociation of $\text{PdCl}_2(\text{TOPO})_2$ could be shown to be Eqs. 6–8, with Eq. 7 being the rate-limiting step [9]:

$$\text{PdCl}_2(\text{TOPO})_2 \underset{\text{fast}}{\overset{K}{\rightleftarrows}} \text{PdCl}_2(\text{TOPO}) + \text{TOPO} \tag{6}$$

$$\text{PdCl}_2(\text{TOPO}) + \text{Cl}^- \underset{\text{slow}}{\overset{k_2}{\rightleftarrows}} \text{PdCl}_3^- + \text{TOPO} \tag{7}$$

$$\text{PdCl}_3^- + \text{Cl}^- \underset{\text{fast}}{\rightarrow} \text{PdCl}_4^- \tag{8}$$

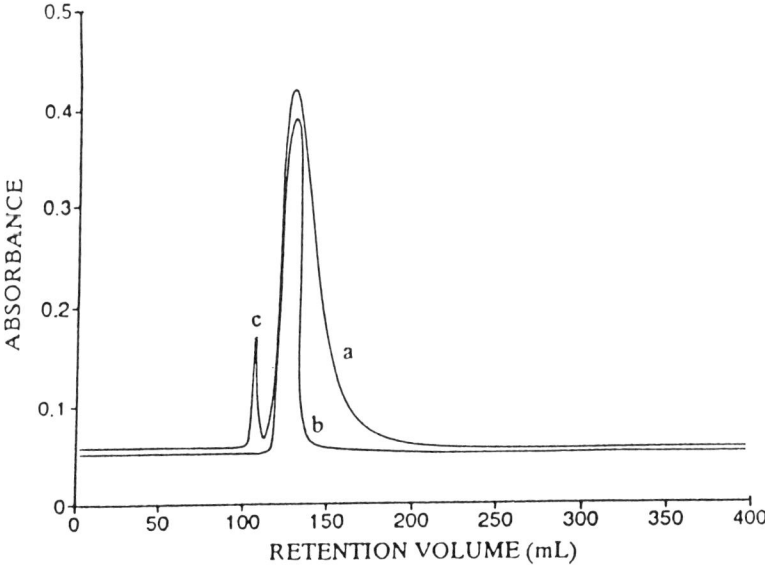

Figure 6 CPC chromatograms of 3-picoline and $\text{PdCl}_2(\text{TOPO})_2$ at identical distribution ratios in the heptane–H_2O phase pair; $V_s/V_m = 0.2$, flow rate 2 ml/min. (a) $10^{-3}\ M$ Pd(II), $0.3\ M$ TOPO, $0.1\ M$ Cl$^-$, pH = 3. (b) 3-picoline, pH = 6.1. (c) dead volume peak.

This was independently verified by studying the dissociation of $PdCl_2(TOPO)_2$ in Triton X-100 micelles used as a model for the two-phase system using the stopped flow technique, as this reaction is too fast for conventional spectrophotometric kinetic measurements. Thus, dissociation reactions with half-lives in the millisecond–second range can adversely affect the CPC efficiencies. A very significant finding of this work is that a direct linear correlation exists between $CETP_{ck}$ and $t_{1/2}$, as shown in Fig. 7 for the Pd(II)–TOPO system. *Since the $CETP_{ck}$ values are a measure of the half-lives of the slow dissociation steps in metal complex dissociation reactions, CPC is a useful tool for examining such reactions.* This analysis of the bandwidths of the CPC chromatograms indicates that extraction and back-extraction reactions that appear to be rapid in single-stage equilibrations may still be slow enough to reduce the efficiencies of multistage separations.

B. The Tervalent Lanthanides

The $CETP_{ck}$ values of the lanthanides with Cyanex 272 as the extractant were also measured as a function of the flow rate of the mobile phase, the pH, and the

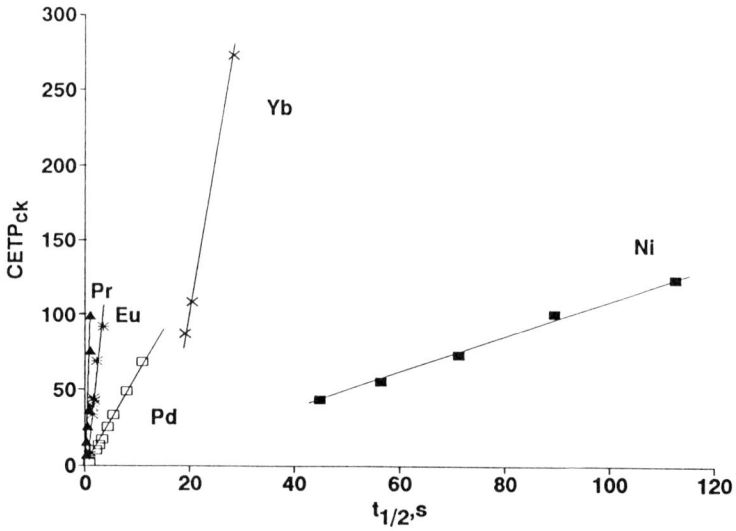

Figure 7 $CETP_{ck}$ versus $t_{1/2}$ determined in Triton X-100 or Brij 35 micelles. Pr, Eu, and Yb in heptane–H_2O phase pair, 0.1 M Cyanex 272, $V_s/V_m = 0.18$, flow rate 1 ml/min. Pd in heptane–H_2O phase pair, $V_s/V_m = 0.2$, flow rate 2 ml/min, [TOPO] = 0.1–0.5 M, [Cl$^-$] = 0.1–0.3 M, pH = 3. Ni in $CHCl_3$–H_2O phase pair, $V_s/V_m = 0.4$, flow rate 0.4 ml/min, pH = 5.8–6.2, [HPMBP] = 0.0045–0.01 M.

concentration of Cyanex 272 in the heptane phase. The dependence of $CETP_{ck}$ on these parameters was more complex than in the case of the Pd(II)–TOPO system. Independent measurement of the kinetics of dissociation of the lanthanide–Cyanex 272 complexes was necessary to understand the behavior of $CETP_{ck}$. Such measurements in Triton X-100 micelles are not straightforward as in the case of $PdCl_2(TOPO)_2$, as the former complexes lack the UV-visible spectral handle available in the latter complex. We developed a *metallochromic indicator method* to overcome this difficulty [10]. This method relies on the rapid complexation of the tervalent lanthanide upon the dissociation of the lanthanide–Cyanex 272 complex by H^+, by a metallochromic indicator like arsenazo III:

$$M(HL_2)_{3(m)} + 3H^+ \underset{slow}{\rightleftarrows} M^{3+} + 3(HL)_{2(m)} \tag{9}$$

$$2M^{3+} + 2\,\text{arsenazo III} \underset{fast}{\rightleftarrows} (M-\text{arsenazo III})_2 \tag{10}$$

where the subscript m represents the species in the bulk micellar pseudophase. The ligand Cyanex 272 chelates the tervalent lanthanides in its dimeric form. The formation of the lanthanide–arsenazo III complex yields the mechanism of dissociation of the lanthanide–Cyanex 272 complex, the rate constants, and the extraction equilibrium constants for the extraction of the lanthanide complexes into the bulk micellar phase [11]. The rate-limiting step from such kinetic studies was determined to be the attack of the complex by H^+:

$$M(HL_2)_{3(m)} + H^+ \underset{slow}{\overset{k_{3m}}{\rightarrow}} M(HL_2)_2^+ + 2HL_{(m)} \tag{11}$$

These kinetic studies also indicated that in addition to the 1:3 lanthanide–Cyanex 272 dimer complex, the 1:2.5 dimer complex was also formed, depending on the pH and concentration of Cyanex 272. The rate of interconversion of the 1:3 and 1:2.5 lanthanide–Cyanex 272 dimer complexes could be determined from these kinetic measurements [11].

The correlation of $CETP_{ck}$ with $t_{1/2}$ determined by the metallochromic indicator method is shown in Fig. 7 for Pr, Eu, and Yb, indicating that this is not as good as in the case of Pd(II)–TOPO. The dependence of the distribution ratios of these metals on pH and concentration of Cyanex 272 indicates that the equilibrium in Eq. 1 is predominant in the heptane–H_2O system. This would indicate that the dissociation mechanism in Eq. 11 would govern the $CETP_{ck}$ values, which should not vary with the concentration of Cyanex 272. It was found, however, that the $CETP_{ck}$ values of Eu vary with the concentration of Cyanex 272. The complexity of the behavior of $CETP_{ck}$ of the lanthanides could be attributed to several factors, namely, the dimeric nature of the ligand, the higher interfacial activities of Cyanex 272 and the lanthanide complexes compared with the Pd(II)–TOPO system, and the formation of different types of complexes such as the 2.5 dimer and 3 dimer complexes and possibly even higher aggregates in solvents like heptane. Thus, the lack of a good $CETP_{ck}$–$t_{1/2}$ correlation is an indication of a more complex chemistry than that suggested by equilibrium measurements. This again reinforces the value of CPC in uncovering complex chemistries associated with apparently simple systems.

C. The Ni(II) System

The generality of the $CETP_{ck}$–$t_{1/2}$ relation was further tested by examining the CPC behavior of Ni(II) with 1-phenyl-3-methyl-4-benzoyl-5-pyrazolone (HPMBP) [12]. The ligand HPMBP, which is similar to β-diketones and exists predominantly in its enol form due to intramolecular and/or intermolecular hydrogen bonding, is capable of extracting metal ions at relatively low pH values. The Ni(II)–HPMBP system in the $CHCl_3$–H_2O phase pair was used to understand the influence of interfacial activity of the ligand, and hence the liquid–liquid interface, on the $CETP_{ck}$ values. This was accomplished by correlating the $CETP_{ck}$ values with the $t_{1/2}$ values measured in the $CHCl_3$–H_2O phase pair using the microporous Teflon phase separator (MTPS) as well as those measured in Brij 35 micelles. The mechanism of the dissociation of the $Ni(PMBP)_2$ complex from the CPC and kinetic experiments was determined to be

$$Ni(PMBP)_2 + H^+ \underset{fast}{\overset{K_{-2}}{\rightleftarrows}} Ni(PMBP)^+ + HPMBP \quad (12)$$

$$Ni(PMBP)^+ + H^+ \underset{slow}{\overset{k_{-1}}{\rightleftarrows}} Ni(II) + HPMBP \quad (13)$$

with the rate-limiting step being the dissociation of the 1:1 metal–ligand complex.

The experimental second-order rate constant k_{-1} has a bulk aqueous and an interfacial component:

$$k_{-1} = k^b + k^i K_H^i K_{DC}^i da \quad (14)$$

Here k^b and k^i are the second-order dissociation rate constants in the bulk aqueous and aqueous–organic interfaces, K_H^i is the distribution constant of proton between the aqueous phase and the interface, taken to be 1, K_{DC}^i the distribution constant of the metal complex between the bulk organic phase and interface, d the interfacial thickness, and a the specific interfacial area (area per unit volume). The product $K_{DC}^i d$ and a values were independently determined from equilibrium measurements using MTPS. Thus, the k^b and k^i values could be obtained from the MTPS kinetic experiments, and these values were essentially the same, namely, 8511 $M^{-1} s^{-1}$. The half-life for the dissociation is related to the overall dissociation rate constant k_{-1} as

$$t_{1/2} = \frac{0.693}{k_{-1}[H^+]} \quad (15)$$

A linear correlation between $CETP_{ck}$ and $t_{1/2}$ values determined in Brij 35 micelles was found (Fig. 7). The dependence of $CETP_{ck}$ on the pH of the mobile phase was used to derive k_{-1} using Eq. 15. This composite k_{-1}, and the k^b and k^i from MTPS measurements were used to obtain the specific interfacial area a from Eq. 14. This specific interfacial area, for a stationary to mobile phase volume ratio of 0.39 and flow rate of the mobile phase of 0.4 ml/min, was determined to be 207.7 cm^{-1}. This value is similar to the specific interfacial areas generated at a stirring rate of 3500 rpm in the MTPS experiments for the Ni(II)–HPMBP system at the $CHCl_3$–H_2O interface. We have thus been able to obtain for the first time an estimate of the specific interfacial areas in the CPC experiments by a comparison of the CPC results with independent MTPS kinetic and equilibrium studies. This analysis also shows that the CPC setup, wherein aqueous mobile phase droplets move through the stationary organic phase held in discrete channels by a centrifugal force, is capable of generating interfacial areas similar to that of highly stirred system such as MTPS.

The specific interfacial area 207.7 cm^{-1} corresponds to an average mobile phase droplet radius of 144 μm ($r = 3/a$). This may be compared with 450 μm, the droplet radius calculated by equating the surface tension strength of the drop to the interfacial tension at the HPMBP concentration of 0.01 M employed in the pH dependence studies in CPC, and to 42 μm, the radius of droplets calculated from the data by A. Foucault et al. (see Chapter 2), for the chloroform/water system (no HPMBP) in the ascending mode, according to the Stokes law. Thus, the drop size is roughly $r_{\text{surface tension strength}} \approx 3 \times r_{\text{specific interfacial area}} \approx 3 \times r_{\text{Stokes model}}$. This suggests that the motion of the mobile phase droplets in CPC, through the ducts and channels under a centrifugal force, is more complex than that predicted by conventional equations normally applicable to the motion of fluids under gravity.

It is evident from Fig. 7 that the $CETP_{ck}$–$t_{1/2}$ correlations for the different metal–ligand systems examined so far are not identical; that is, the same CPC bandwidth represents a different $t_{1/2}$ value for each different system. In many metal–ligand systems, we are forced to correlate $CETP_{ck}$ values with $t_{1/2}$ values from micelles, as the dissociation reactions are too fast to be studied by the MTPS setup. This in turn results in correlating a system with a combination of bulk aqueous and interfacial kinetics (CPC) with a system that has predominantly interfacial kinetics, namely, micelles. Thus, the slope of the $CETP_{ck}$–$t_{1/2}$ plot will vary depending on the degrees of participation of the bulk aqueous and interfacial reactions in the CPC experiments and the magnitudes of the bulk and interfacial rate constants.

IV. SUMMARY

The significant results from the work performed to date in our research group are

1. Several families of metal ions can be completely separated using CPC by complex formation or ion pair formation.
2. The oxidation states of metals and species arising due to the hydrolytic equilibria of metals could be separated as well.
3. Slow chemical kinetic steps significantly lower the efficiencies of metal separations. The CETP values provide information on the mechanisms of metal complex dissociation reactions as well as the rate constants for the rate-limiting steps.
4. The CPC efficiencies are influenced by the interfacial activities of the ligands and metal complexes. Correlation of the CETP values with independent kinetic and equilibrium measurements provides a way of determining the average mobile phase droplet size in CPC experiments.

REFERENCES

1. G. Morrison and H. Freiser, *Solvent Extraction in Analytical Chemistry*, John Wiley, New York (1957).
2. T. Braun and G. Ghersini, *Extraction Chromatography*, Elsevier, New York (1975).
3. S. Motomizu and H. Freiser, *Solvent Extr. Ion Exch.*, *3*: 637 (1985).
4. S. Muralidharan, R. Cai, and H. Freiser, *J. Liq. Chromatogr.*, *13*: 3651 (1990).
5. Y. Surakitbanharn, S. Muralidharan, and H. Freiser, *Solvent Extr. Ion Exch.*, *9*: 45 (1991).
6. Y. Surakitbanharn, H. Freiser, and S. Muralidharan, *Sep. Sci. Technol.* (submitted).
7. E. Ma, H. Freiser, and S. Muralidharan, *Anal. Chem.* (submitted).
8. M. J. Cleare, P. Charlesworth, and D. J. Bryson, *J. Chem. Tech. Biotech.*, *29*: 210 (1979).
9. Y. Surakitbanharn, S. Muralidharan, and H. Freiser, *Anal. Chem.*, *63*: 2642 (1991).
10. K. Inaba, S. Muralidharan, and H. Freiser, *Anal. Chem.*, *65*: 1510 (1993).
11. K. Inaba, H. Freiser, and S. Muralidharan, *Solvent Extr. Res. Develop.* (in press).
12. F. Chen, H. Freiser, and S. Muralidharan, *Langmuir* (in press).

12
Preparative Centrifugal Partition Chromatography

Rodolphe Margraff
Rhône-Poulenc Rorer, S.A., Vitry-sur-Seine, France

I. INTRODUCTION

Chromatographic purifications of low-molecular-weight pharmaceuticals are based mainly on two different modes: adsorption and partition.

Adsorption chromatography exploits the interaction of one part of the solute with a solid support such as silicagel. In normal phase this interaction is restricted to the polar side of the molecule and presents two risks: chemical degradation and insufficient selectivity in the case of natural products, which are often biosynthesized together with homologs and analogs.

In such cases liquid–liquid partition chromatography is a better choice, because this mode is amenable to sufficient selectivity under nondegrading conditions.

Liquid–liquid partition chromatography follows Nernst's distribution law [1] which states:

> A solute will partition between two partially miscible solvent layers in a constant and reproducible ratio K of its concentration in one phase to its concentration in the other. Under ideal conditions K is only dependent on the solvent system, temperature, and pressure. It is independent on the solute concentration. In the presence of several molecular species, each solute is distributed as if the others were not present.

In practice this distribution has been extrapolated from the separatory funnel and made "chromatographic" in two different ways.

Martin and Synge, often referred to as the inventors of partition chromatography owing to their famous paper published in 1941 [2], after an unsuccessful attempt to build a countercurrent apparatus for the separation of amino acids from silk hydrolysates, realized that it is easier to immobilize the aqueous phase on a column filled with porous silicagel and to percolate through the organic phase composed of chloroform stabilized with 1% ethanol. This system is easy to operate and can generate a relatively large number of theoretical plates within practical column heights.

However, silicagel is acidic and can promote hydrolysis of sensitive solutes, as was demonstrated for the penicillins. The replacement of silicagel by cellulose in 1944 had the problem of very low flow rates, leading to the same result.

At the same time (1944), Craig [3] proposed a support-free system based on a set of separatory funnels assembled in series along an axis, allowing a semi-automatic operation of what was improperly called a countercurrent distribution. A true countercurrent was not achieved until 1963 when Post and Craig modified the system to allow simultaneous transfer of both phases in opposite directions from a central cell. In this case the partition coefficient of the compound to be purified is adjusted to 1 and polar impurities are removed with the polar phase, whereas less polar impurities are withdrawn in the opposite direction with the less polar phase; in the meantime, the purified compound remains immobile at the injection point.

Even though only one phase is withdrawn, it can be shown that it is advantageous to have about equal volumes of both phases in the different cells and to choose a composition for the solvent mixture that will yield K~1 for the compound to be purified.

In packed-column partition chromatography (PCPC) (where K is defined as the ratio of the solute concentration in the stationary phase to its concentration in the mobile phase ($K = C_s/C_m$)), this is not feasible, because the ratio of the two liquid phases is fixed within certain limits imposed by the presence of the solid support (~20%), the volume of the impregnated liquid stationary phase (V_s~5%) and the porous interstitial volume accessible to the mobile phase (V_m~75%). Hence, with $K = 1$, the capacity factor $k' = K$ (V_s/V_m) would be too low and the solute would elute too fast, without being separated from impurities. Thus, in PCPC, K must be increased by 15× in order to secure a k' of around 1.

This remains true when the filling material has been made hydrophobic for immobilization of the apolar phase, as is true for reversed-phase partition chromatography, first conceived by Howard and Martin in 1950 [4]. Among the ways of achieving this is the covalent bonding by alkylation of silicagel with a functionalized silane, for example, an octadecylisane. This method is used in the most popular chromatographic technique today, namely, reversed-phase high-performance liquid chromatography (HPLC). The practical outcome for solvent consumption and peak dilution will be discussed later.

Although Craig's apparatuses gained great success, especially in the antibiotics and other natural products fields, they have fallen into disuse—with some exceptions. For peptide production the Swiss firm Labortec AG (a subsidiary of Bachem AG) still manufactures and distributes a series of models in various sizes with typically 200 extraction tubes of 20 and 2000 ml capacity each for the laboratory and industrial versions, respectively. Their advertising claim that countercurrent methods can be used successfully in the isolation of natural substances, e.g., fermentation products and plant extracts; the purification of peptides; and in industrial production, seems to us very accurate, with the possible objection that there is no reason to limit their use to natural products; synthetic organics and even minerals can be purified as well. Countercurrent methods remain particularly suitable for peptides because the very low solubility of those peptides rich in hydrophobic residues renders them intractable by HPLC in both modes.

Natural products are often isolated as a complex of very similar molecular species whose separation is a more difficult challenge than in the case of synthetics. An interesting comparison between countercurrent distribution and preparative HPLC for the resolution of a kitasamycin complex and a xanthophyll fatty acid mixture led researchers from Bayer [5] to conclude that it is for the user to decide when to use liquid–liquid distribution and when to use preparative HPLC, since the two methods should be regarded as equivalent.

Although modern Craig apparatuses are fully automated and easy to operate, they remain time consuming when compared with actual centrifugal partition chromatographs.

The purpose of this chapter is to highlight the differences between HPLC and centrifugal partition chromatography (CPC) and to show how approximately equivalent performances at laboratory scale diverge profoundly at an industrial scale. Only the hydrostatic equilibrium–based systems produced by Sanki Engineering, LTD., (Kyoto, Japan) will be considered here.

II. CHROMATOGRAPHIC PARAMETERS

To start with, it may be advisable to review some chromatographic definitions in order to prevent the frequent confusion between *efficiency* (the plate number N) and *efficiency of the separation* (the resolution Rs).

It is well known that countercurrent chromatography in general (and CPC in particular) is a poorly efficient process because of the limited diffusion rate of the solutes between the liquid phases. Typically a CPC chromatograph can generate a few hundred theoretical plates, roughly one order of magnitude less than HPLC. The confusion between N and Rs is maintained by the manufacturers of HPLC columns, who focus their marketing on "column efficiency."

The efficiency of a chromatographic separation or, better, its resolution Rs is expressed in the Knox equation as the product of three terms related respectively to the efficiency N, the selectivity α, and the capacity k'.

$$Rs = \frac{1}{4}\sqrt{N}\frac{\alpha-1}{\alpha}\frac{k_2'}{1+k_2'} \quad \text{or} \quad Rs = \frac{1}{4}\sqrt{N}(\alpha-1)\frac{k_1'}{1+k_1'}$$

This approximate equation is obtained from the four fundamental chromatographic equations:

$$Rs = 2\frac{t_{R2} - t_{R1}}{w_{b1} + w_{b2}} \quad \text{or} \quad Rs = 2\frac{V_{R2} - V_{R1}}{w_{b1} + w_{b2}}$$

$$N = 16\left(\frac{t_R}{w_b}\right)^2$$

$$t_R = t_o(1 + k')$$

$$\alpha = \frac{k_2'}{k_1'}$$

In CPC, where elution peaks are gaussian, these equations remain valid. However, K, becoming now accessible, is more practically used than k'.

Substituting $V_R = V_m + KV_s$ into the Rs equation gives:

$$Rs = 2V_s\frac{K_2 - K_1}{w_{b1} + w_{b2}}$$

From this equation it is obvious that resolution increases with the volume V_s of the stationary phase (capacity), with the selectivity (K_2/K_1), and with the reduction in band width.

In practice laboratory preparative scale HPLC and CPC offer similar resolutions using, respectively, a 2" I.D.–column and a 250-ml rotor for 5-g injections. That is to say that HPLC's better efficiency ($\sqrt{N}$ is 3:1 to 5:1 vs. CPC) is offset by CPC's higher capacity and selectivity. In particular selectivity in CPC, providing that the solvent system has been properly chosen, has always been found to be higher than 2—and mostly in the range 3–4—something very uncommon in HPLC.

In fact, it is in the optimization of selectivity that CPC has the most to offer, although this optimization may appear obscure to the beginner. Figure 1 compares how CPC and HPLC are used to separate (or resolve) two unresolved peaks. In CPC resolution is achieved by improving selectivity: the peaks are moved away each other but their shapes remain unaffected any variation in efficiency. The situation is reversed in HPLC, where the stress is made on efficiency in order to slim the peaks. In other words, CPC focuses on the peak of interest and its vicinity, like a zoom lens, whereas HPLC is useful in lining up several thin peaks, for instance, as in an amino acid analysis. What happens upon scaleup is discussed in the following example.

PREPARATIVE CPC

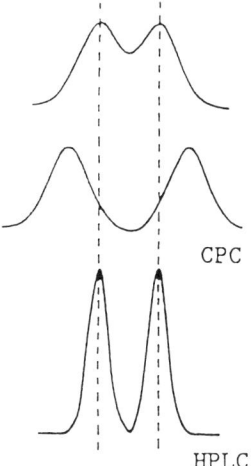

Figure 1 Two ways of improving resolution.

III. PURIFICATION OF 10-DEACETYL-BACCATIN III

10-Deacetyl-baccatin III, or "10-DAB," the starting material for docetaxel (Taxotere®) semisynthesis, is produced by the European yew *Taxus baccata* together with its 19-hydroxy analog, or "19-OH-10-DAB" (Fig. 2). A chromatographic separation step was judged necessary to separate 10-DAB from 19-OH-10-DAB.

Using the polystyrene-divinylbenzene resin Diaion HP 20 (Mitsubishi), it has been shown that 1 kg of 10-DAB can be purified in a single run on a 100 × 40 cm column in 8 hours with 2500 to 3500 L of aqueous (50 to 60%) methanol.

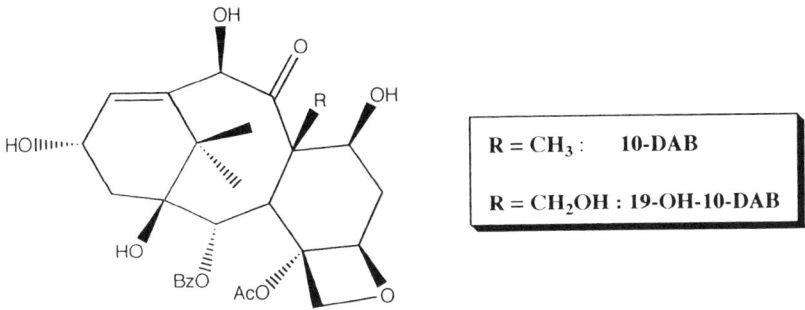

Figure 2 10-DAB (10-deacetyl-baccatin III) and its 19-hydroxy analog, 19-OH-10-DAB.

The difficulty and expense of this procedure for a large-scale production of 10-DAB has led us to evaluate CPC for an alternative isolation procedure.

We describe here the use of the new Sanki HPCPC chromatograph with stacked disks of a total volume of 250 ml. Our objective was to purify up to 5 g per run on this laboratory chromatograph to ensure a reasonable throughput with a larger centrifuge. Our strategy consisted of seven successive steps dealing with the: (1) choice of solvents, (2) choice of solvent system, (3) adjustment of solvent system coefficients for $K \sim 1$, (4) dynamic checking of the solvent system in the centrifuge, (5) optimization of the injection, (6) analytical-scale run, and (7) scaleup.

A. Solvents

A large majority of systems described in the literature contain halogenated solvents, as a consequence of the broad use in the past of the droplet countercurrent chromatograph (DCCC), the use of which requires stable droplets over a period of 2 to 3 days. The system $CHCl_3$-MeOH-H_2O meets this requirement, and the mixture 7/13/8 was perhaps the most widely used system in DCCC.

For preparative CPC we have abandoned the use of halogenated and aromatic solvents as well as acetonitrile for health hazard and environmental reasons. We considered only the following set: heptane (or isooctane), ethyl acetate, MtBE (methyl *tert*-butyl ether), acetone, MIBK (methyl isobutyl ketone), methanol, butanol and water, all of which share more general acceptance in industry.

Although chlorinated solvents exhibit unique chromatographic selectivities among individual solvents, we are convinced that mixtures of 3 to 5 of the above-cited solvents are able to reproduce the same behaviors.

B. Solvent System

In order to simplify the selection of the composition (the partition coefficient K has to be adjusted close to 1) we developed a set of 6 biphasic systems, A to G (the letter E has been skipped in order to avoid any confusion with F), of continuously varying polarity, and for which K of most solutes cross the unit value when going from A to G (Table 1).

For each of these 6 systems we measured the partition coefficient of both taxoids by dissolving a small amount, typically 5 mg, in a mixture made from 2 ml of each liquid phase. After dissolution under agitation, the concentration was measured in each layer by HPLC. K is here defined as the concentration in the upper organic phase divided by the concentration in the lower aqueous phase (Table 2).

Ideally, the partition coefficient K has to be close to 1. In practice, a K value

Table 1

System	EtOAc	Heptane	MeOH	Water
A	6	6	6	6
B	6	5	5	6
C	6	4	4	6
D	6	3	3	6
F	6	2	2	6
G	6	1	1	6

between 0.2 and 5 can be used without excessive elution time associated with band broadening. On the other hand, the separation factor, α, which is related to solvent system selectivity and is defined as the ratio of the K values of the two compounds (>1), has to be as high as possible in order to counterbalance the low efficiency, expressed in the number of theoretical plates.

System G has a high selectivity (4.6) but $K_{10\text{-}DAB}$ is too high. System F or system D would be a better compromise. Indeed, with system D we obtained very pure 10-DAB:$[\alpha]_D = -42°$ (c = 0.25; MeOH).

Unfortunately, the solubility of 10-DAB in all these systems is far too low for preparative work and, in the case of a poorly soluble compound like 10-DAB, the first parameter to be examined is solubility. 10-DAB solubility was measured by heating for 1 hour at 80°C in a closed autosampler vial 1 ml of solvent and 50 mg 10-DAB (about 99% pure). After cooling to room temperature and filtration of undissolved and recrystallized 10-DAB through a 0.22 μm membrane, the resulting solution was assayed by HPLC (Table 3).

In addition, DMF dissolves 400 mg per ml and ethylene glycol monomethyl-ether 110 mg/ml (at room temperature), but these two solvents were not considered

Table 2

System	$K_{10\text{-}DAB}{}^a$	$K_{19\text{-}OH\text{-}10\text{-}DAB}{}^a$	Selectivity (α)
A	0.035	0.088	2.50
B	0.085	0.100	1.19
C	0.295	0.236	1.25
D	0.700	0.400	1.75
F	2.000	0.660	3.00
G	5.460	1.180	4.60

$^aK = C_{\text{upper phase}}/C_{\text{lower phase}}$

Table 3

Solvent	Solubility in mg/ml at 20°C[a]
THF	45
MeOH	39.8
EtOH	28
Acetone	24.8
2-PrOH	19
EtOAc	10
Water	2.6
MtBE	1.0
Acetonitrile	0.67
Toluene	0.14

[a]Crystallized 10-DAB containing one molecule acetonitrile. Unsolvated 10-DAB (from ethylacetate) is less soluble.

for our initial experiments. THF is toxic and peroxidizes easily. Acetone is a better candidate although seldom used in CPC, with the exception of the system MIBK-acetone-water, which has been described for the purification of the macrolide antibiotic erythromycin [6] and has been known since 1939 [7].

C. System Coefficients

Starting with the system MIBK-acetone-water 20/1/20, we measured $K_{10\text{-DAB}}$ = 10.8 and $K_{19\text{-OH-}10\text{-DAB}}$ = 1.58. These make interesting values for the selectivity α = 6.8 but $K_{10\text{-DAB}}$ is too high. To lower it, we decided to add more acetone because acetone will partition slightly in favor of the aqueous phase and consequently will lower the K coefficients.

For that purpose we prepared the mixtures 2/0.1/2, 2/1/2, 2/2/2, 2/3/2, and 2/4/2. The system 2/5/2 is monophasic, as shown in Figure 3, where the binodial curve delimitates the biphasic area in which a mixture of composition A separates into an upper layer of composition C and a lower layer of composition B. Segment BC is a tie line. All solvent systems with composition points lying on the same tie line yield the same conjugate phases, the composition of which are found at the intersections of the tie line and the binodial curve. The respective volumes of the phases are inversely proportional to the lengths of the tie line segments on either side of the system composition point. Thus:

$$\frac{V_{\text{aqueous}}}{V_{\text{organic}}} = \frac{\text{Volume of B}}{\text{Volume of C}} = \frac{\overline{AC}}{\overline{BA}}$$

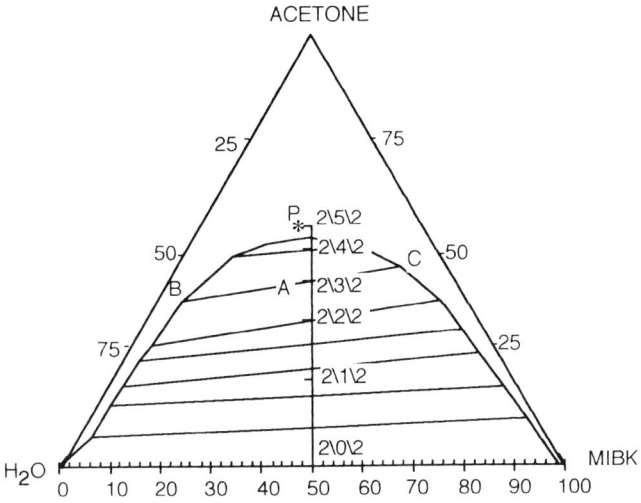

Figure 3 Phase diagram for system MIBK-acetone-water.

As expected both K values decrease with an increase of the acetone content (Table 4).

These partition coefficients were measured, as explained previously, by HPLC quantitation of the solutes partitioned between the two liquid phases (shake-flask method). The experimental error for such a single partition step is very high and more accurate values can be calculated from the elution profiles on the chromatograms obtained using system 2/3/2 (see Section III.D). The greater discrepancy observed for 19-OH-10-DAB ($K = 0.5$ by the shake-flask method and slightly above 1 by chromatography) may be explained by the stability of the solvate (19-OH-10-DAB, acetonitrile) during the shake-flask experiment.

System 2/4/2 (or 1/2/1) is very close to the Plait point, P (where the two phases become mutually soluble, giving a single phase of composition P; see

Table 4

System	2/0.1/2	2/1/02	2/2/02	2/3/02	2/4/02
$K_{10\text{-DAB}}$	10.8			3	
$K_{19\text{-OH-10-DAB}}$	1.58			0.5	
α	6.8			6	

Fig. 3), and separates into two phases with similar polarities and probably decreased interfacial tension. As a consequence settling time is increased and retention of the stationary phase in the chromatograph becomes more difficult, making this system unsuitable for CPC. Consequently, we selected the system MIBK-acetone-water: 2/3/2.

D. Dynamic Checking of the Solvent System in the Centrifuge

To maximize resolution, the retention of the stationary phase (either upper or lower) $S_F = V_s/V_c$ has to be as high as possible and preferably above 50% [S_F is the fraction of the total column volume (V_c) occupied by the stationary phase (V_s)]. This is an unpredictable behavior for a given system and has to be checked for every new one. Retention is favored by increasing the speed of the rotor with the correlating benefit of limiting a potential bleeding of the stationary phase. However, there is a limitation introduced by the accompanying back-pressure increase, which reduces the allowed flow rate with respect to the maximum pressure (60 bars) imposed by construction of the rotor. Reduction of flow rate is detrimental to throughput. We therefore, selected the conditions given in Table 5. By setting the pressure limit at 55 bars and with the use of an HPLC piston pump, both systems can probably be run at 1400 rpm to eliminate bleeding in the descending mode.

E. Injection

Injection is a crucial step in CPC. Improper injection may affect resolution and even can result in a complete loss of the stationary phase retention, a phenomenon called flooding. It is commonly accepted that the injection volume can reach 10% of the total volume V_c and that the injection solvent may be either the upper or the lower phase, or even a mixture of both.

We have observed that viscosity and interfacial tension are important parameters: above a given threshold, flooding will result from emulsification

Table 5

Mode	Flow rate (ml/min)	Speed (rpm)	S_F	Pressure (bars)	Bleeding
Ascending	3	1300	75%	50[a]	—
(upper phase mobile)	4	1400	75%	50[a]	—
Descending	3	1300	73%	50[a]	2.5% (75 μl/min)
(lower phase mobile)					

[a]Pressure limit of Pharmacia P 500 syringe pump.

following injection. This can be foreseen by mixing the injection solution with the opposite phase and measuring the settling time: a large increase predicts flooding.

Viscosity and interfacial tension are not a problem for 10-DAB CPC, whereas solubility is. To reach our goal of 5 g per run, we had to cross the 10% injection volume limit and to choose the upper (organic) phase for the injection solvent. Surprisingly, solubility of 10-DAB in this phase (point C on Fig. 3, with a composition of MIBK 43%, acetone 47%, and water 10%) is at least 3 times higher than in pure acetone.

Thus, solubilizing 5 g of crude 10-DAB (the sample used contained 74.5% 10-DAB and 3.8% 19-OH-10-DAB) needs 60 ml and 1.5 h heating in an oil bath at 80°C. No degradation was observed even without the addition of inert atmosphere, and the resulting solution remained stable for several hours at room temperature, provided the use of a piston pump was avoided (shocks from piston and from check valves initiate crystallization of the taxoids). We used an FMI® valveless rotating piston pump, which has a limitation of 7 bars. To overcome the back pressure in the rotor, it was necessary to reduce its speed to 100 rpm during the injection (which is better than injection in a motionless rotor).

In an attempt to substitute DMF for acetone (the system MIBK-DMF-H_2O: 1/1/1 can be used in CPC), we were very surprised to observe how a small amount of MIBK precipitates 10-DAB from a DMF solution. This system could therefore not be used.

F. Analytical Chromatography

Two analytical runs operated at 1300 rpm and 3 ml/min in the normal phase mode (organic phase mobile, ascending mode $S_F = 75\%$) and in the reversed-phase mode (aqueous phase mobile, descending mode $S_F = 73\%$) are shown on Figure 4. 10-DAB (10 mg) in organic phase solution was injected through a 500 µl loop in the rotor full of stationary phase.

Splitless detection was made using an evaporative light-scattering detector (ELSD) (Sedex 45) from Sedere, Alfortville (France).

Peaks A and B are mixtures of several components: in A (less polar than 10-DAB) the most abundant is 10-deacetyl-taxol (1.7% in the above batch); B corresponds to a mixture of 14-hydroxy-10-DAB and 13-*epi*-10-DAB, respectively, 0.2 and 0.8% in the same batch.

From the partition coefficient equation ($K = C_{\text{upper phase}}/C_{\text{lower phase}}$) and from the elution profiles, we calculated the following chromatographic parameters:

Descending mode: $$K = \frac{V_R - V_M}{V_S}$$

Ascending mode: $$K = \frac{V_S}{V_R - V_M}$$

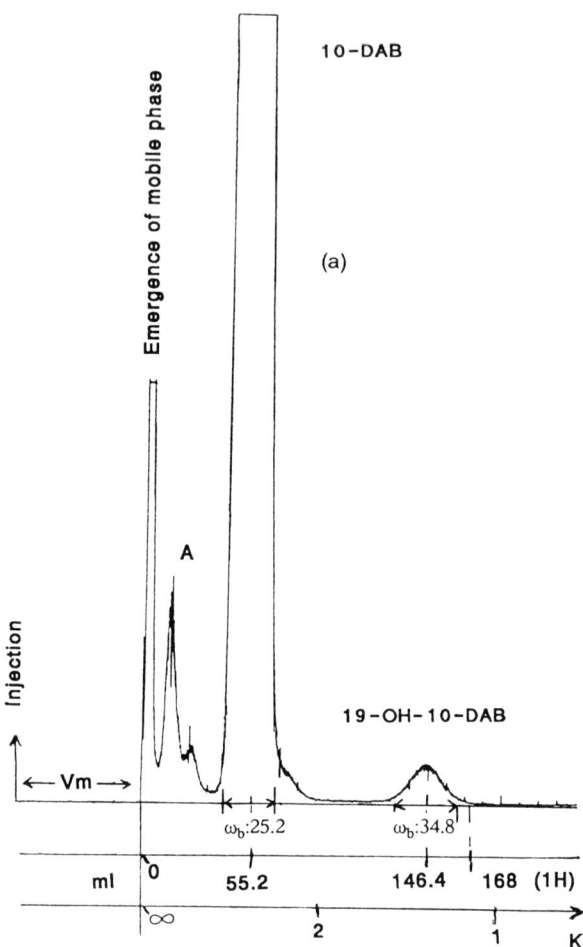

Figure 4 CPC chromatogram with ELSD detection after injection of 10 mg in 500-μl upper phase. (a) ascending mode, organic phase mobile, normal phase; (b) descending mode, aqueous phase mobile, reversed phase; (c) dual mode elution, first descending, then ascending after 371 ml.

PREPARATIVE CPC

343

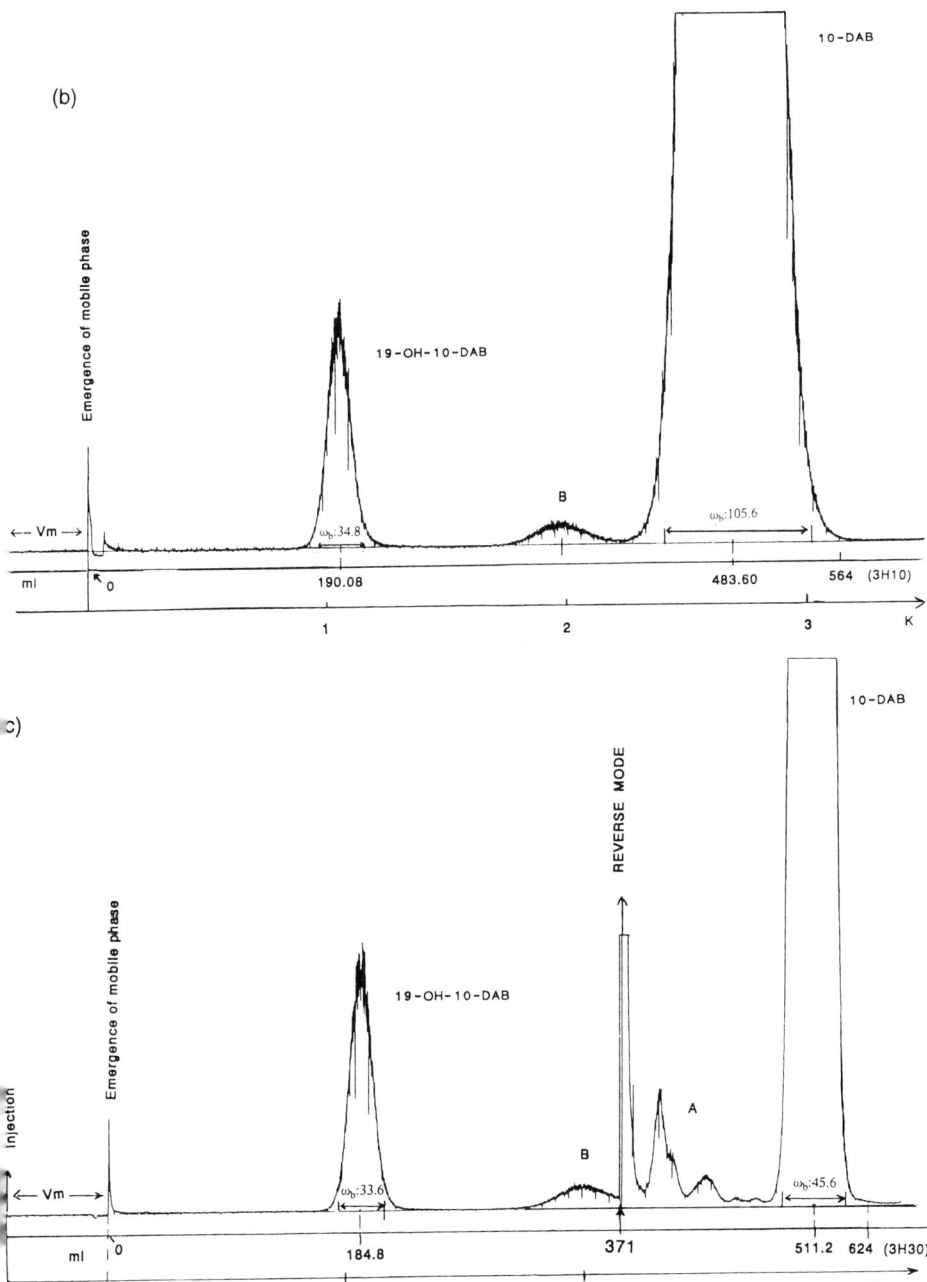

V_R = elution volume
V_M = mobile phase volume
V_S = stationary phase volume
$N = 16(V_R/w_b)^2$
$$Rs = 2\frac{V_{R2} - V_{R1}}{w_{b1} - w_{b2}}$$
$\alpha = \dfrac{K_2}{K_1}$ with $\alpha > 1$
N = Number of theoretical plates (efficiency)
w_b = peak width at baseline
Rs = resolution
α = selectivity factor

Table 6 shows the various experimental parameters. In the ascending mode elution was completed in 1 h, whereas in the descending mode after elution of 10-DAB for 3 h, impurities of A still remained in the apparatus. However, upon scale-up using an injection volume overload, the 10-DAB peak in the ascending mode rapidly overlapped the 19-OH-10-DAB peak.

The descending mode with dual-mode elution has a higher resolution and is a better choice for preparative separations. In order to avoid excessive elution time for A, we reversed the elution mode after the elution of B, just before the emergence of 10-DAB. Then using the organic phase as the mobile phase, A, which traveled poorly, eluted first and 10-DAB, which had to turn back the entire column volume, eluted last (Fig. 4c).

Interestingly the 10-DAB peak slimmed and w_b was reduced from 105.6 ml to 45.6 ml, thus the slight increase in total elution time—from 3 h 10 min to 3 h 30

Table 6

Parameter	Ascending mode		Descending mode	
	10-DAB	19-OH-10-DAB	10-DAB	19-OH-10-DAB
V_R (ml)	55.2	146.4	483.6	190.1
w_b (ml)	25.2	34.8	105.6	34.8
N	77	283	335	497
K	3.2	1.2	2.66	1.04
α		2.6		2.55
Rs		3		4.2

min—was compensated by a reduced peak volume and a corresponding reduced evaporation time for recovery.

In dual-mode CPC efficiency remains unchanged for 19-OH-10-DAB: $N = 484$ and rises to a theoretical* value of 2010 for 10-DAB. At the same time resolution doubles to a theoretical 8.2, too.

G. Scaleup

For scaleup we no longer considered impurities A and B, which were eliminated during semisynthesis of docetaxel (Taxotere®); the only aim was the complete separation of 19-OH-10-DAB from 10-DAB at injection up to 5 g per run.

Increasing the injected amount and volume up to 2 g in 25 ml (10% of V_c) affects the chromatogram only slightly. Thereafter, injection volume overload becomes increasingly evident in peak broadening.

The 19-OH-10-DAB peak reached the "foot" of the 10-DAB peak for somewhat less than 60 ml. At 60 ml the overlapping remained negligible, as can be seen from the injection of 5 g in 60 ml organic phase (described below). Injection was 10 ml/min in a rotor full of organic phase and rotating at 100 rpm. Speed was then increased to 1300 rpm and elution started with aqueous phase at 3 ml/min; 12 ml was collected every 4 min. After emergence of 73 ml organic phase (V_m = dead volume) corresponding to a retention $S_F = 70\%$, pressure rose to 52 bars and bleeding was 0.3 ml per fraction (2.5%). 19-OH-10-DAB eluted in fractions 13 to 20 (0.14 g). The mode was then reversed for a back development with the organic phase in the ascending mode. Taxoids A and B were still only partly resolved and the last fractions (36–42) contained a mixture of 10-DAB and 19-OH-10-DAB, although in a negligible amount. Purified 10-DAB was collected in fractions 27–35. Finally, the whole contents of the column were extruded by reducing the speed to 100 rpm, reversing the mode to ascending, and increasing the organic phase flow rate to 8 ml/min. After 260 ml the system was ready for the next purification cycle.

Recovered amounts were as shown in Table 7. The injected 5 g contained 3.725 g 10-DAB (74.5%) and 0.19 g 19-OH-10-DAB (3.8%). The remaining 21.7% were taxoids A and B (about 8.9%), solvents 5.85%, and uncharacterized impurities.

Crystallization in acetonitrile (systematically used for recovery from the intractable evaporation residue) yields 10-DAB solvated with one molecule of acetonitrile. Thus 3.725 g is equivalent to 4.005 g of solvate, and the corrected yield for purified 10-DAB is 98.1%. Surprisingly the volume yielding 98.1% was

---

*Equations expressing the elution parameters for initially retained solutes when recovered by reversed development have not yet been developed.

Table 7

Fractions	Amount (g)	Fraction content
13–20	0.14	19-OH-10-DAB
21–26	0.80	Yellow gum
27–35	3.93[a]	10-DAB
36–42	0.03	10-DAB (min.) + 19-OH-10-DAB (maj.)
Extrusion	0.04	10-DAB (maj.)

[a]After crystallization in acetonitrile; an added 0.07 g was found in the mother liquor.

obtained from only 9 fractions (27–35), that is to say, 108 ml and a dilution factor of 1.8 from the injection volume of 60 ml.

If one considers the analytical dual-mode run, where the 10-DAB peak had a width of 45.6 ml following an injection of 500 µl (dilution factor 90), much understanding is still needed of the mechanisms that govern volume overload, especially in the dual mode. Nevertheless, the above scaleup procedures are effective and the results are highly reproducible (as a result of Nernst's law) and suitable for automation. For that purpose, the steps are summarized in Table 8.

Since the preparation of mixture 2/3/2 furnishes an excess of aqueous phase (42.65%), this wasting can be avoided by shifting from point A toward point C on the tie line BC (refer back to Fig. 3).

H. Purified 10-DAB

Starting from a sample titrating 74.5% 10-DAB and 3.8% 19-OH-10-DAB, we have obtained in two identical runs 19-OH-10-DAB–free 10-DAB titrating 89% and 92.8%, respectively, the only difference being crystallization in acetonitrile for the former, and in ethyl acetate for the latter. Upon correction for solvent content, these titers were, respectively, 93.8% and 94.5%.

Our objectives were fulfilled (quantitatively: 5 g, and qualitatively: 19-OH-10-DAB not detectable). The remaining impurities are discarded during semisynthesis (of docetaxel) (although the same is not true for 19-OH-10-DAB).

In conclusion, it has been shown that 10-DAB can be separated from 19-OH-10-DAB with an almost quantitative recovery in 3 h 14 min in using only 0.8 liter solvent for 5 g (160 L/kg). In comparison to packed liquid chromatography, the savings in solvent is 15- to 20-fold, and further improvement is not ruled out (the extrusion step may be shortened or even omitted). This example clearly demonstrates the performance capability of CPC—and it should be noted that this is not the only such example. It should also be emphasized that although it was not our objective to purify 10-DAB at >99%, 99% purification *has* been done in

Table 8

Step	Operation	Duration (min)	Mode	Speed (rpm)	P (bars)	Flow rate (ml/min)	Volume (ml) upper phase	Volume (ml) lower phase
1	Injection: 5 g in 60 ml upper phase	6	desc.	100	5	10	60[a]	
2	Acceleration	0.5	desc.	1,300	10	0		
3	Elution: lower phase mobile up to fraction 20 (12-ml fractions)	80	desc.	1,300	52	3		240[a]
4	Mode reversal: upper phase mobile up to fraction 45 (12-ml fractions)	75	asc.	1,400	25	4	300	
5	Extrusion with upper phase	32.5	desc.	100	5	8	260	
	Total	194					560[a]	240

[a]The first eluted 73 ml is upper phase in excess corresponding to dead volume (V_m) for $S_F = 70\%$. This can be reused for dissolution and injection (60 ml) in the next cycle, and the corresponding 60-ml volume does not have to be added to the 560 ml of upper phase necessary for one cycle. One cycle will have a duration of 3 h 14 min and will use up to 560 ml upper phase (70%) and 240 ml lower phase (30%).

other cases, including docetaxel (Taxotere®). CPC is not restricted to the pre-purification of complex mixtures, but, like HPLC, it *can* yield highly pure compounds.

IV. PREPARATIVE CPC VERSUS PREPARATIVE HPLC

Before entering into the details of this comparison it is noteworthy to recall that if, on an industrial scale, preparative HPLC is best performed on a single large column, CPC can be performed with the use of several small units working in parallel. In the case of 10-DAB, an alternative to a huge 40-cm–I.D. HPLC column would be three 5-liter units instead of one 30-liter unit.

Keeping industrial applications in mind, the differences between HPLC and CPC can be classified by three main topics: (1) solid support, (2) solvents, and (3) automation. We shall examine these points, with an emphasis on the production of valuable chemicals at low-to-medium scales.

A. Solid Support

Besides the commonly evoked cost and chemical degradation problems, it is important to point out that recovery is never quantitative when a solid support is used. A loss of several percent on underivatized silicagel is not uncommon, and this drawback in the case of a very expensive sample becomes serious, especially if the sample is highly toxic, such as paclitaxel (Taxol®) or docetaxel (Taxotere®). Disposal of the solid waste is always a serious problem. Nevertheless, faced with the problem of cross-contamination, the temptation to use a "one-way" cheap matrix is high for the pharmaceutical and chemical industries. This makes CPC especially attractive for these "at-the-edge" purifications.

B. Solvents

Reduction in solvent consumption by an average factor of 10, which can reach 15–20 in the case of a poorly soluble compound such as 10-DAB, is a decisive advantage of CPC over HPLC. This difference results from the dissymmetry in the ratio of the stationary to mobile phase introduced by the adsorbent, as already mentioned in the introduction of this chapter. In CPC this ratio is close to 1, and consequently $K = k'$ and the solute is equally soluble in both phases. In HPLC K is about 10 to 15 times k', the solute concentration in the mobile phase is 10 to 15 times lower, and the wasting of solvents becomes inevitable.

This solvent-wasting feature of the HPLC has been recognized by several groups, such as the leading preparative column manufacturer Prochrom, which proposes a remedy with the so-called simulated countercurrent moving-bed chromatography, now referred to as continuous chromatography. Indeed, continuous chromatography is able to solve the problem of excess solvent consumption

PREPARATIVE CPC 349

but the number of extra columns needed, though smaller, still boosts operating complexity and investment costs, thereby reducing any real benefit for the user.

C. Automation

Continuous operation under full automation, the dream of the process engineer and the plant manager, is surely the future for preparative CPC. In addition to no risk of cross-contamination, the dual mode, which is practically equivalent to two successive HPLC runs in direct and reversed-phase modes on two separate columns, contributes to peak sharpening and solvent saving.

Automation is, in addition, greatly facilitated by the exclusion of any displacement effects, which are ruled out as a consequence of Nernst's law, which states that in the presence of several molecular species, each solute is distributed as if the others were not present.

In conclusion, besides the savings of a solid support, CPC features some additional advantages, such as huge reduction in solvent consumption and automation possibilities that indicate higher throughputs than for HPLC. The economics of the process have not been considered here, but it is apparent that operating costs are dramatically lower for CPC.

V. DISADVANTAGES OF CPC

The disadvantages of CPC may be classified as "software" and "hardware." The principal "software" disadvantage of CPC is precisely the absence of any CPC optimization software of the sort that would help in selecting a solvent system. Chapter 4 in this volume brings an important contribution to the solution of this problem, which looks more psychological than really technical.

Concerning the hardware, present commercial chromatographs are limited to an operating pressure of 60 bars. This limit restricts the number of usable solvent systems on a preparative scale because flow rates that are too low are detrimental to throughput and are also detrimental to the efficiency, N, which increases with the pressure as the drop size decreases. The possibility of performing separations at higher pressure would probably improve performance in a similar manner as the reduction in granulometry improves the plate number in HPLC.

VI. CONCLUSION

The two ways of performing partition chromatography with (reversed-phase HPLC) and without (CPC) a solid support show paradoxically similar performances in term of resolution at a laboratory preparative scale. However, they differ profoundly at the extremes: HPLC is more suited to analytical chromatography and CPC seems to be tailor made for industrial chromatography. For several reasons, the chromatography community has spent—and continues to spend—much effort on the development of preparative chromatography, whereas CPC has been not only neglected but was, to some extent, discredited too. The prestige of countercurrent chromatography can now be considered restored, but the actual situation remains very uncomfortable for the practitioner because the available CPC centrifuges, though reliable, suffer from an Achilles' heel: the customer service and commercial practices of the sole manufacturer, Sanki, are not necessarily compatible with the introduction CPC for industrial processes outside Japan. There is an urgent need to change this situation, because CPC has to be considered as a very useful and promising tool for industrial chromatography.

REFERENCES

1. W. Z. Nernst, *Phys. Chem.* (1891).
2. A. J. P. Martin and R. L. M. Synge, *Biochem. J.*, *35*: 1358 (1941).
3. L. C. Craig, *J. Biol. Chem.*, *155*: 519 (1944).
4. G. A. Howard and A. J. P. Martin, *Biochem. J.*, *46*: 532 (1950).
5. W. Gau and H. J. Ploschke, *Abstracts PREP '92 Symposium*, Nancy (France), 77–82.
6. T. Y. Zhang, *J. Chromatogr.*, *315*:287–297 (1984).
7. D. F. Othmer et al., *Ind. Eng. Chem.*, *33*: 1240 (1939).

Appendix I
Various Ways to Fill a CPC

Filling a CPC column is always easy, much easier than filling an HPLC column, and it can be done very quickly, once we are familiar with the nature of the solvent, miscibilities, and relative densities. Except when the instrument is new, or when it has been intentionally emptied by pushing out the two liquid phases with air or nitrogen, the CPC column should always be filled with liquids. They can be the two phases that were used for a separation, or a cleaning solvent, such as methanol or acetone, used to remove the contents of the column because the column will be stored for a while or we want to start with a new biphasic solvent system containing different solvents.

Following are the most common ways to fill or refill a CPC column. Data are given for a laboratory apparatus, that is, a column of 110–240 ml and a rotor of an average radius of 8–12 cm (this corresponds to the CPC LLN and to the HPCPC).

I. THE CPC IS EMPTY, THAT IS, FILLED WITH AIR OR NITROGEN

A correct filling procedure will prevent gas bubbles from being trapped in the channels, which could create some trouble for on-line detection, or for the stability of the stationary phase.

The best way is to use a rotational speed of 300–600 rpm, and to fill the column in the *ascending mode* with the stationary phase, whether it is the lower phase or the upper phase. A flow rate F of about 5 ml/min can be used, so that the gas is gently pushed out by the liquid, which appears at the outlet of the column after a time $t = V_c/F$, V_c the total volume of the column; the column effluent may then be charged with gas bubbles for a few milliliters and generally becomes clear in a very short time.

Once this has been done, the operational rotational speed, correct mode, and flow rate will be set up, and the mobile phase pumped into the column; a constant increase of the back pressure to an equilibrium value in the column will reflect a correct equilibration of the two phases in the channels. The back pressure increases until the mobile phase starts to elute (see Chapter 1); then it should stay constant, or it can be slowly decreasing, due to some bleeding of the stationary phase, mechanically dragged out by the mobile phase.

Flushing out the two liquid phases with air or nitrogen is the reverse procedure. The gas outlet is connected at the inlet of the CPC column, which is rotated at 300–600 rpm, and *descending mode* is chosen. A pressure of 5–20 bars (70–300 psi) is enough to allow a fast flow of the liquids. The total volume of the collected liquids is generally less than the volume of the column, V_c; and this is due to the wettability of the inner surface of the CPC column (the total inner surface of the HPCPC is around 44 m^2).

II. THE CPC IS FILLED WITH ONE OR TWO LIQUID PHASES

The mode that will be adopted (ascending or descending) to refill the CPC column will depend upon the density difference between the new liquid phase and the phases that are in the column. The *ascending* mode will be selected if the *new* phase is *heavier*, or the *descending* mode if the *new* phase is *lighter*; a rotational speed of 500–700 rpm will be chosen, and higher flow rate, that is, 7–10 ml/min or even more, since no hydrostatic pressure is developed in this opposite mode; if liquids are too viscous, the flow rate should be lower. After a time, $t = V_c/F$, the new phase appears at the outlet of the CPC column, and it takes a few additional milliliters before it becomes clear. The column can then be operated and the mobile phase pumped in the appropriate mode (ascending if lighter, descending if heavier).

III. FILLING WITH BOTH STATIONARY AND MOBILE PHASES

This is the fastest way to refill a CPC column with fresh stationary and mobile phases. It only requires a pump equipped with a multisolvent delivery device, such as any HPLC pump with gradient ability.

One reservoir, A, will be filled with the upper phase, the other one, B, with the lower phase, the two phases being mutually saturated; it is always advantageous to keep a small layer of the lower phase in the reservoir that contains the upper phase, and vice versa.

The pump is then programmed to deliver a %B that corresponds to the ratio of mobile to stationary phase that will assure a correct equilibrium between the two phases when the chromatographic conditions are set up. This %B is known from previous runs or from the physical parameters of the two phases (see Chapter 2).

Then, the CPC column being not rotating, the mixture $A + B$ is pumped at highest flow rate, 10–20 ml/min, in order to minimize the demixing of the two solvents. To pump this mixture of two immiscible liquids, experiments were performed with various HPLC pumps; pumping problems were never encountered. Systems with premixing before the pump head, and with mixing after two pump heads, operated well; if one phase is too viscous, the flow rate has to be decreased.

After a time, $t = V_c/F$, the two phases elute together and, after a few additional milliliters, using a graduated cylinder, we can check that the ratio of the two phases coming out is the same as they are upon entering the CPC column.

The column is now ready, the mobile phase is selected, the operational rotational speed, the flow direction, and the desired flow rate are set up, and the chromatographic run begins.

Appendix II
CPC Instrumentation

A centrifugal partition chromatograph is, according to the classification of Y. Ito [1] a hydrostatic equilibrium system with rotary seals. Centrifugal partition chromatography is a type of countercurrent chromatography, which uses both hydrodynamic and hydrostatic systems. The hydrodynamic apparatus has been extensively described in books by Y. Ito [1] and W. Conway [2].

Hydrostatic apparatuses are not numerous; they are manufactured by one company in Japan [3], and distributed in various ways [4]. They include analytical- or semiprep-scale apparatuses, and large industrial-scale installations. Following is some information concerning these apparatuses.

I. THE CPC LLN

This instrument was introduced in 1984 and sold until 1992; it is now replaced by the HPCPC (*vide infra*). Figure 1 shows the instrument with its accessories, a power supply, a pump, a valve module with the injection valve and the elution mode valve, a UV detector, and a fraction collector. The CPC main frame is a 44 × 67 × 70 cm centrifuge operating in the range 200–2000 rpm that can be thermostated from 15 to 35°C in an ambient temperature of 25°C. Both the power supply and the centrifuge are equipped with a 24-hour timer for automatic unattended system operation and shutdown. The rotor is a 30 cm (diameter) by 15 cm (height) cylinder made with compartments that receive cartridges of two kinds (12-cartridge capacity). These cartridges consist of engraved polychlorotrifluoroethylene (PCTFE) plates separated by Teflon seal sheets and maintained with two stainless steel plates (see Fig. 2).

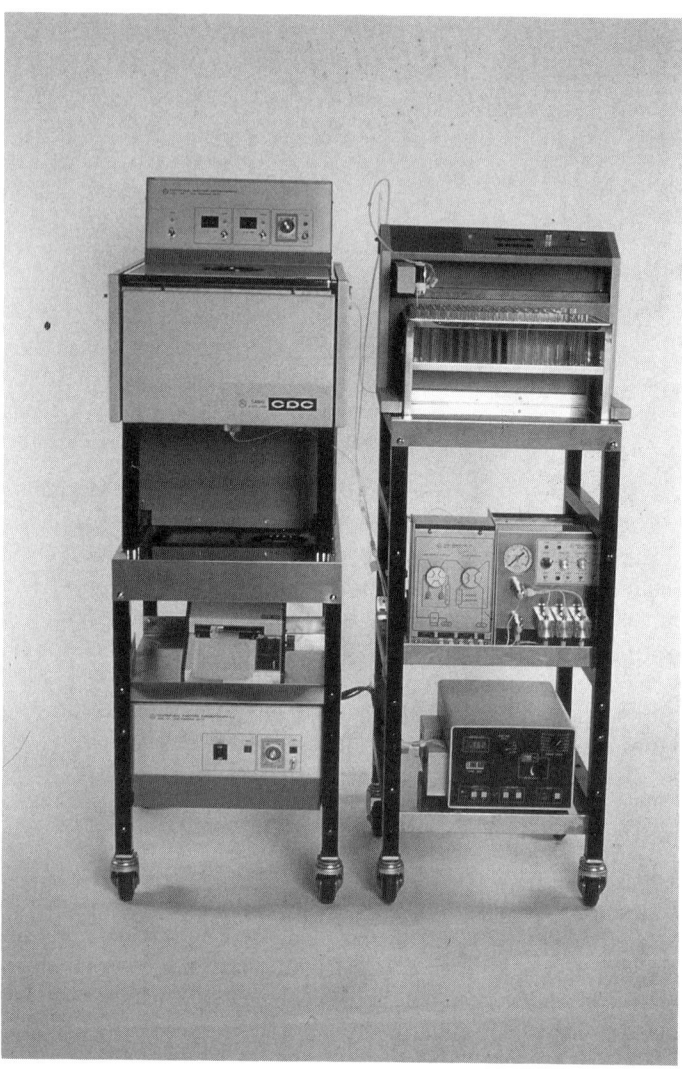

Figure 1 General view of the CPC LLN and its accessories.

CPC INSTRUMENTATION

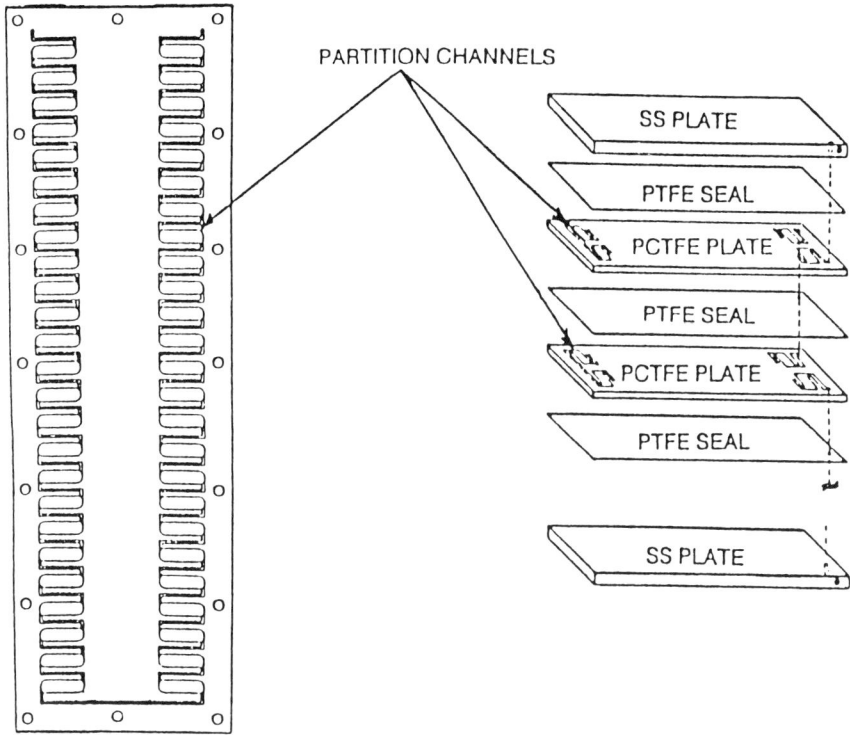

Figure 2 Expanded view of a PCTFE plate found in the cartridges of the CPC LLN.

A. Partition Cartridge–Type 250W

There are 400 channels per cartridge, with a total internal volume of about 18 ml. Each cartridge consists of 4 PCTFE plates, with 50 channels on each side of each PCTFE plate (see Fig. 2). A channel is 12.4 mm long by 2.4 mm wide, and a duct is 12.4 mm long by 0.9 mm wide; they are both 1.1 mm deep. The total internal volume is then the sum of ≈13 ml for the 400 channels, and ≈5 ml of the ducts, that is, 73% and 27% of the total volume, respectively. Up to 12 cartridges can be connected in series, but most of the applications found in the literature were performed using 3 or 6 cartridges (see Fig. 3). Figure 4 is a schematic of the rotor equipped with 12 cartridges, and a zoom view of the flow pattern in a cartridge.

B. Partition Cartridge–Type 1000 E

There are 40 channels per cartridge. The cartridges are similar to the type 250W, except that each channel is larger, which results in fewer channels per cartridge.

Figure 3 Upper view of the rotor of the CPC LLN arranged with six 250W cartridges.

They are called *preparative cartridges*, with a total nominal volume of 75 ml, which results in a "CPC column" of ≈900 ml for 12 cartridges.

C. The Rotary Seals

The rotary seals of the CPC LLN consist of two drilled rods terminated with carefully polished disks, which are maintained in close contact by a spring. One rod is made of a hard ceramic while the other is made with soft graphite and acts as a dry lubricating face. This assembly is able to maintain a leak-free connection up to 60 bars, provided that the two-faced disks are extremely clean. Special attention must be paid if the liquid phases contain some dissolved solids, such as buffers. They are, however, easy to dismantle and wash by sonication in hot water and solvents (do not rotate the sealing disks against each other if you suspect some particles to be present, particularly after a period of inactivity).

II. THE HPCPC, OR SERIES 1000, CHROMATOGRAPHIC SYSTEM

This new instrument was introduced at the 1992 Pittsburgh Conference. Figure 5 shows the instrument; the four-way valve, which controls the elution mode, is mounted on the HPCPC, so that the instrument can be connected between the

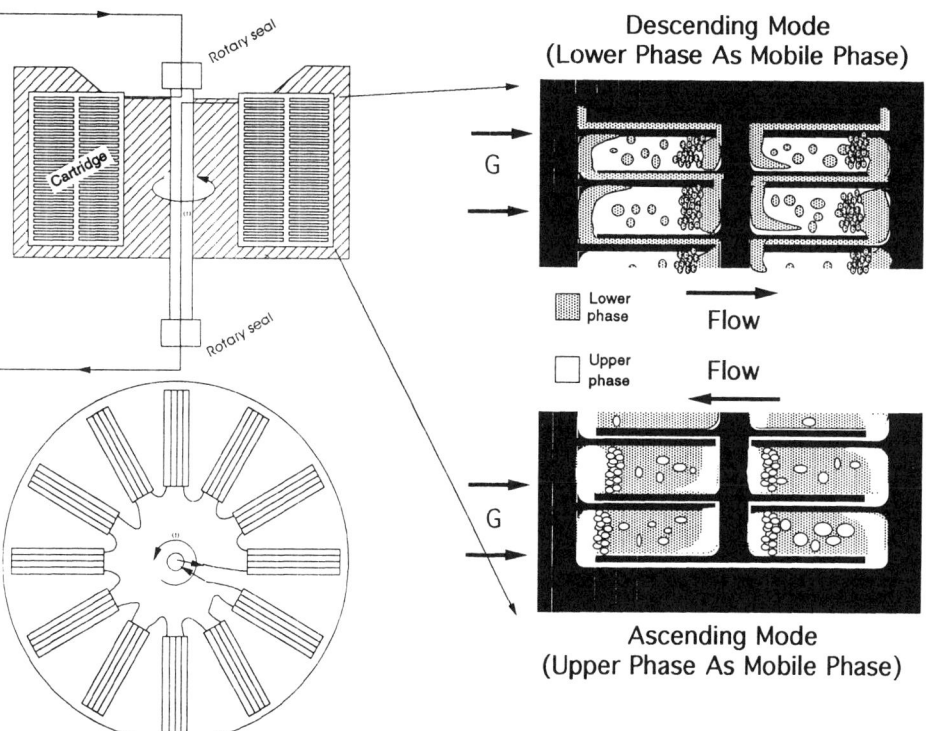

Figure 4 Schematic of the CPC LLN showing the rotor, the configuration of the cartridges, and the flow pattern in the channels that constitute the CPC column.

outlet of a sample injector and the inlet of a detector, in a manner similar to a standard HPLC column.

The HPCPC main frame is a 31 × 47 × 50 cm centrifuge weighting around 43 kg, and operating in the range 0–2000 rpm; it cannot be thermoregulated, and it has no automatic shutdown facilities (a remote control will be adapted later). The rotor consists of two packs of six disks each, connected through a 1/16" tubing, and easily removable. The upper disk pack is shown on Fig. 6. A disk is made of polyphenylene sulfide (PPS) and consists of 89 channels and ducts on each side, which have been obtained by injection molding. A detail of a disk is shown in Fig. 7.

A channel is 15 × 2.8 × 2.1 mm and a duct is 15 × 1 × 1 mm, which results in an internal volume of ≈88 μl per channel, and the total volume for two disk packs is around 220 ml [5], 85% of that volume being for the channels and 15% for the ducts.

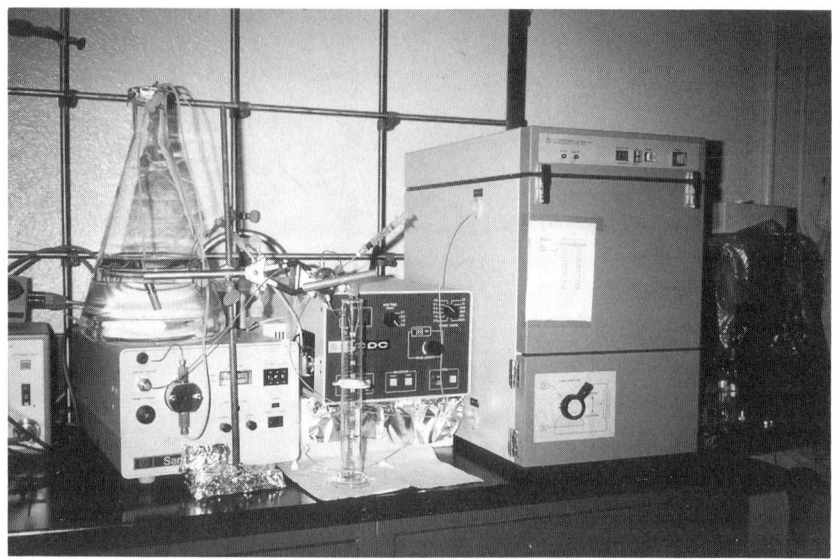

Figure 5 General view of the HPCPC.

Figure 6 Upper view of the rotor of the HPCPC.

CPC INSTRUMENTATION

Figure 7 Detail of a disk of the HPCPC.

A. The Rotary Seals

The rotary seals of the HPCPC (the upper one can be seen in Fig. 6) are made of a drilled sapphire rod going through two toroidal seals similar to those used with HPLC pump pistons. They need much less maintenance than the previous seals used with the CPC LLN. Only one toroidal seal is in movement with respect to the sapphire rod, and it is very easy to replace, with a seal insertion tool comparable with that used for HPLC pump.

III. LARGER INSTRUMENTS

Very little detailed information is available concerning larger CPC instruments. This is because they are custom designed for specific separation processes. Their basic design and operation are similar to the smaller laboratory preparative units. The lack of detailed information concerning these instruments is probably due to the proprietary separations they perform.

Table 1 shows the main characteristics of large-scale CPC. The interested reader will have to contact the manufacturer for additional detailed information.

Table 1 Specifications of Large-Scale Centrifugal Partition Chromatographs

Feature	Series 2000	LLI-005	LLI-0030
Internal volume, liters	1.4	5	27–30
Capacity (approx.), gm	30	150	1000
Flow rate (approx.), ml/min	20–40	50–140	400–700
Max rpm	1500	1300	1000
Rotor radius	15	15	25
Number of channels	1332	1144	792
Channel length, cm	2.9	4.6	8
Number of rotor packs	1	2	2
Number of disks per rotor pack	18	13	11

IV. FUTURE TRENDS

Hydrostatic apparatuses are easier to scale up than their parent hydrodynamic apparatuses. It is then easy to imagine, in the near future, bigger instruments, with a very high throughput. We can also imagine that future rotors will be able to work at higher pressure, leading to more efficient systems and making CPC technology more attractive.

It is also desirable for the scientific community that other companies develop new instruments. Imagine what HPLC might be today if only one company was making the HPLC instruments. It is desirable, too, that new groups of research get involved in centrifugal partition chromatography. It is time now for CPC to grow up faster and to prove its usefulness.

REFERENCES

1. N. B. Mandava and Y. Ito, eds., *Countercurrent Chromatography*, Marcel Dekker, Inc., New York (1988).
2. W. D. Conway, ed., *Countercurrent Chromatography*, VCH Publishers, New York (1990).
3. Sanki Engineering Limited, 2-16-10, Imazato, Nagaokakyo, Kyoto 617, Japan.
4. At the time this book was printed, there was a complete reorganization of the international distribution of the Sanki manufactured instruments, so the reader is advised to ask Sanki Japan for the best way to get information and order instruments.
5. A large French company received recently two HPCPC units with a total volume of only 180 ml for each unit; it is strongly suggested that one verify the total volume of a new instrument before starting any experiment or calculation.

Appendix III
Ternary Diagrams

From the numerous ternary diagrams found in the collection compiled by Sørensen and Arlt,* we have selected some systems that have been used, or that could be useful, for CPC applications. The mole % data have been transformed into volume % diagrams, which we know to be theoretically inexact because we are neglecting contractions normally observed when mixing different solvents. This can be perceived, for example, when making up a mixture corresponding to the midpoint of a tie line, which generally lead to two phases that do not have exactly the same volume. But these small discrepancies have no consequences for CPC applications and are insignificant when compared with the advantages of using these diagrams.

The ternary diagrams are classified according to the best solvent (see Chapter 4, "Solvent Systems in Centrifugal Partition Chromatography"), as shown in Table 1.

---

*Ternary diagrams in this appendix adapted from J. M. Sørensen and W. Arlt, *Liquid–Liquid Equilibrium Data Collection*, Dechema, Frankfurt/Main, Germany (distributed by Scholium International, Great Neck, N.Y.) (1980). The ternary diagram for methyl *tert*-butyl ether/acetonitrile/water adapted from S. J. Gluck and M. P. Wingeier, *J. Chromatogr.*, 69: 547 (1991). The ternary diagrams for heptane/methanol/water and 1,2-dichloroethane/methanol/water adapted from W. D. Conway, *Countercurrent Chromatography*, VCH, New York (1990).

Table 1 Ternary Diagrams Classified According to Best Solvent

Best solvent	Solvent system	Page
Carboxylic acid	Benzene/formic acid/water	367
	1,2-Dichloroethane/formic acid/water	367
	Chloroform/formic acid/water	368
	Cyclohexane/acetic acid/water	368
	Hexane/acetic acid/water	369
	Toluene/acetic acid/water	369
	Diethyl ether/acetic acid/water	370
	Chloroform/acetic acid/water	370
	1,2-Dichloroethane/acetic acid/water	371
	Isoamyl alcohol/acetic acid/water	371
	1-Butanol/acetic acid/water	372
	Methyl isobutyl ketone/acetic acid/water	372
	Ethyl acetate/acetic acid/water	373
Aromatic solvent	Heptane/benzene/dimethyl formamide	373
	Hexane/benzene/ethylene glycol	374
	Cyclohexane/toluene/acetonitrile	374
	Heptane/toluene/dimethyl sulfoxide	375
	Heptane/toluene/sulfolane	375
Chlorinated solvent	Cyclohexane/chloroform/acetonitrile	376
Ether	Cyclohexane/diethyl ether/methanol	376
Nitrile	Toluene/acetonitrile/water	377
	Chlorobenzene/acetonitrile/water	377
	Methyl *tert*-butyl ether/acetonitrile/water	378
	Methyl isobutyl ketone/acetonitrile/water	378
	Ethyl acetate/acetonitrile/water	379
Ketone	Toluene/acetone/water	379
	Toluene/acetone/ethylene glycol	380
	Diethyl ether/acetone/water	380
	Chloroform/acetone/water	381
	Methyl isobutyl ketone/acetone/water	381
	Cyclohexane/acetone/ethylene glycol	382
	Hexane/acetone/water	382
	Heptane/acetone/water	383
	Ethyl acetate/acetone/water	383
	Ethyl acetate/acetone/ethylene glycol	384
	Cyclohexane/methyl ethyl ketone/water	384
	Hexane/methyl ethyl ketone/water	385
	Heptane/methyl ethyl ketone/water	385
Alcohol	Heptane/1-butanol/acetonitrile	386
	Heptane/1-butanol/ethylene glycol	386
	Hexane/1-butanol/water	387
	Heptane/1-butanol/water	387
	Carbon tetrachloride/1-butanol/water	388
	Chloroform/1-butanol/water	388
	Ethyl acetate/1-butanol/water	389

Table 1 Continued

Best solvent	Solvent system	Page
Alcohol	Benzene/1-butanol/water	389
	Ethyl acetate/2-butanol/water	390
	Benzene/2-butanol/water	390
	Toluene/2-butanol/water	391
	Toluene/1-propanol/water	391
	Benzene/1-propanol/water	392
	1-Butanol/1-propanol/water	392
	Benzene/2-propanol/water	393
	Heptane/2-propanol/water	393
	Chloroform/2-propanol/water	394
	Diisopropyl ether/2-propanol/water	394
	Hexane/ethanol/water	395
	Hexane/ethanol/acetonitrile	395
	Heptane/ethanol/acetonitrile	396
	Chloroform/ethanol/water	396
	Diethyl ether/ethanol/water	397
	Ethyl acetate/ethanol/water	397
	Hexane/methanol/acetonitrile	398
	Heptane/methanol/acetonitrile	398
	Heptane/methanol/water	399
	Toluene/methanol/water	399
	Chloroform/methanol/water	400
	1,2-Dichloroethane/methanol/water	400
	Ethyl acetate/methanol/water	401
	1-Butanol/methanol/water	401
Amide	Ethyl acetate/N,N-dimethyl acetamide/water	402
	Toluene/dimethyl formamide/water	402
	Chloroform/dimethyl formamide/water	403
Dimethylsulfoxide	Tetrahydrofuran/dimethylsulfoxide/water	403
Pyridine	Toluene/pyridine/water	404
Hexane	Perfluorohexane/hexane/benzene	404
Tetrahydrofuran	Cyclohexane/tetrahydrofuran/methanol	405
	Heptane/tetrahydrofuran/water	405
	1,2-Dichloroethane/tetrahydrofuran/water	406
	Carbon tetrachloride/tetrahydrofuran/water	406
	Chloroform/tetrahydrofuran/water	407
	Methylene chloride/tetrahydrofuran/water	407

Remarks:

Chlorinated solvents, which have been widely used in the past, are often prohibited for industrial applications; but they are still useful for laboratory-scale purifications.

Benzene is no longer "popular," but the ternary diagrams involving benzene have been indicated because benzene can generally be replaced with toluene, which will result in a slightly different, but comparable ternary diagram.

In the same way, heptane can replace hexane in ternary diagrams involving hexane. Note that the binodal will shift in a way that will reflect the difference in the solubility of heptane and hexane in the two other solvents.

I. HOW TO USE A TERNARY DIAGRAM

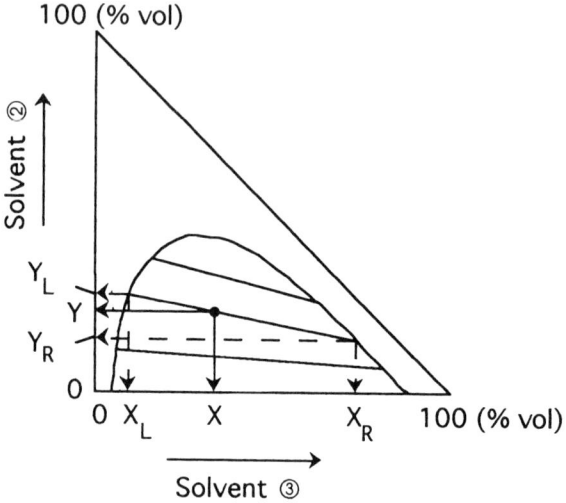

To make up the biphasic system corresponding to S, you have to mix X% of the solvent ③, Y% of the solvent ②, and $100 - X - Y$ % of the solvent ① ("%" = volume %). You will obtain the left phase, S_L, and the right phase, S_R, with the following composition:

	Left phase	Right phase
Solvent ①	$100 - X_L - Y_L$ %	$100 - X_R - Y_R$ %
Solvent ②	Y_L %	Y_R %
Solvent ③	X_L %	X_R %

Each phase can be prepared separately, and, in order to be sure that the equilibrium between the two phases is maintained, it is a good practice to add to a freshly prepared left (right) phase a small volume of the corresponding right (left) phase until a cloud or droplets appear.

TERNARY DIAGRAMS

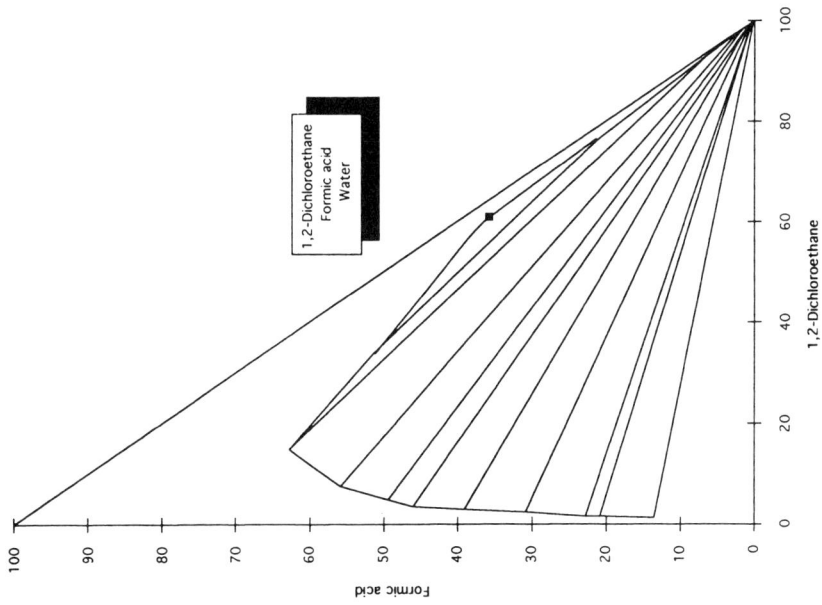

Best solvent: Formic acid. **Comments:** Possibility of gradients in the reversed phase mode.

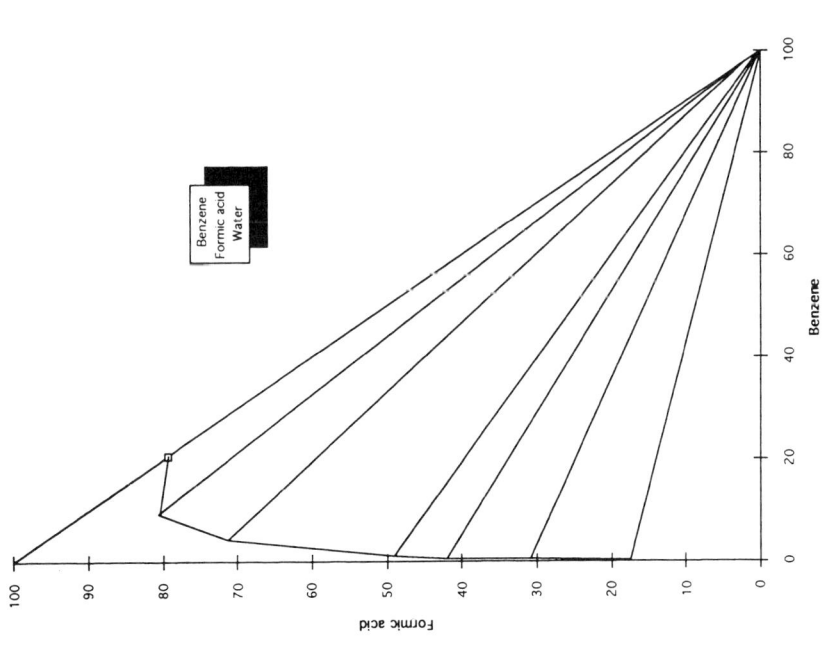

Best solvent: Formic acid. **Comments:** System 2. Using toluene instead of benzene should give a similar diagram. Possibility of gradients in the reversed phase mode.

APPENDIX III

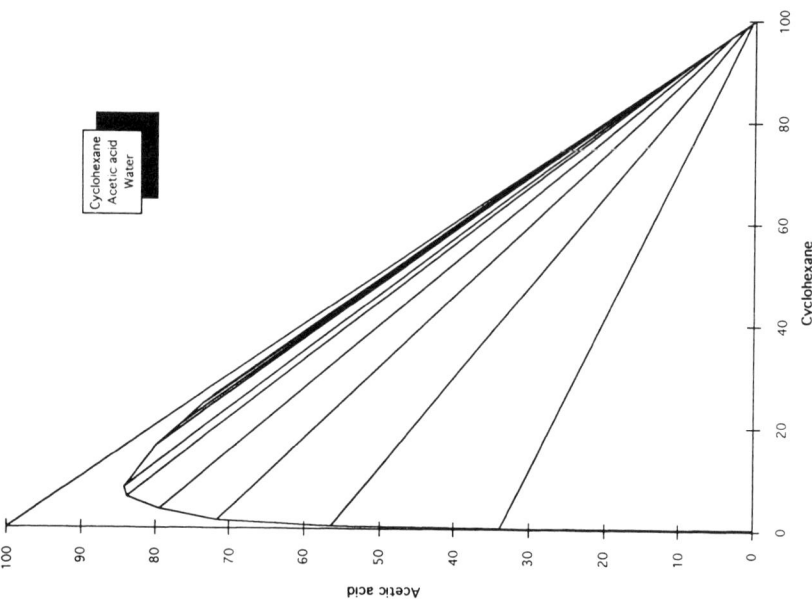

Best solvent: Acetic acid. **Comments:** Possibility of gradients in the reversed phase mode.

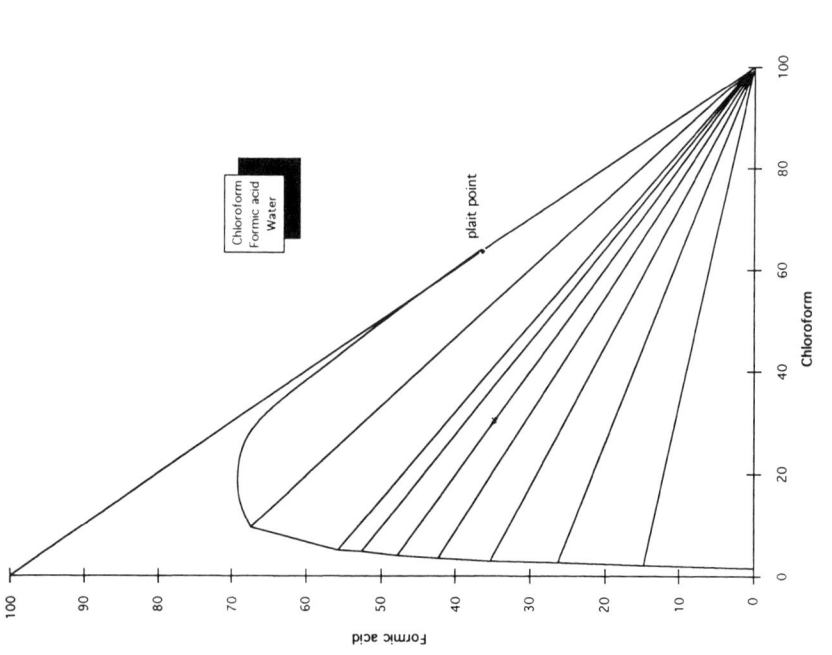

Best solvent: Formic acid. **Comments:** Possibility of gradients in the reversed phase mode.

TERNARY DIAGRAMS

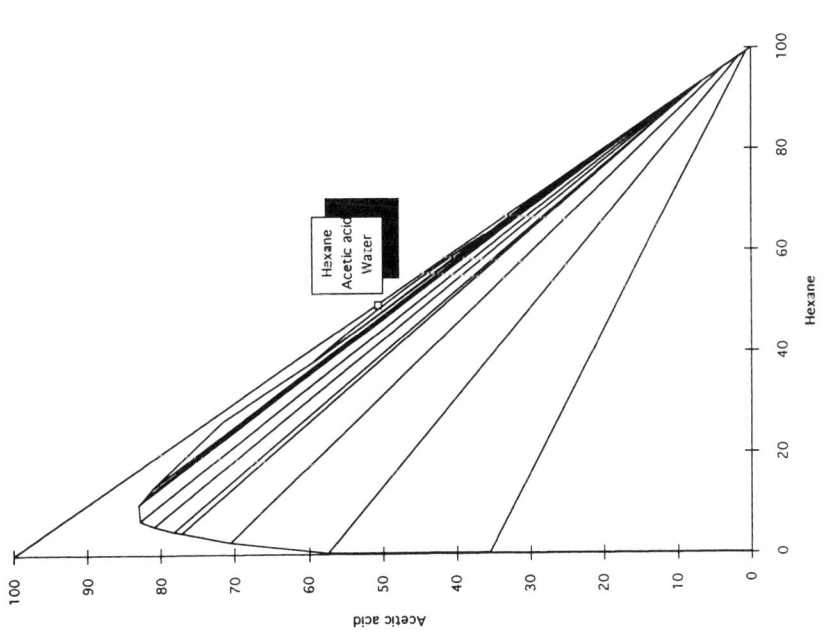

Best solvent: Acetic acid. **Comments:** Gradient in the reversed phase mode; can be used for fractionation of hydrophobic compounds containing some basic moities.

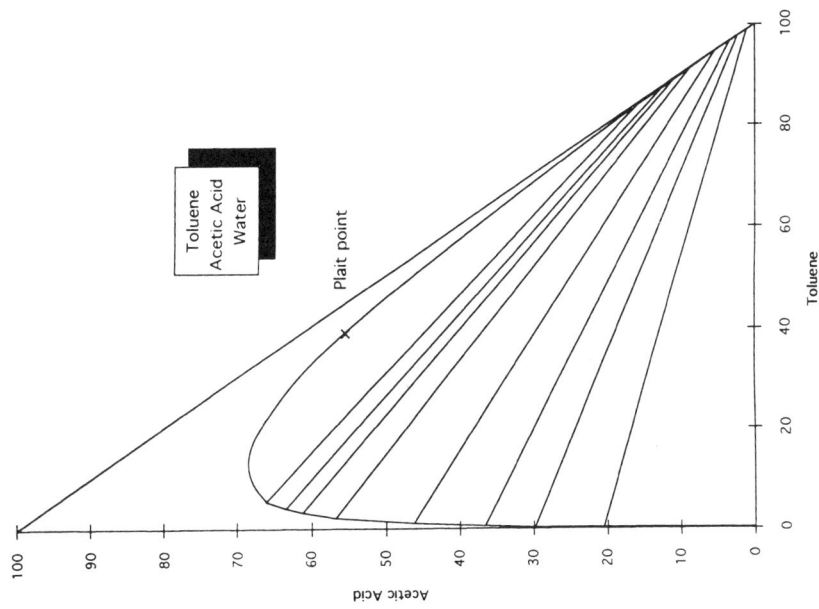

Best Solvent: Acetic acid. **Comments:** Compare with benzene/formic acid/water. Isocratic mode only.

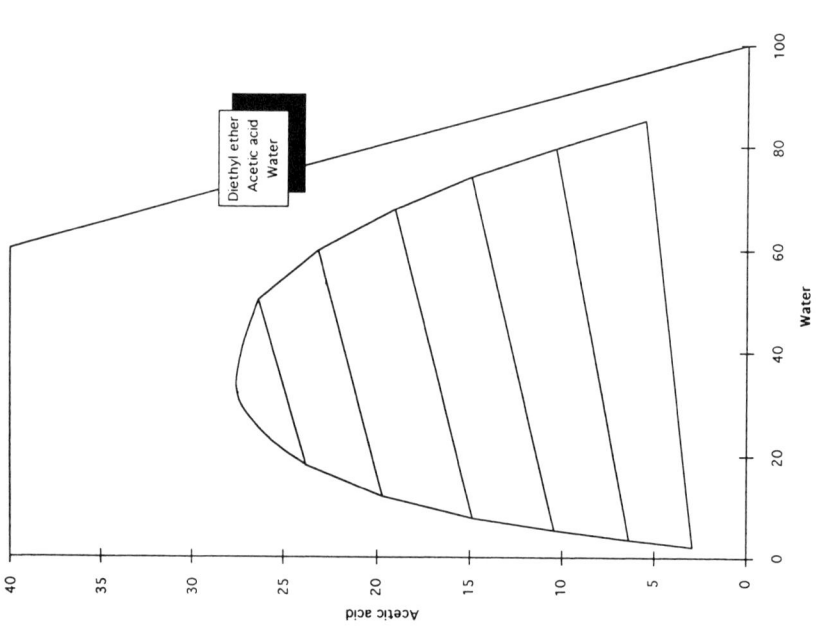

Best solvent: Acetic acid.

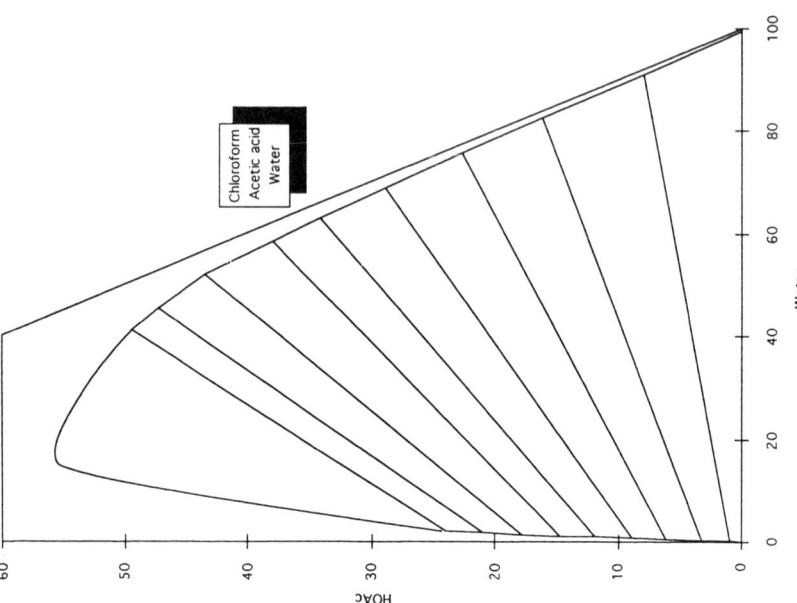

Best solvent: Acetic acid. **Comments:** Compare with chloroform/formic acid/water. The quaternary system chloroform/acetic acid/formic acid/water has been used for purification of large and very hydrophobic peptides.

TERNARY DIAGRAMS

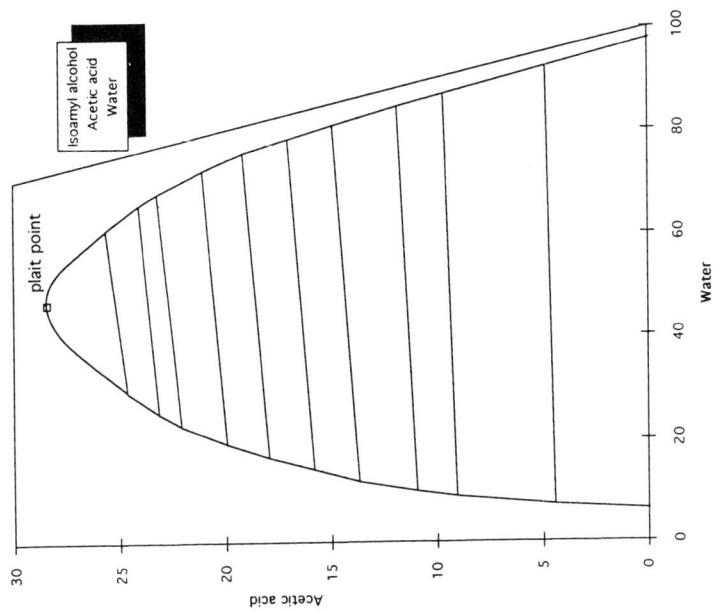

Best solvent: Acetic acid. **Comments:** Polar system, isocratic mode.

Best Solvent: Acetic acid. **Comments:** Compare with 1,2-dichloroethane/formic acid/water. Isocratic mode.

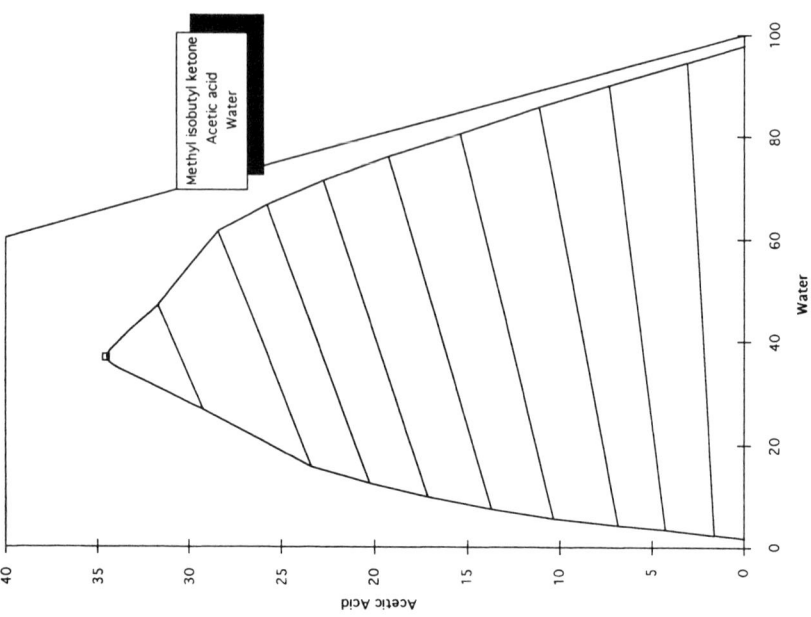

Best solvent: Acetic acid. **Comments:** Polar system, isocratic mode.

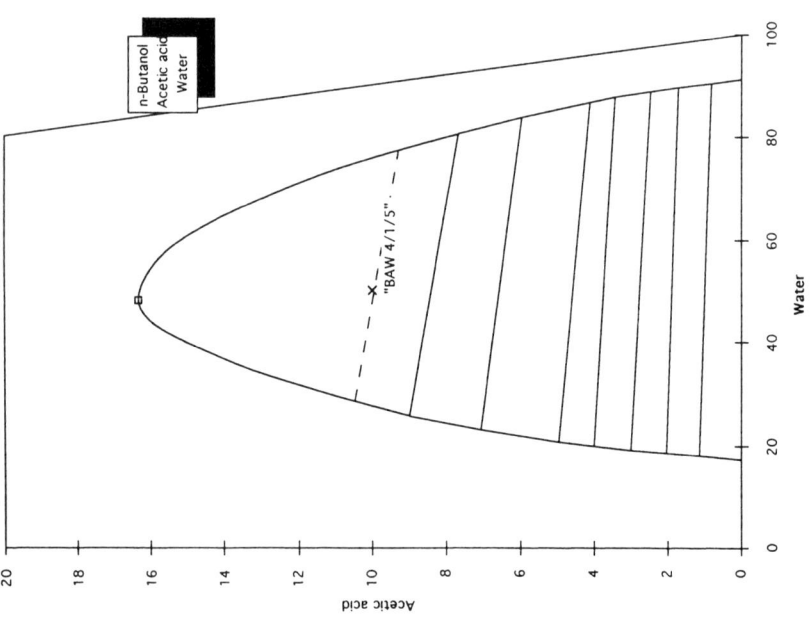

Best solvent: Acetic acid. **Comments:** This system has been extensively used by M. Knight et al. for peptide purification; isocratic mode.

TERNARY DIAGRAMS

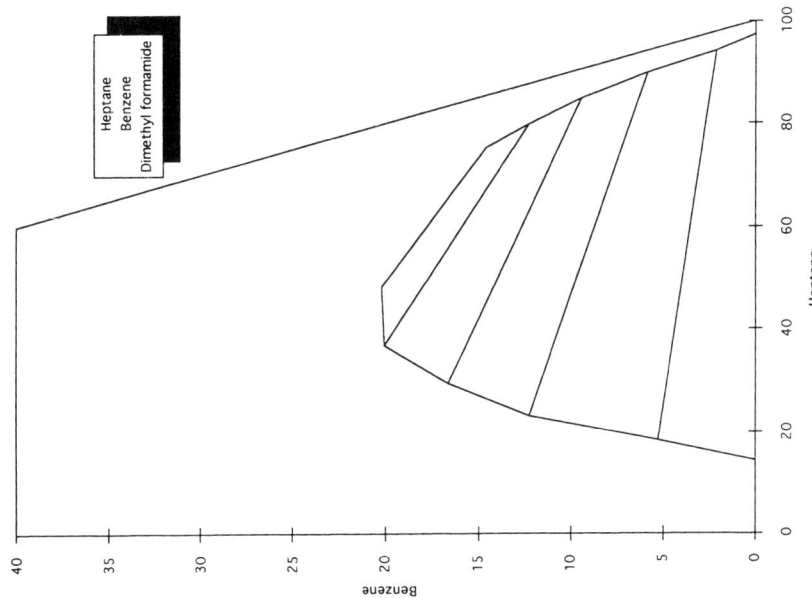

Best solvent: Benzene. **Comments:** Replacement of benzene with toluene has to be tested.

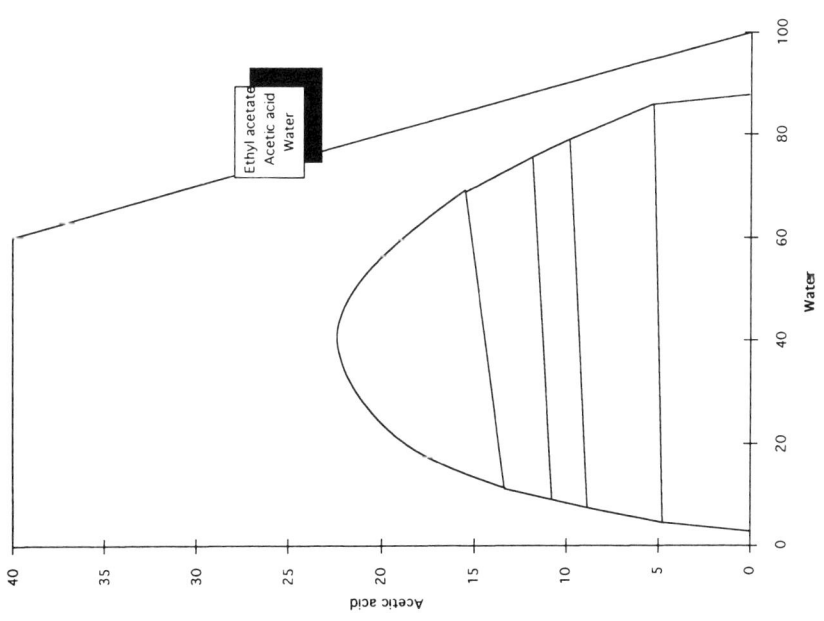

Best solvent: Acetic acid. **Comments:** Isocratic mode.

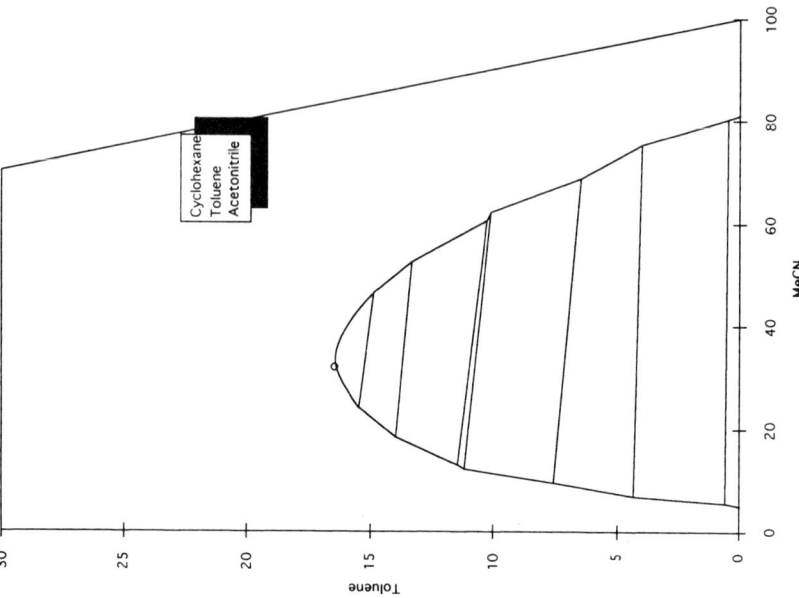

Best solvent: Toluene. **Comments:** Nonaqueous system; hexane or heptane can be tested instead of cyclohexane.

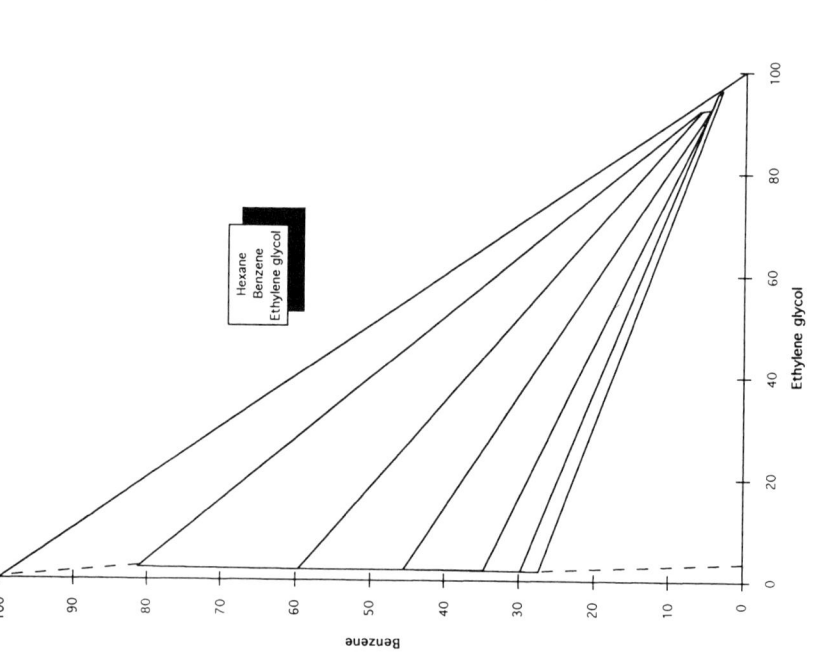

Best solvent: Benzene. **Comments:** System 2. Replacement of benzene with toluene has to be tested.

TERNARY DIAGRAMS

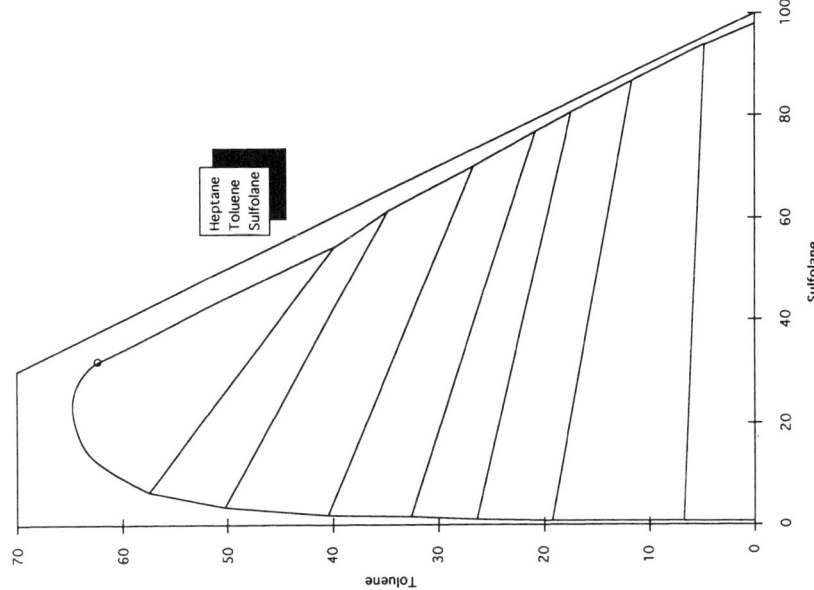

Best solvent: Toluene. **Comments:** This system has to be tested.

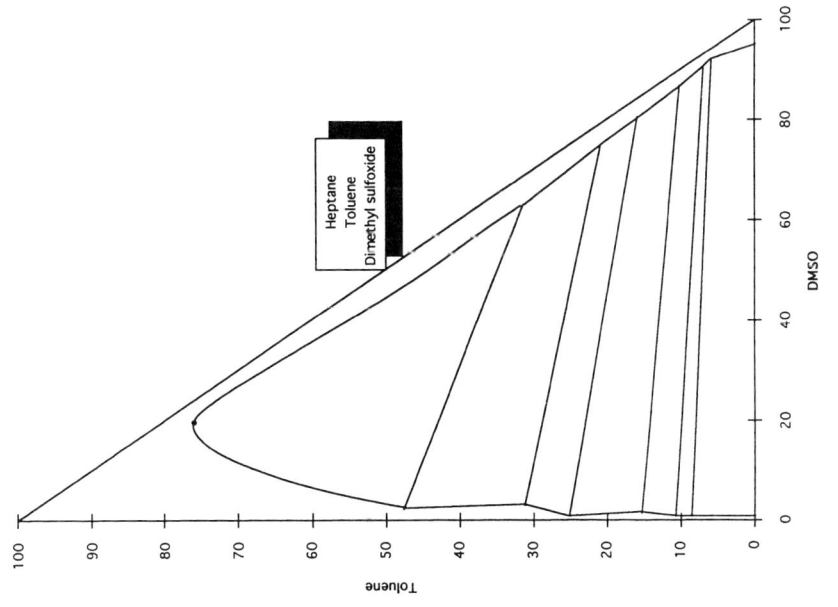

Best solvent: Toluene. **Comments:** Nonaqueous system, isocratic mode.

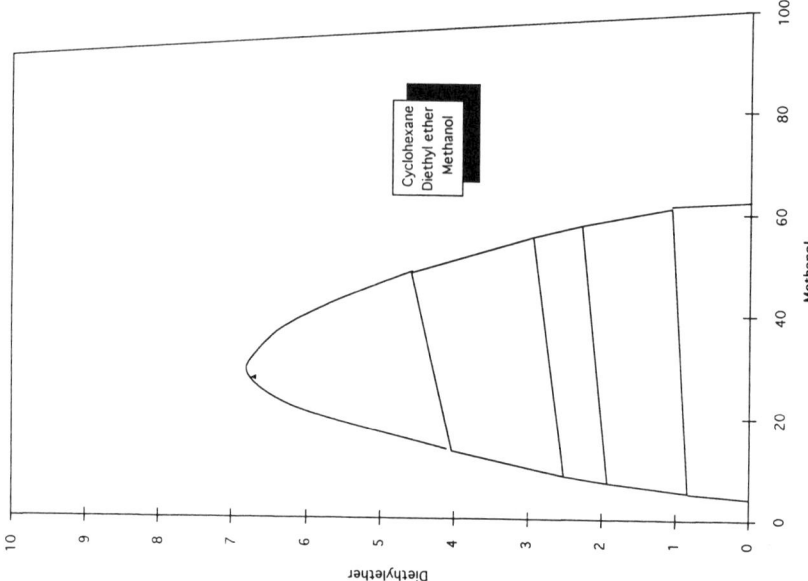

Best solvent: Dimethyl ether. **Comments:** Nonaqueous system.

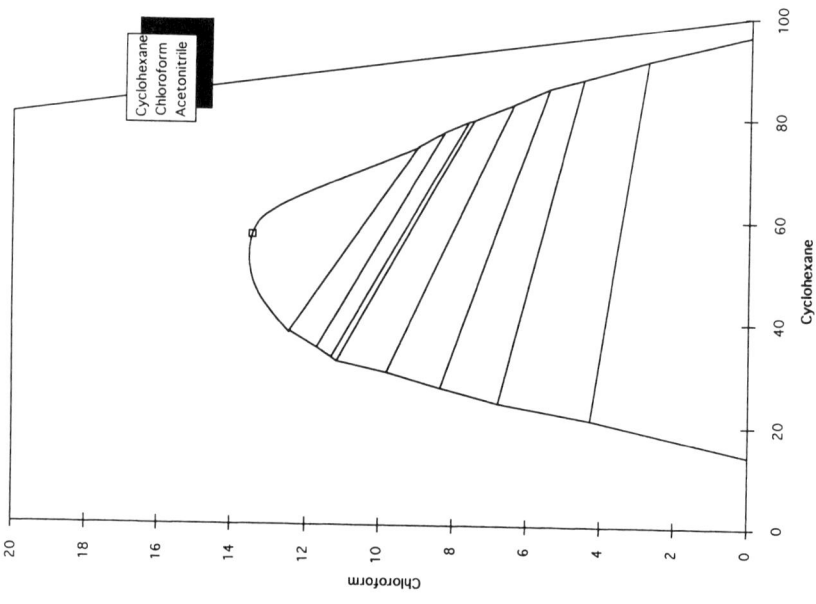

Best solvent: Chloroform.

TERNARY DIAGRAMS

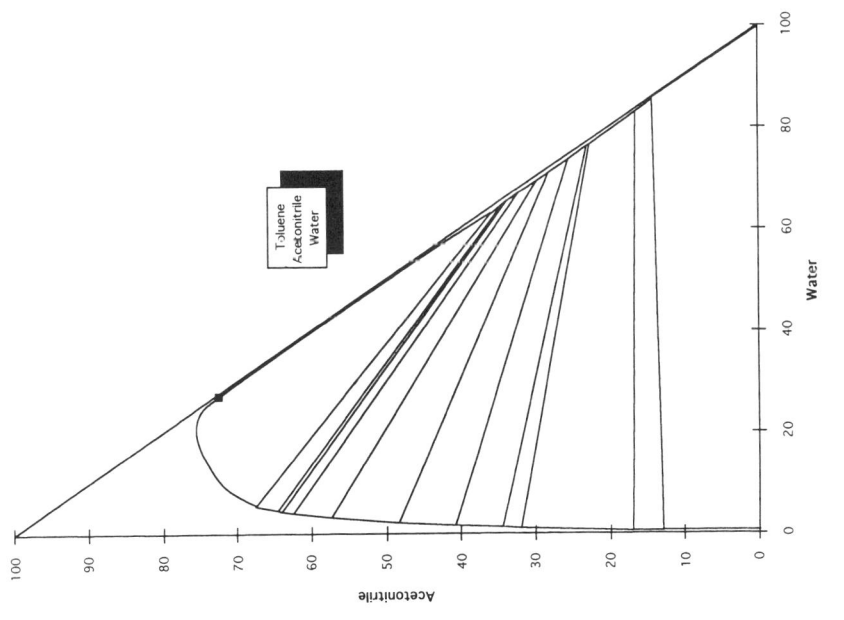

Best solvent: Acetonitrile. **Comments:** Isocratic mode.

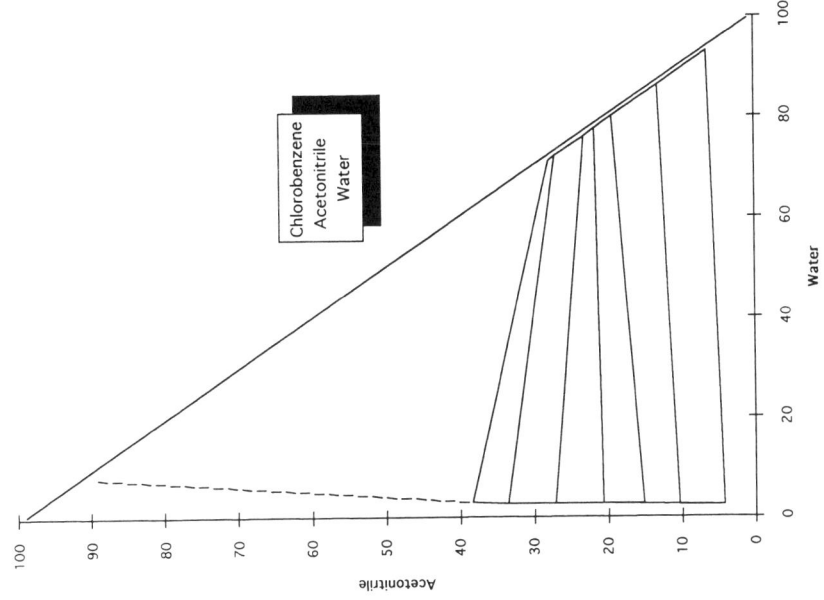

Best solvent: Acetonitrile. **Comments:** Upper part of the diagram is uncertain.

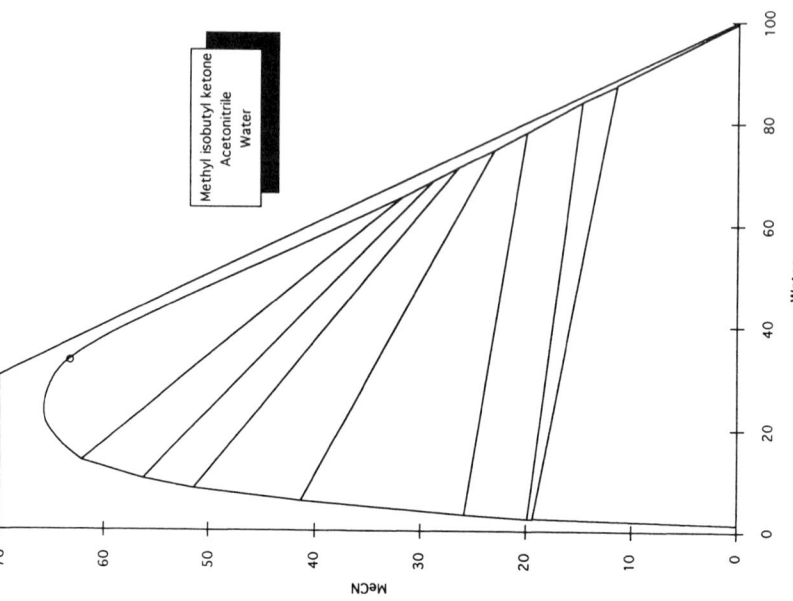

Best solvent: Acetonitrile. **Comments:** Medium polarity system; high content of acetonitrile can be reached.

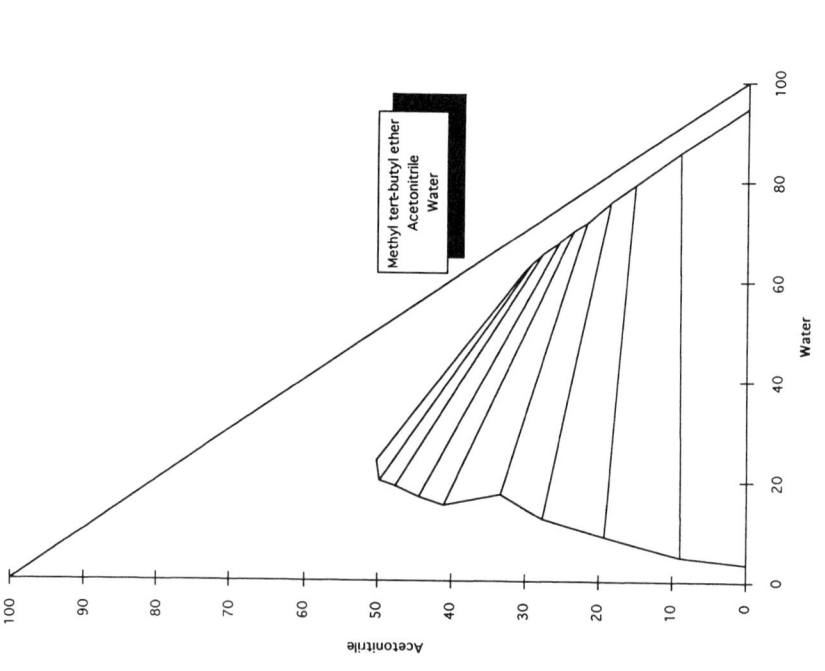

Best solvent: Acetonitrile. **Comments:** Medium polarity system.

TERNARY DIAGRAMS

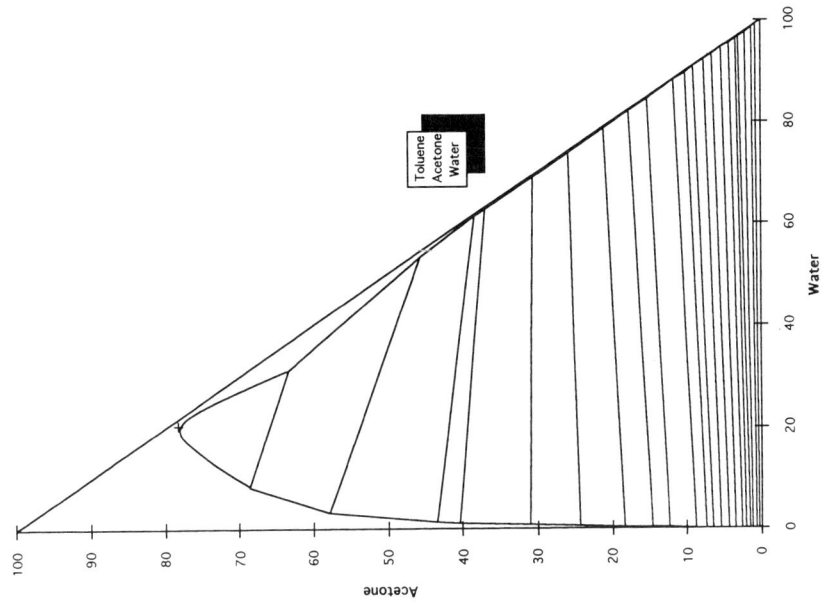

Best solvent: Acetone. **Comments:** Isocratic mode, allowing high content of acetone.

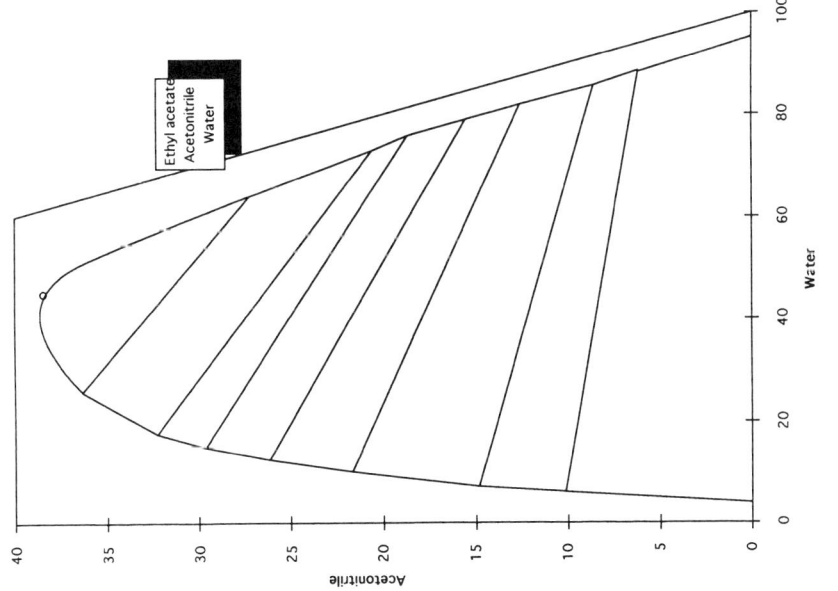

Best solvent: Acetonitrile. **Comments:** Medium polarity system.

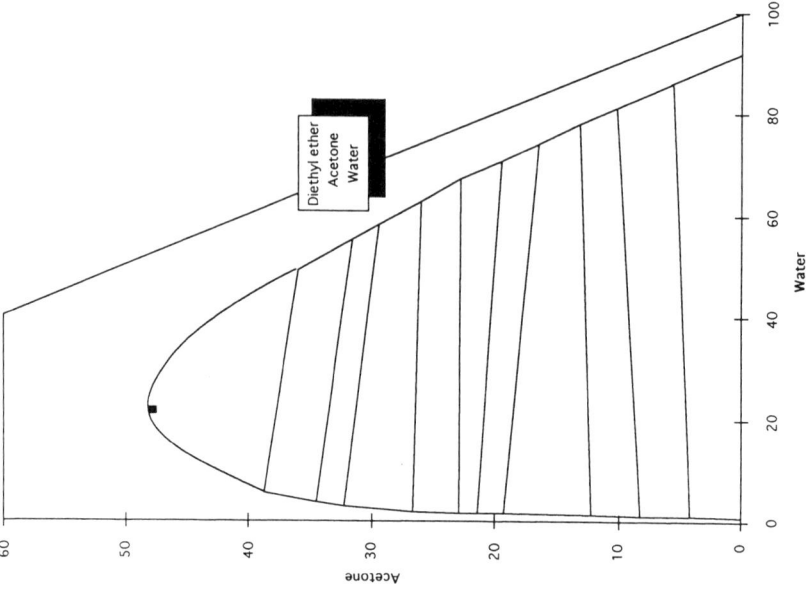

Best solvent: Acetone.

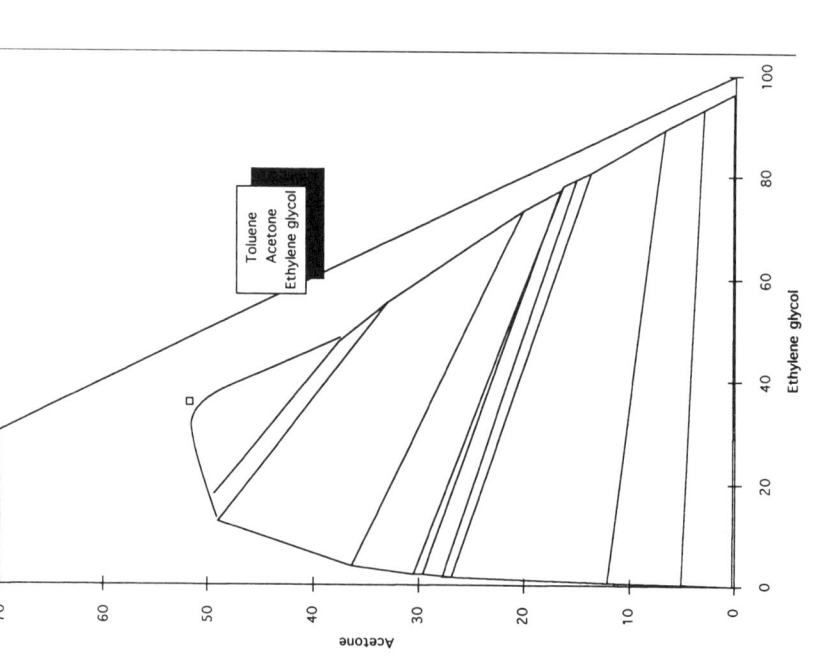

Best solvent: Acetone. **Comments:** Nonaqueous system, allowing high content of acetone.

TERNARY DIAGRAMS

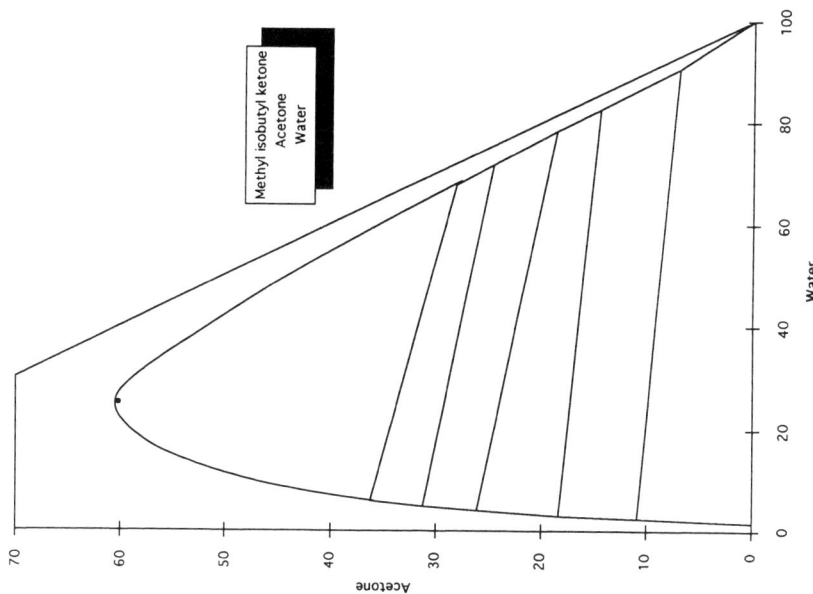

Best solvent: Acetone. **Comments:** A very promising system, medium polarity, isocratic mode.

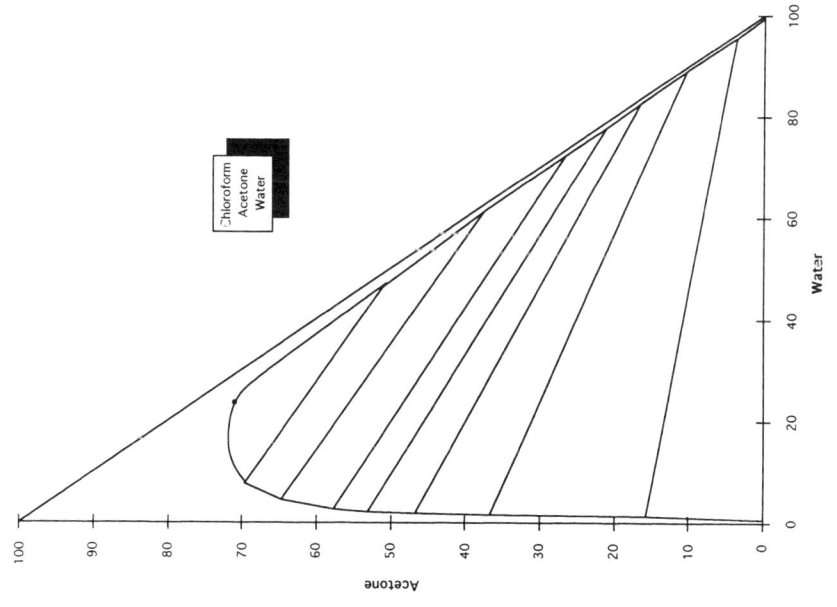

Best solvent: Acetone.

APPENDIX III

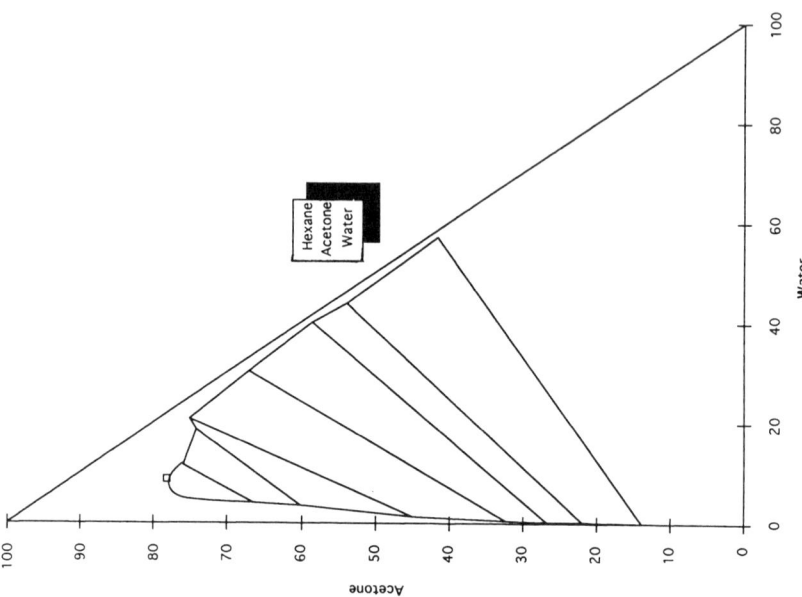

Best solvent: Acetone. **Comments:** Nonaqueous system.

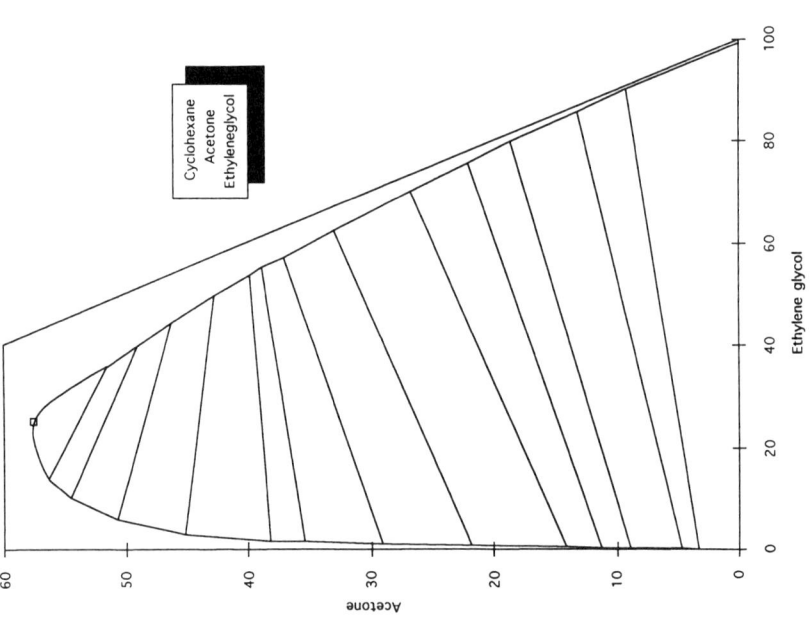

Best solvent: Acetone. **Comments:** Allows high content of acetone.

TERNARY DIAGRAMS

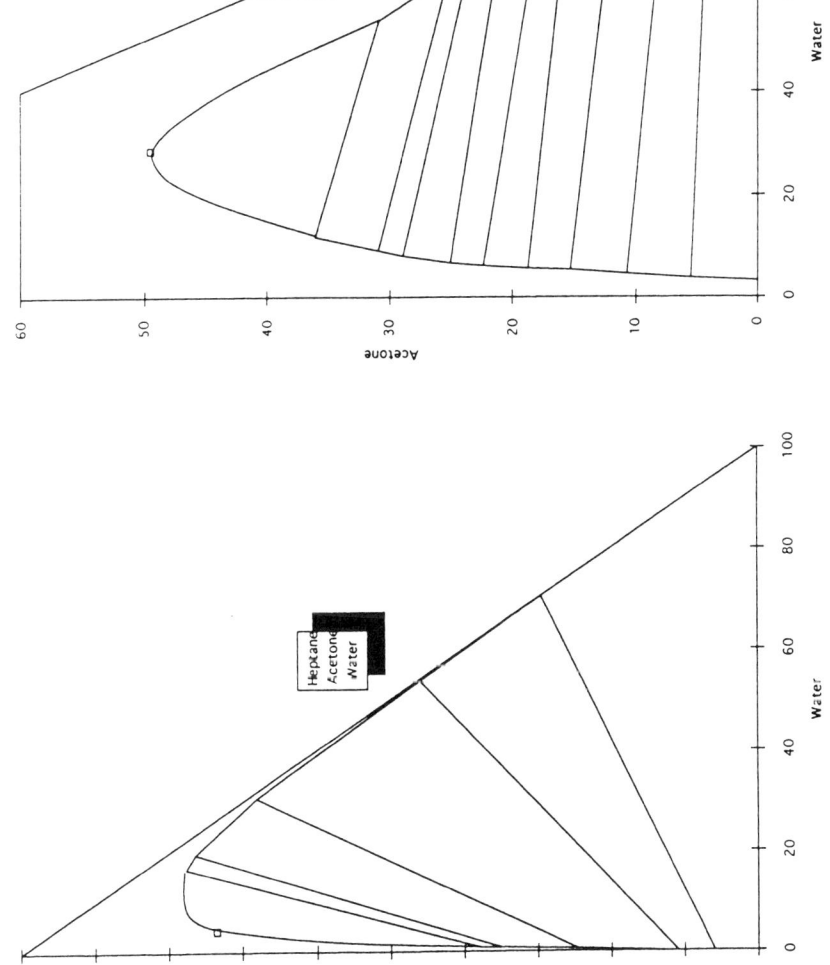

Best solvent: Acetone. **Comments:** Allows high content of acetone.

Best solvent: Acetone. **Comments:** Allows high content of acetone.

APPENDIX III

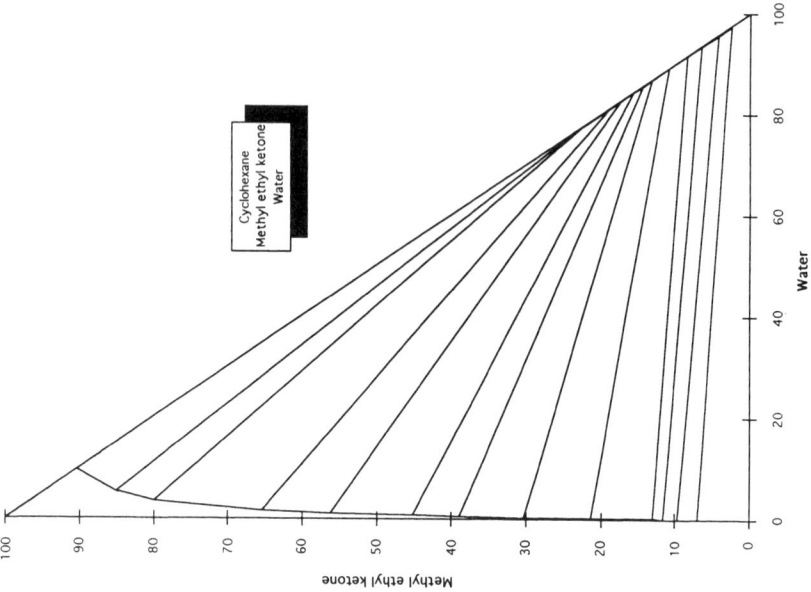

Best solvent: Methyl ethyl ketone. **Comments:** System 2.

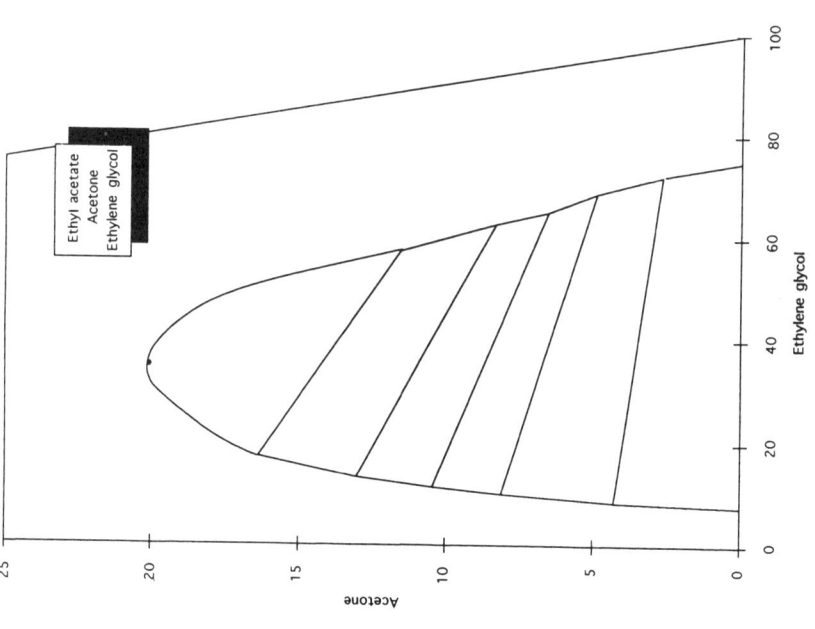

Best solvent: Acetone. **Comments:** Nonaqueous system.

TERNARY DIAGRAMS

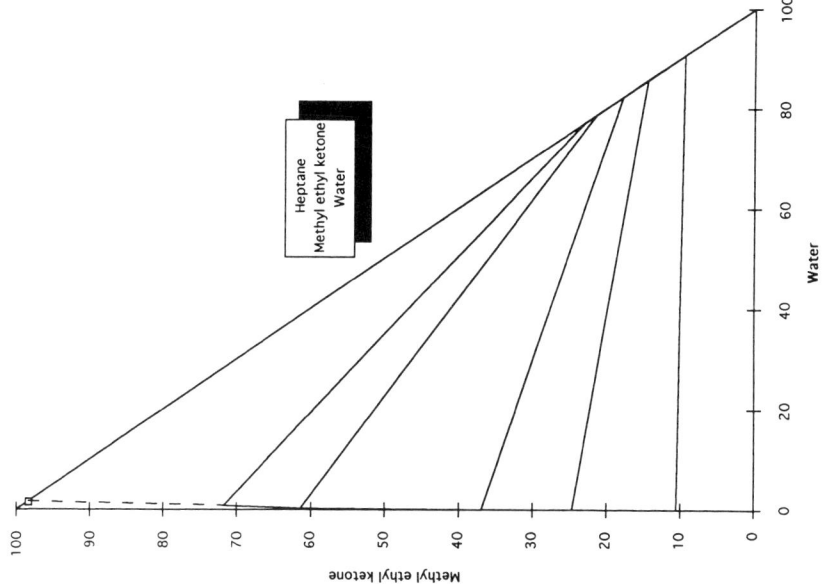

Best solvent: Methyl ethyl ketone. **Comments:** Isocratic mode.

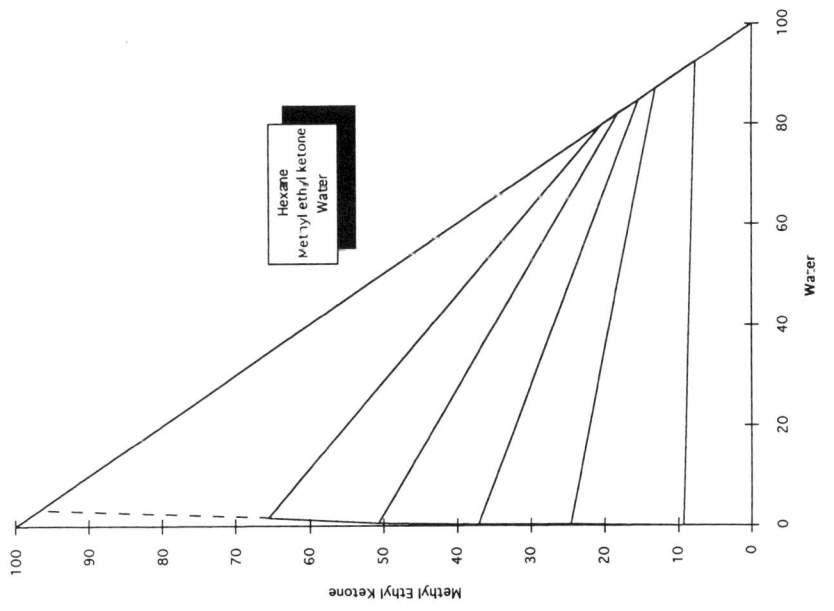

Best solvent: Methyl ethyl ketone.

APPENDIX III

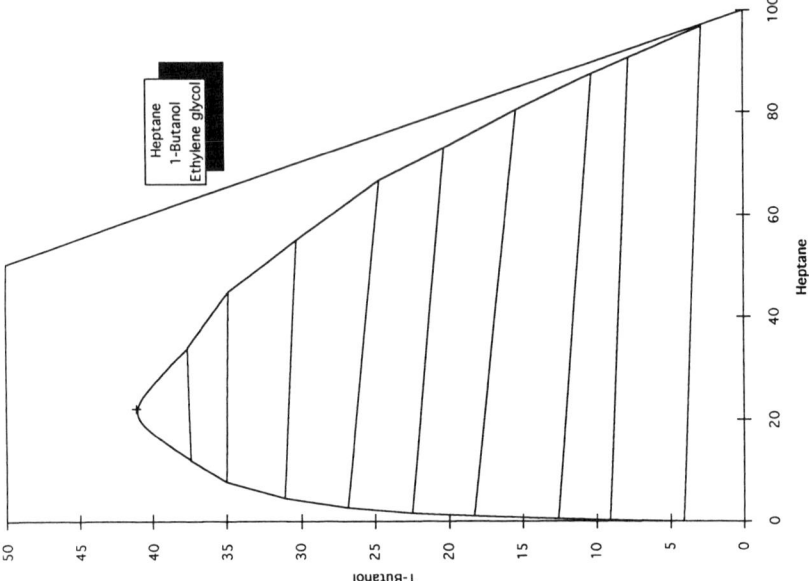

Best solvent: 1-Butanol. **Comments:** Nonaqueous system.

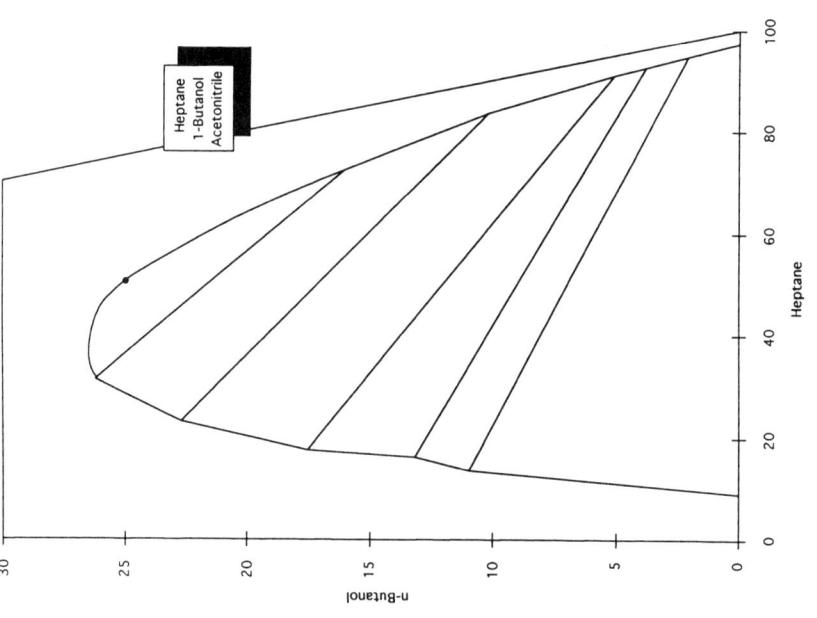

Best solvent: 1-Butanol. **Comments:** Nonaqueous system, low polar system, isocratic mode.

TERNARY DIAGRAMS

387

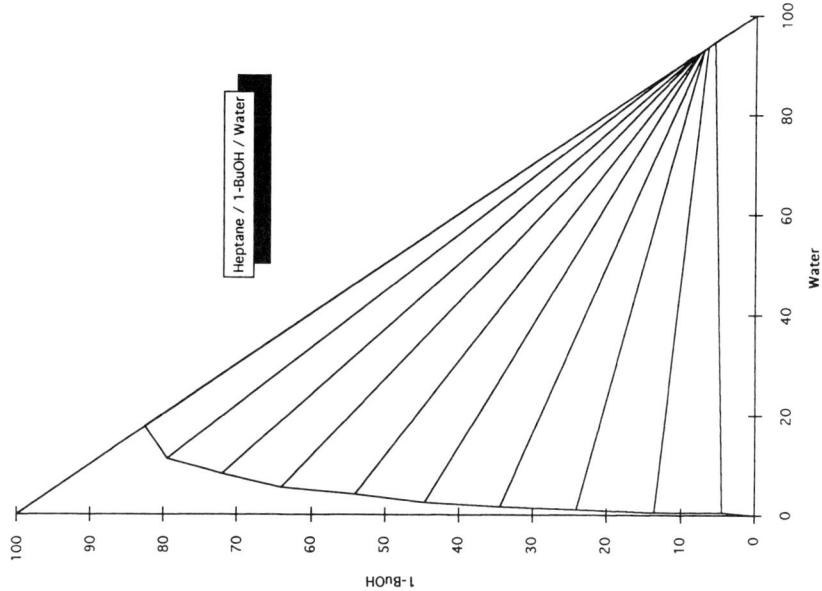

Best solvent: 1-Butanol. **Comments:** Ternary system 2; gradient in the normal phase mode.

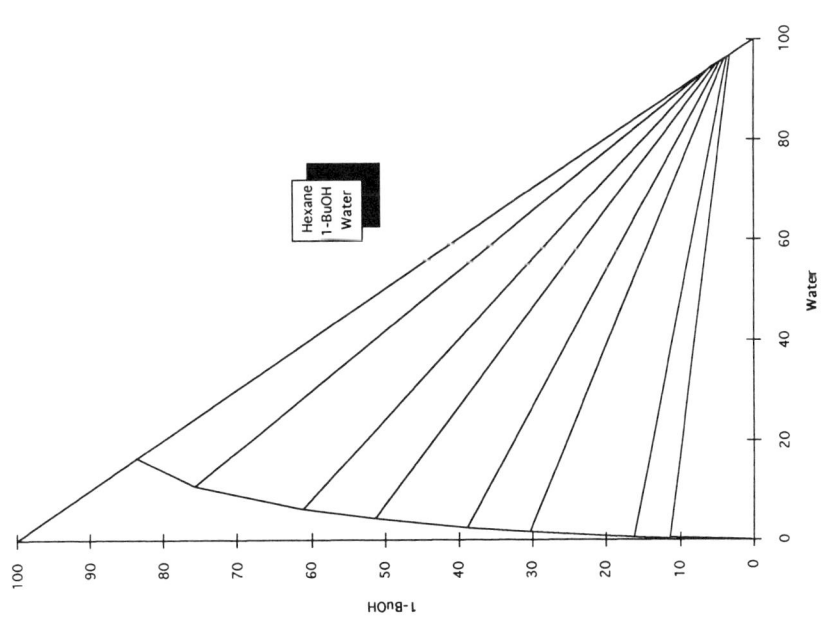

Best solvent: 1-Butanol. **Comments:** Ternary system 2; gradient in the normal phase mode.

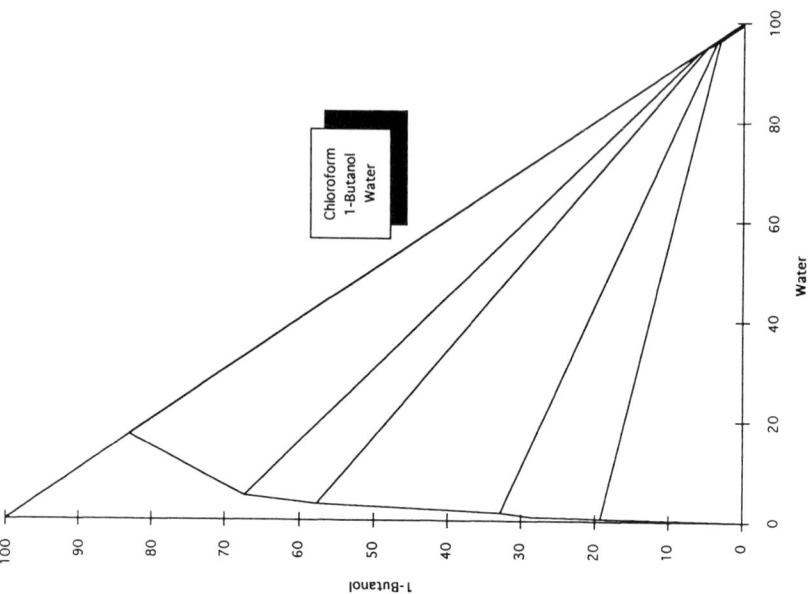

Best solvent: 1-Butanol. **Comments:** Ternary system 2; gradient in the normal phase mode.

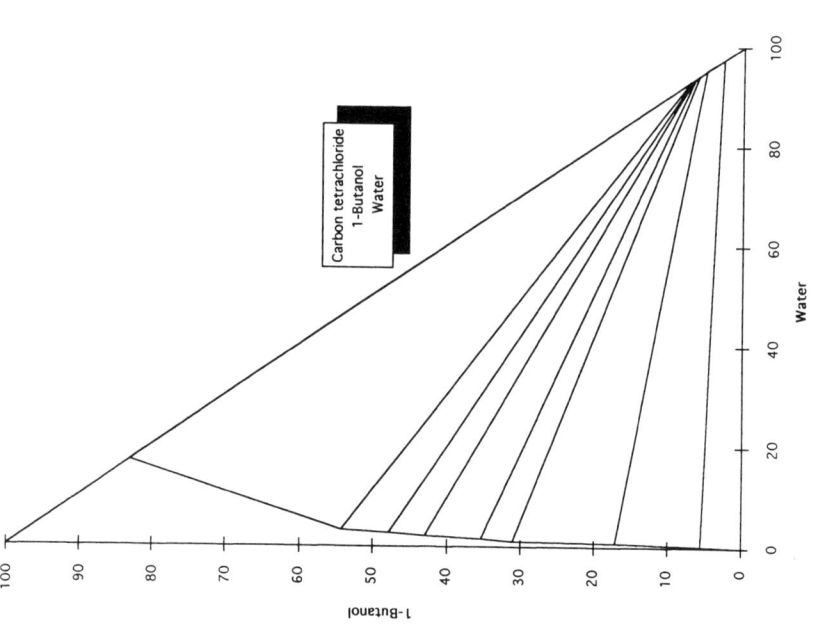

Best solvent: 1-Butanol. **Comments:** Ternary system 2. Carbon tetrachloride is not very popular.

TERNARY DIAGRAMS

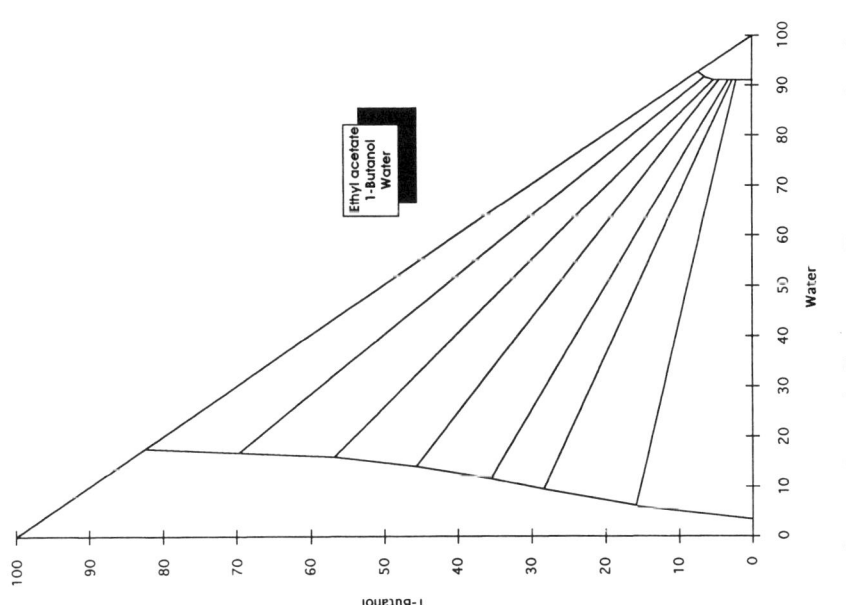

Best solvent: 1-Butanol. **Comments:** Ternary system 2; rather polar system; gradient in the normal phase mode.

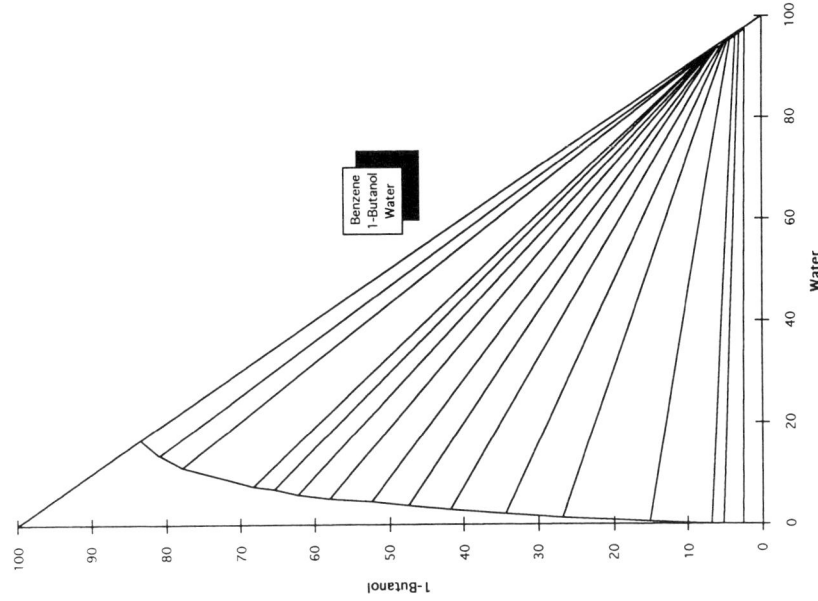

Best solvent: 1-Butanol. **Comments:** Ternary system 2; toluene instead of benzene should give a similar diagram.

APPENDIX III

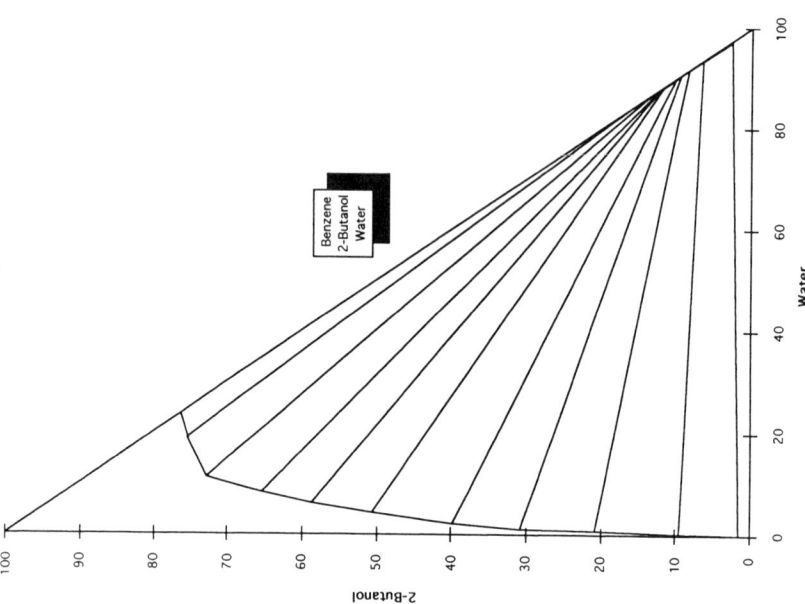

Best solvent: 2-Butanol. **Comments:** Ternary system 2; toluene, instead of benzene, should give a similar diagram; gradient in the normal phase mode.

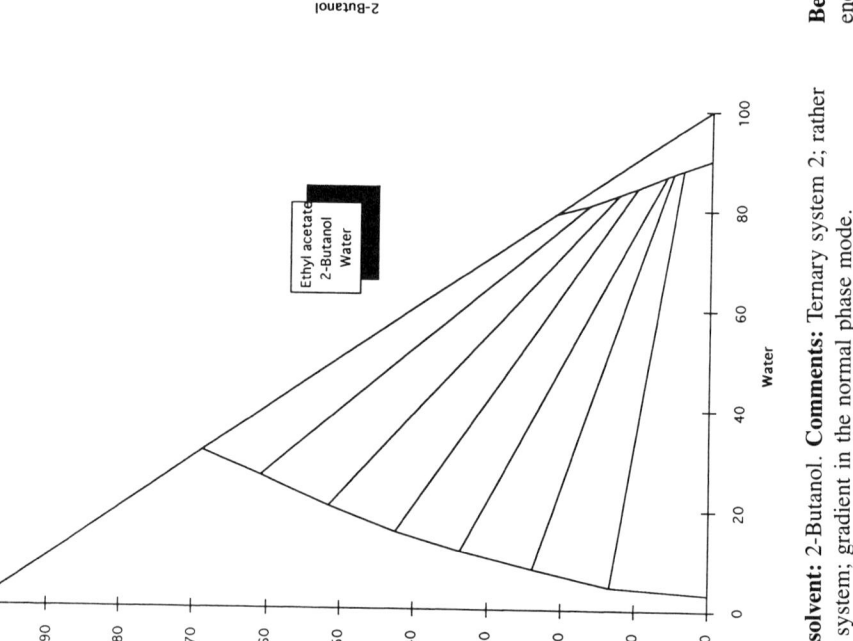

Best solvent: 2-Butanol. **Comments:** Ternary system 2; rather polar system; gradient in the normal phase mode.

TERNARY DIAGRAMS

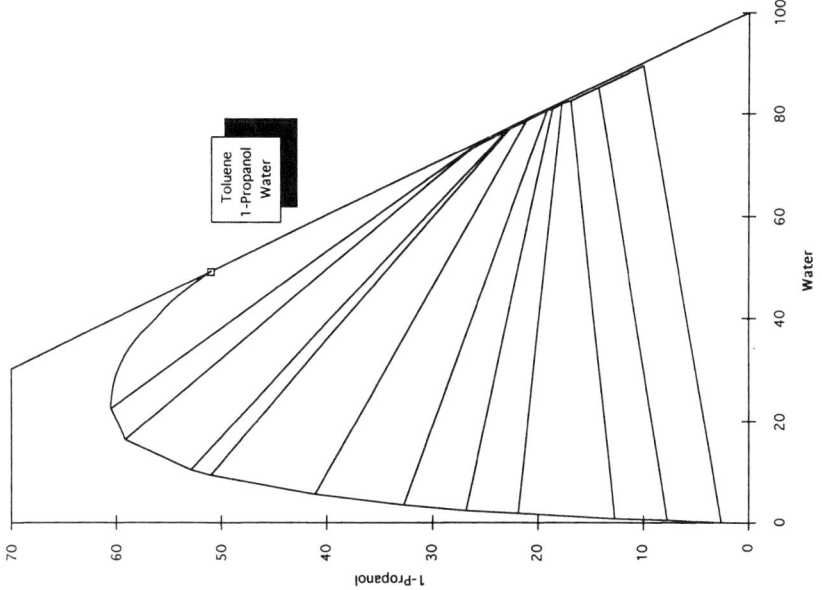

Best solvent: 1-Propanol. **Comments:** Isocratic mode.

Best solvent: 2-Butanol. **Comments:** Isocratic mode.

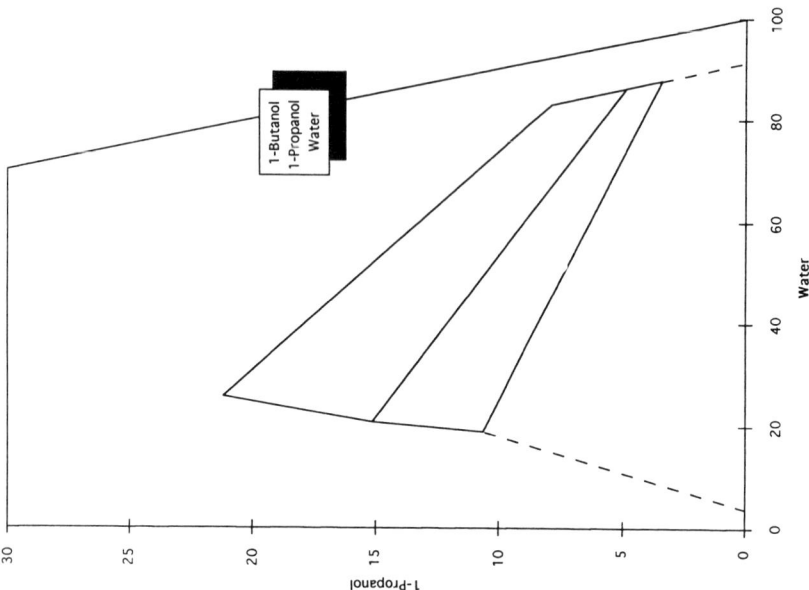

Best solvent: 1-Propanol. **Comments:** Polar system, isocratic mode.

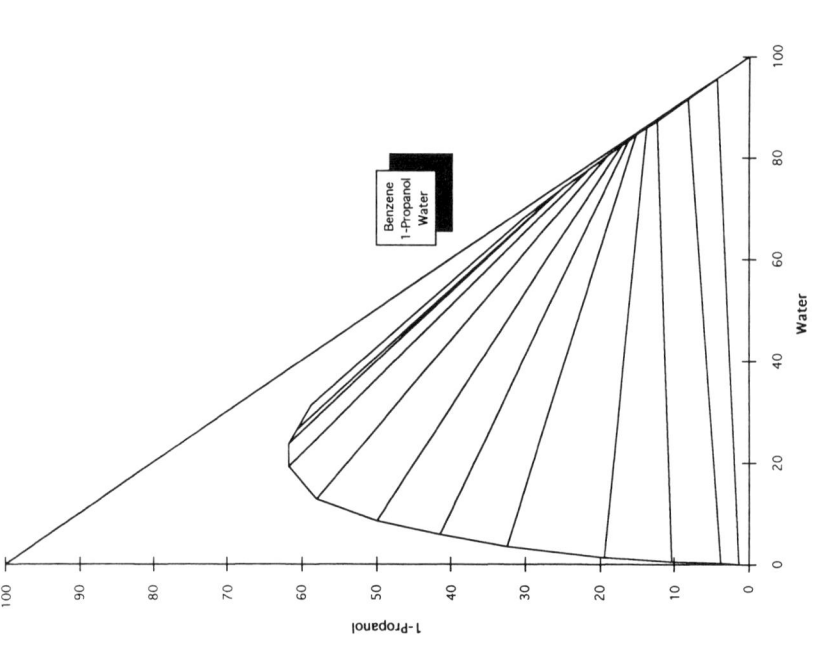

Best solvent: 1-Propanol. **Comments:** Toluene instead of benzene gives a similar diagram; isocratic mode.

TERNARY DIAGRAMS

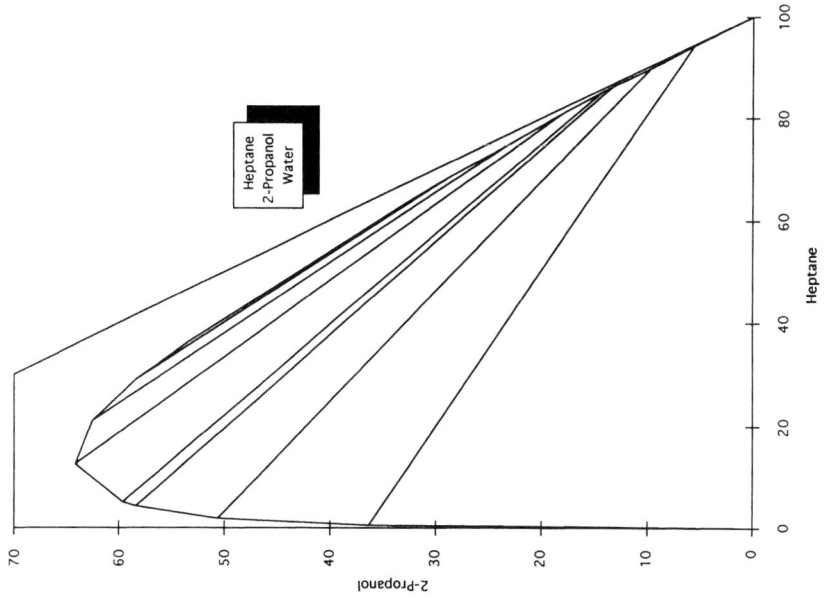

Best solvent: 2-Propanol.

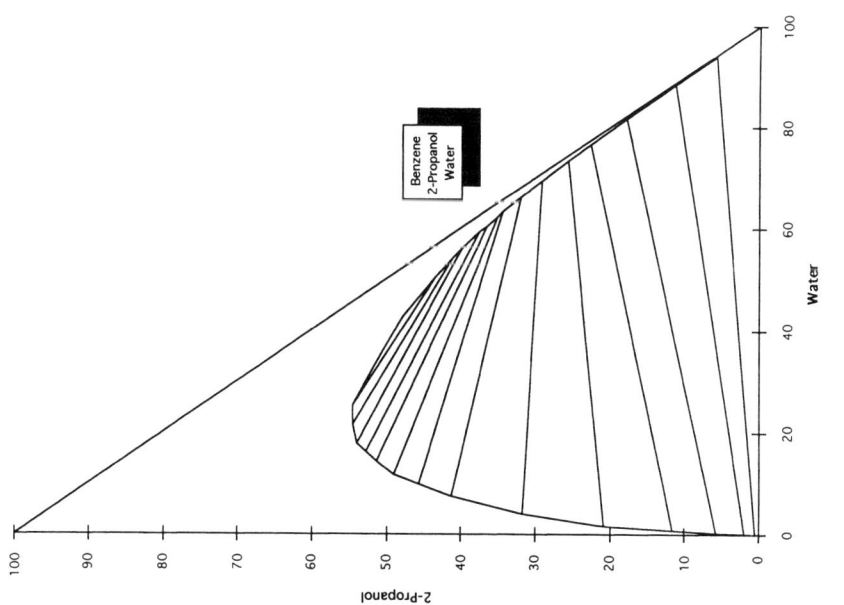

Best solvent: 2-Propanol. **Comments:** Toluene, instead of benzene, should give a similar diagram.

APPENDIX III

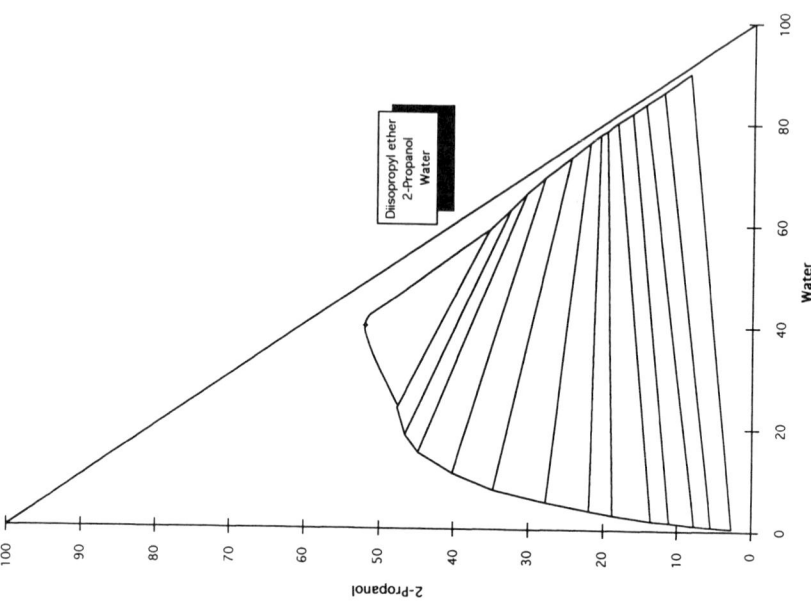

Best solvent: 2-Propanol. **Comments:** Isocratic mode.

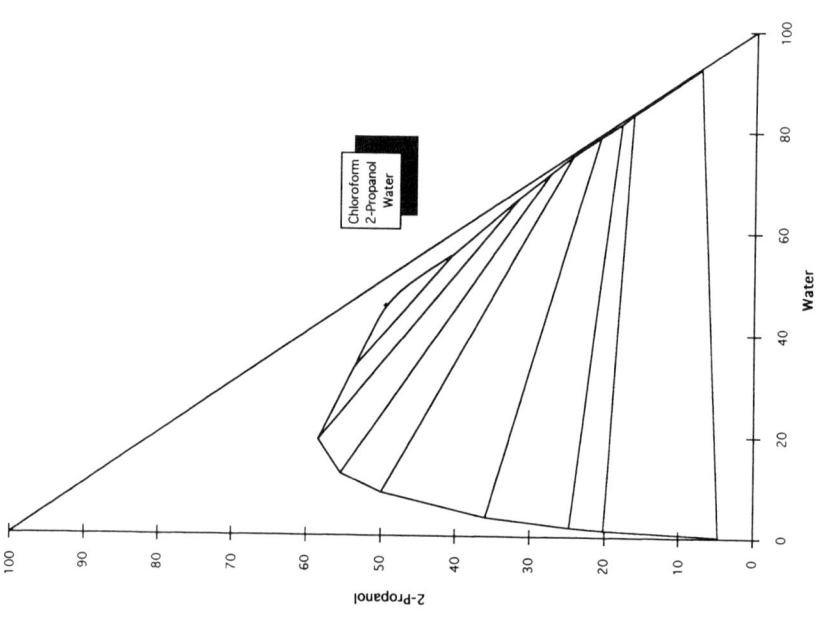

Best solvent: 2-Propanol. **Comments:** Isocratic mode.

TERNARY DIAGRAMS

395

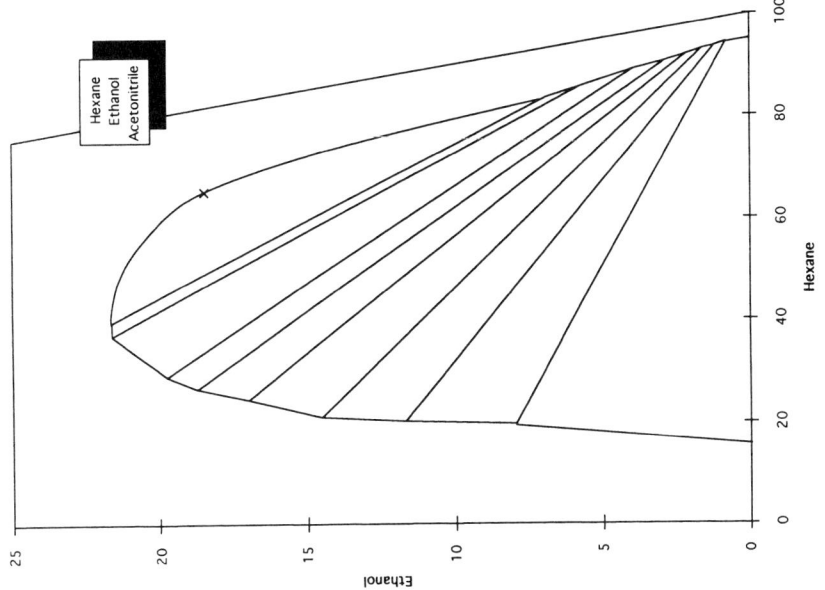

Best solvent: Ethanol. **Comments:** Nonaqueous system.

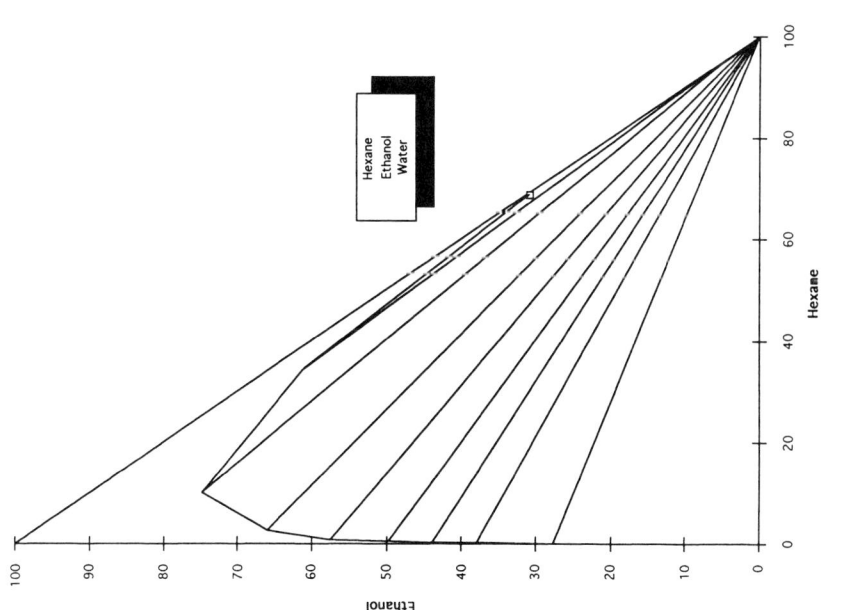

Best solvent: Ethanol. **Comments:** Similar to water/methanol/hexane; gradient in the reversed phase mode.

APPENDIX III

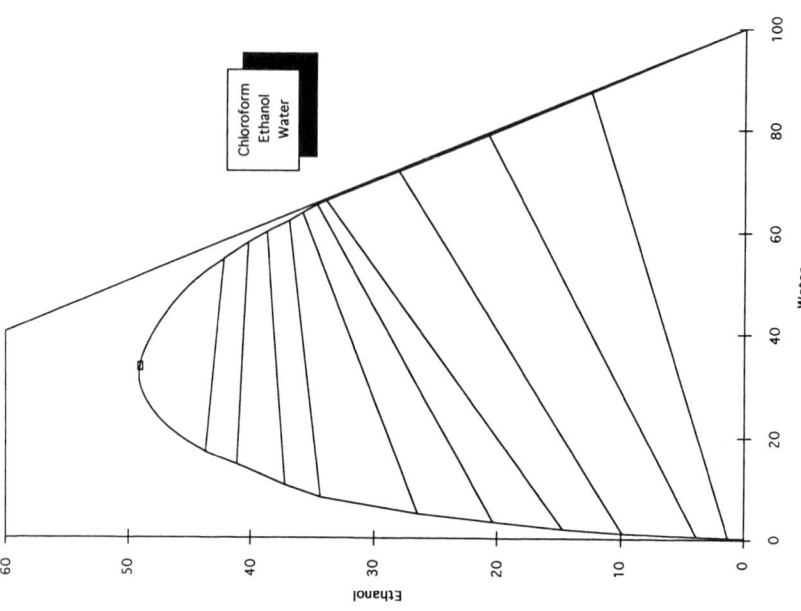

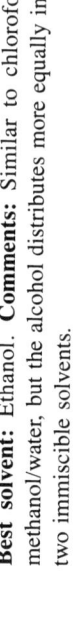

Best solvent: Ethanol. **Comments:** Similar to chloroform/methanol/water, but the alcohol distributes more equally in the two immiscible solvents.

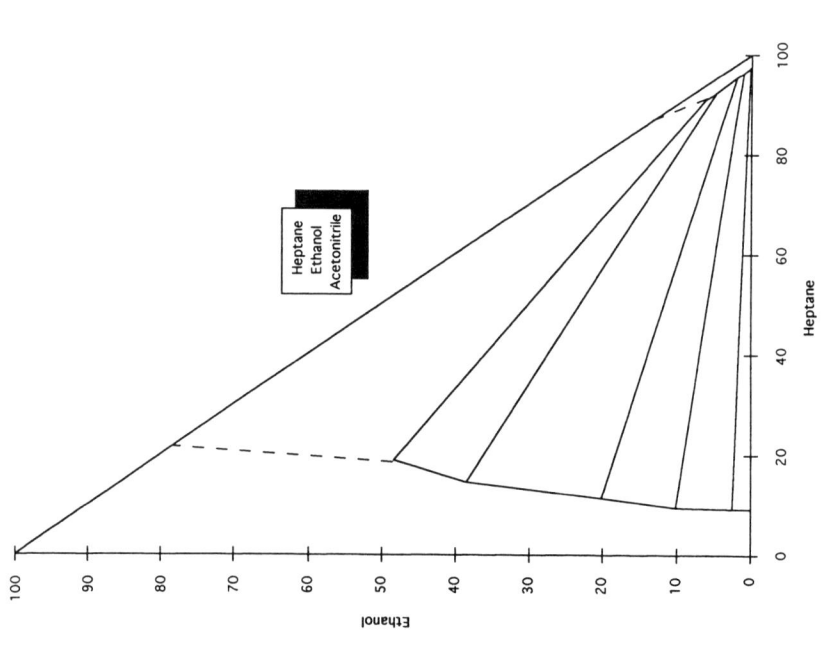

Best solvent: Ethanol. **Comments:** System 2; nonaqueous system.

TERNARY DIAGRAMS

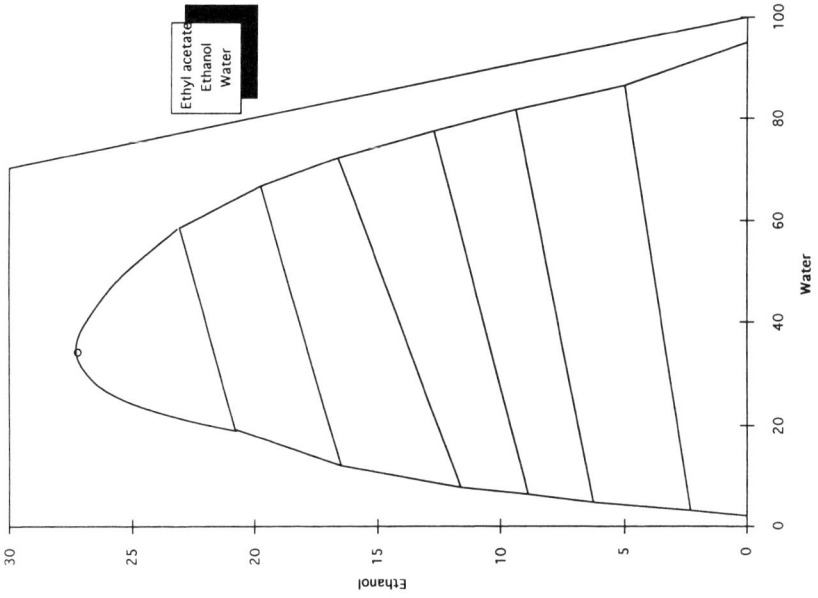

Best solvent: Ethanol. **Comments:** Isocratic mode.

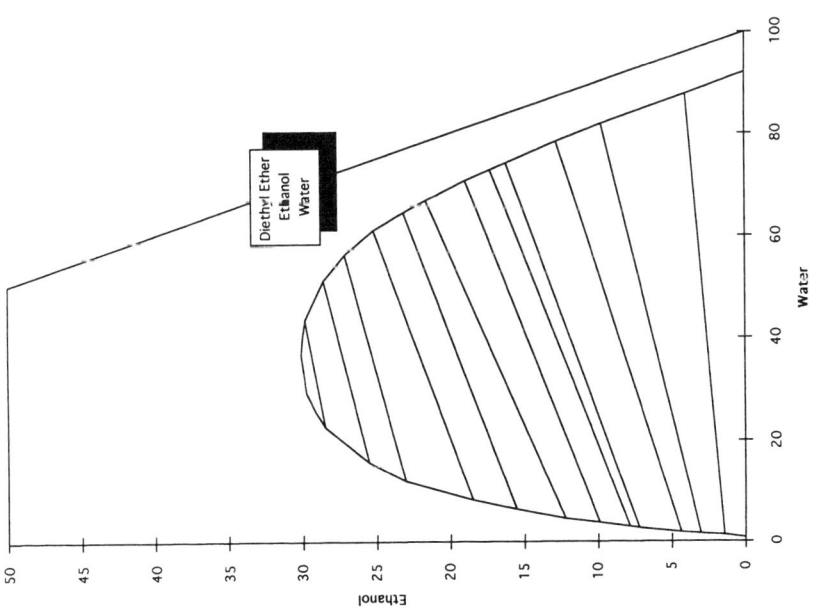

Best solvent: Ethanol. **Comments:** Isocratic mode.

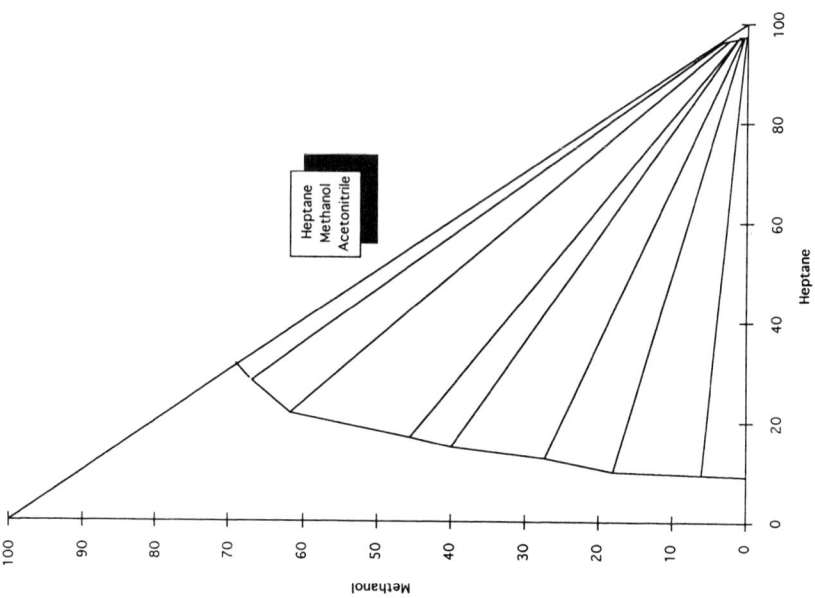

Best solvent: Methanol. **Comments:** System 2; Nonaqueous system; allows gradient runs.

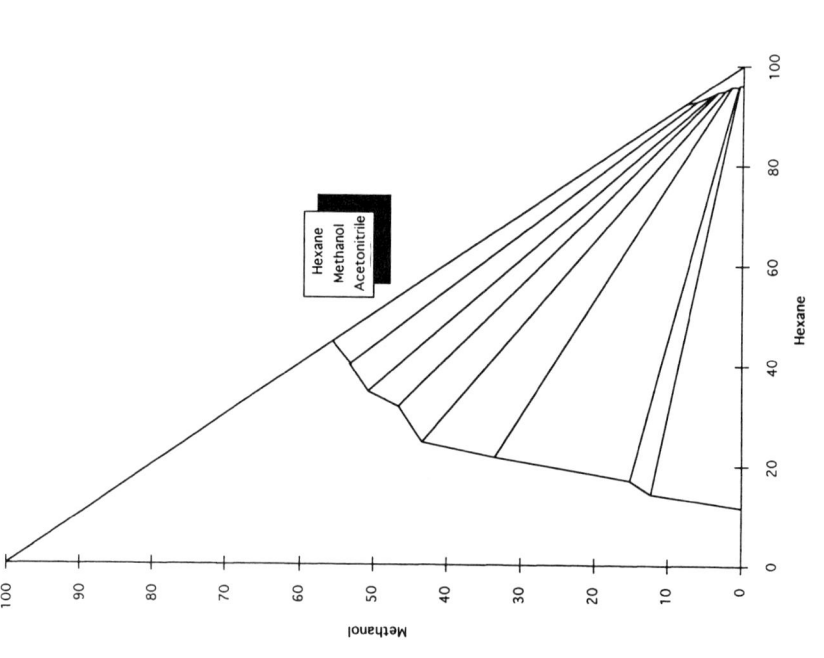

Best solvent: Methanol. **Comments:** System 2; Nonaqueous system (compare with heptane/methanol/acetonitrile).

TERNARY DIAGRAMS

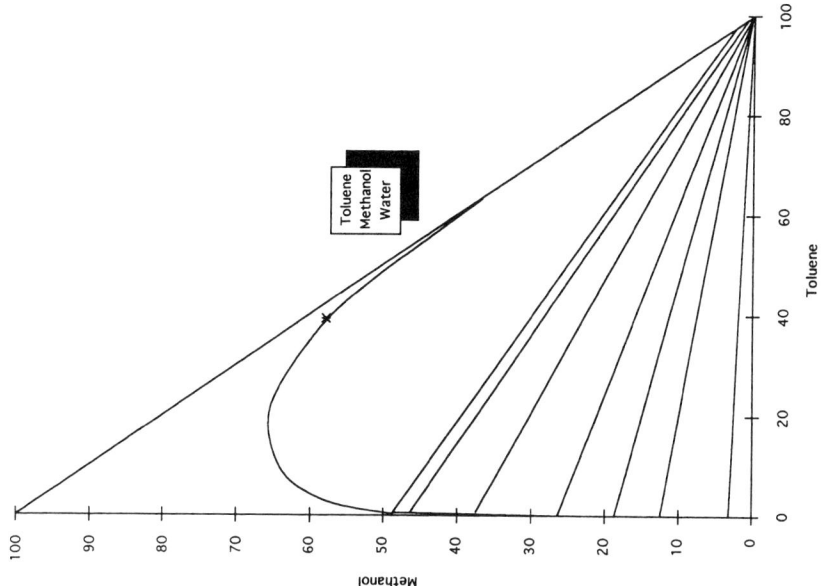

Best solvent: Methanol. **Comments:** Slightly more polar than the system heptane/methanol/water.

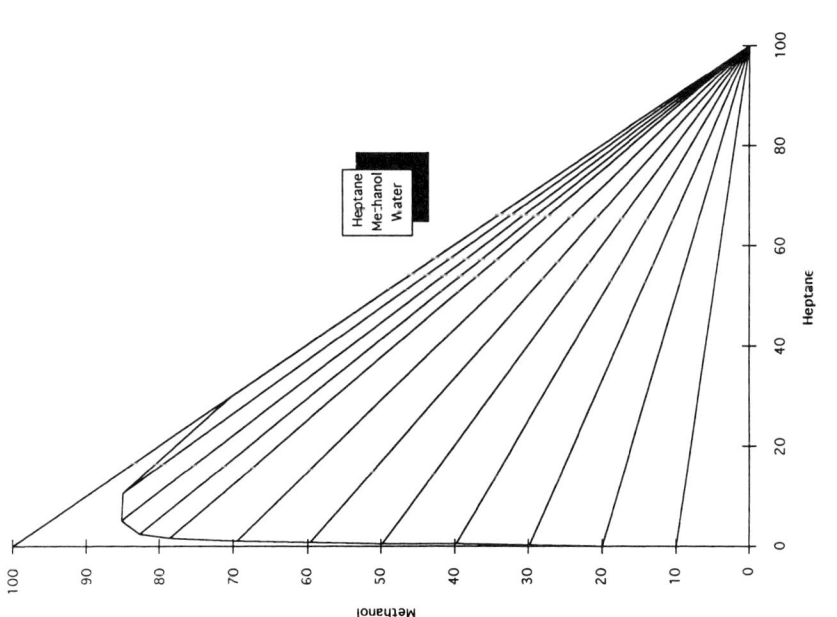

Best solvent: Methanol. **Comments:** Similar to the system hexane/methanol/water.

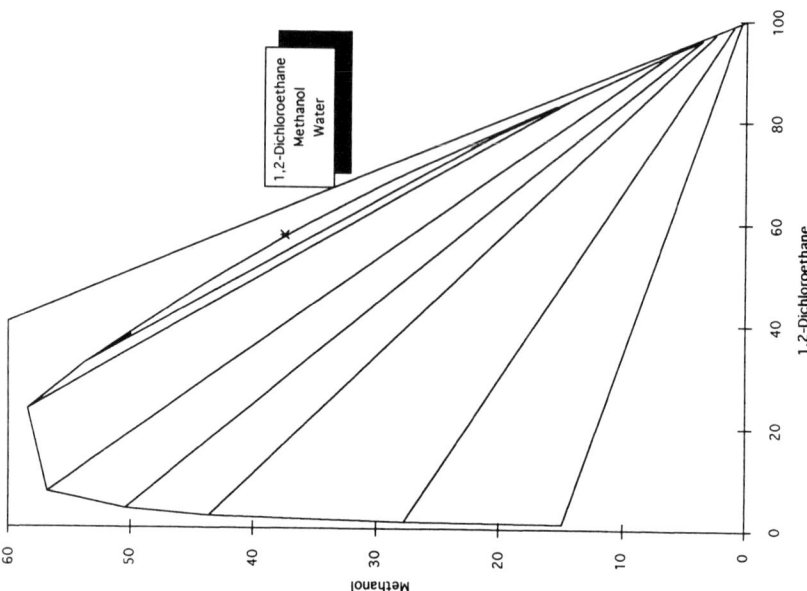

Best solvent: Methanol. **Comments:** Similar to chloroform/methanol/water.

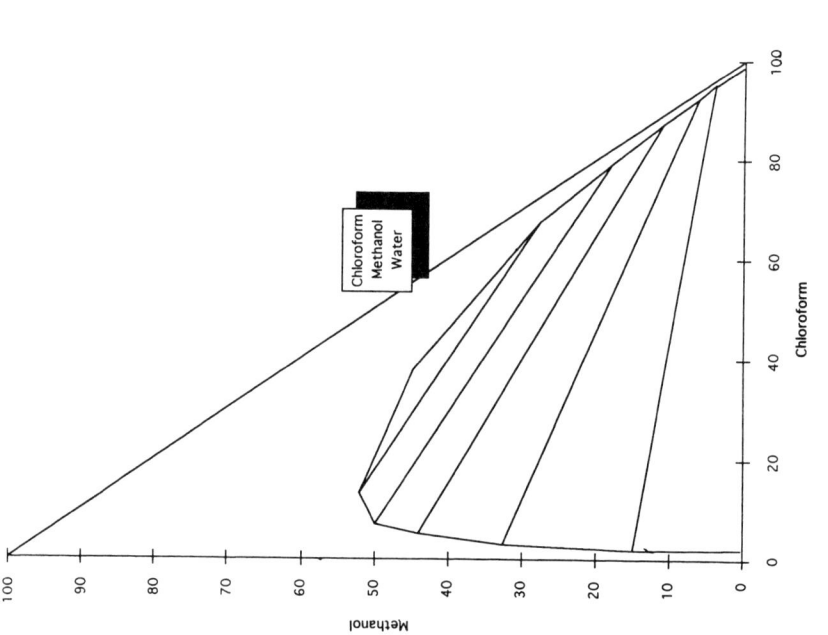

Best solvent: Methanol. **Comments:** Similar to 1,2-dichloroethane/methanol/water.

TERNARY DIAGRAMS

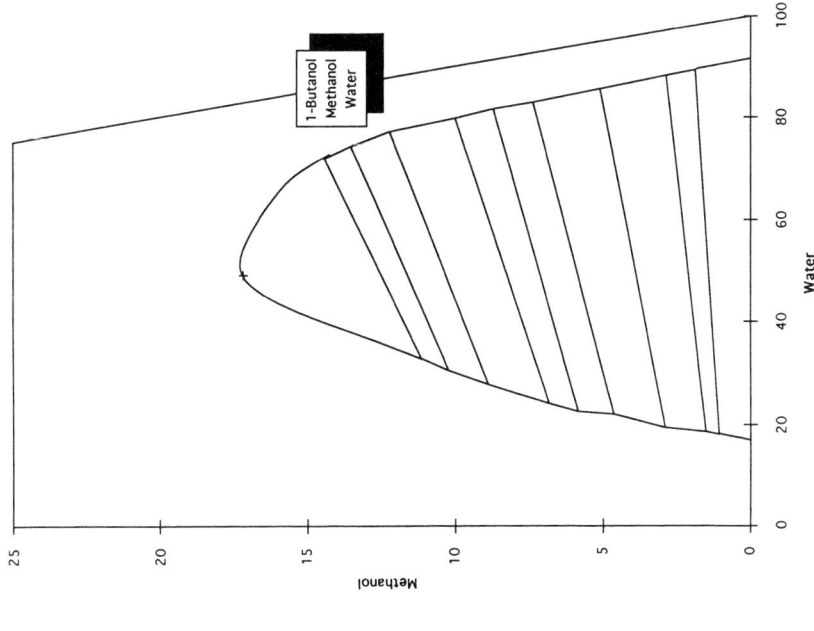

Best solvent: MeOH. **Comments:** Medium polar system, isocratic mode.

Best solvent: Methanol. **Comments:** Polar system, isocratic mode.

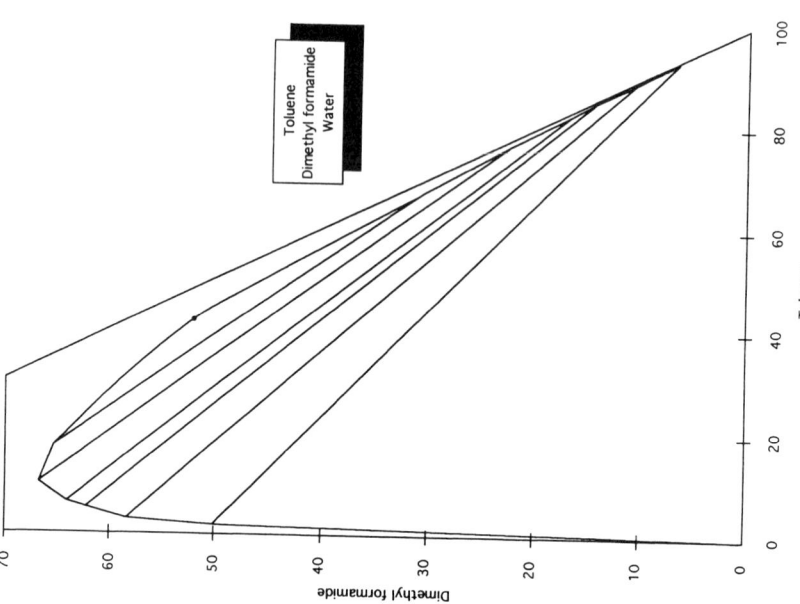

Best solvent: Dimethyl formamide. **Comments:** Isocratic mode; rather high content of DMF can be reached with this system.

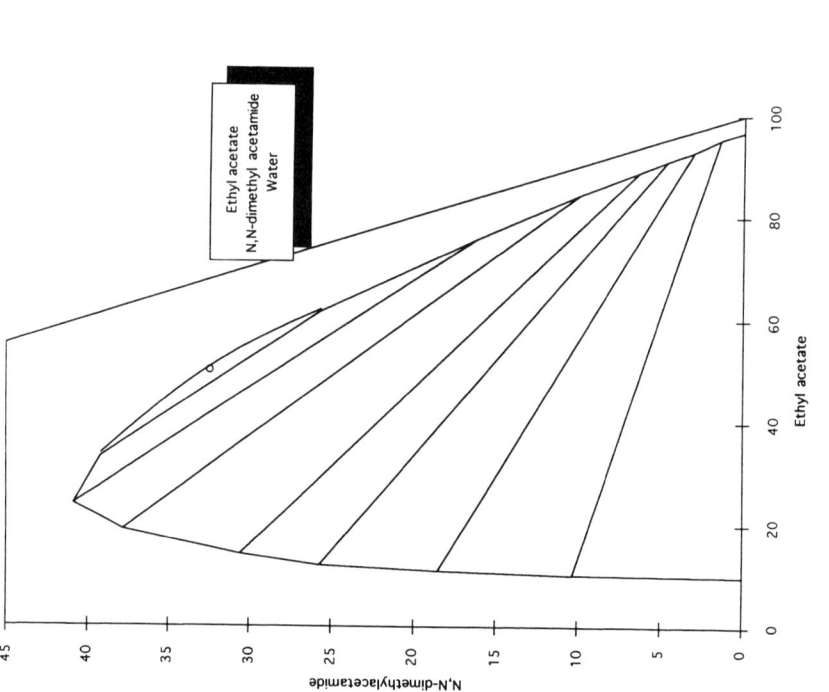

Best solvent: N,N-Dimethyl acetamide. **Comments:** This system has to be tested.

TERNARY DIAGRAMS

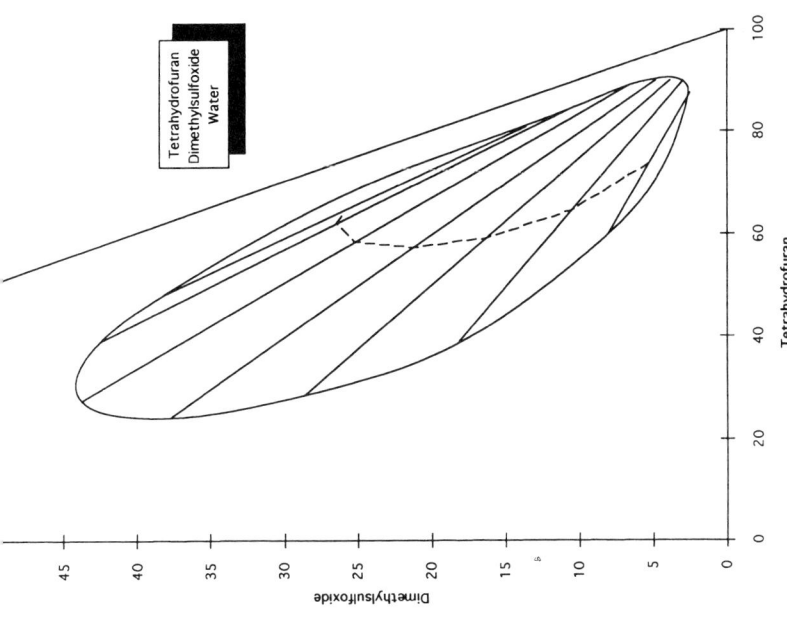

Best solvent: Dimethylsulfoxide. **Comments:** Ternary system 0; unique example in the CPC literature; powerful solvating properties; can be modified by adding some other solvents, e.g., an alkane. The dashed line shows the middle of the tie lines.

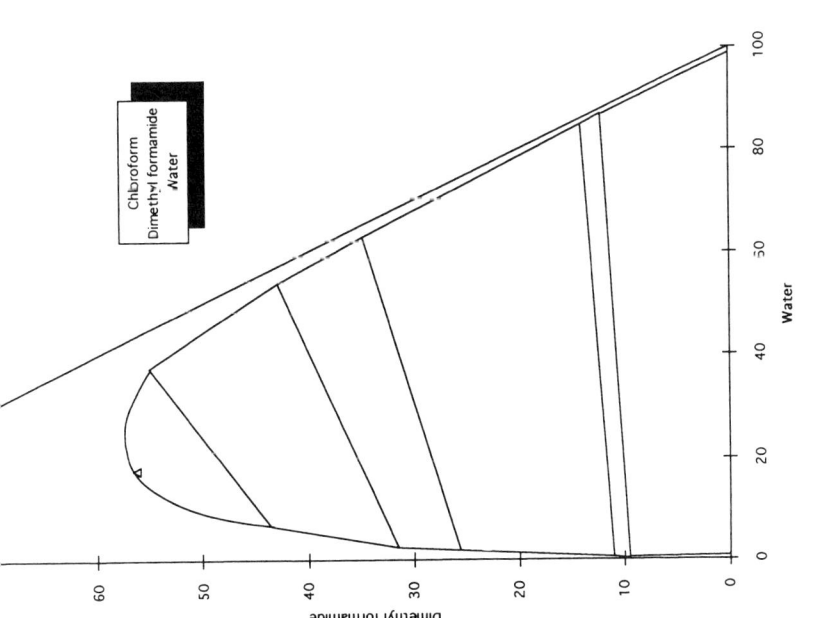

Best solvent: Dimethyl formamide. **Comments:** Isocratic mode.

APPENDIX III

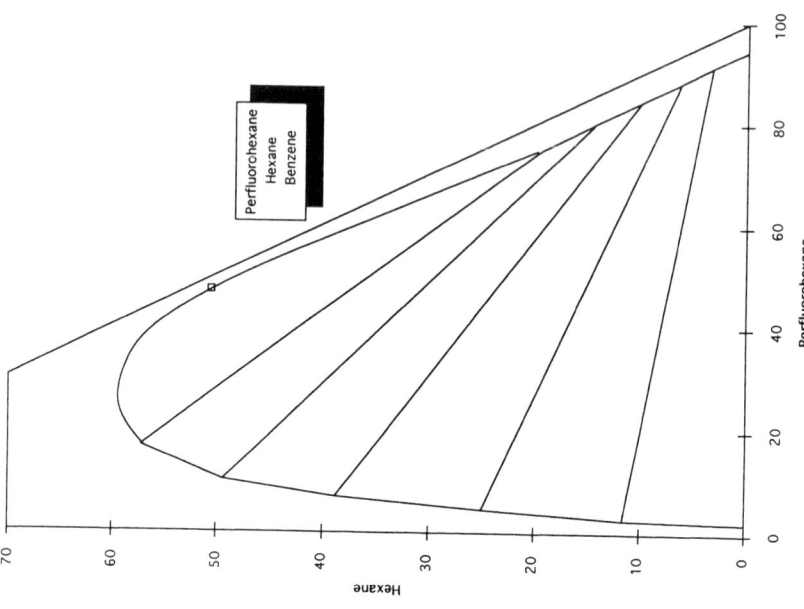

Best solvent: Hexane. **Comments:** Very hydrophobic system; can be useful for very hydrophobic mixtures, such as fullerenes. Has to be tested.

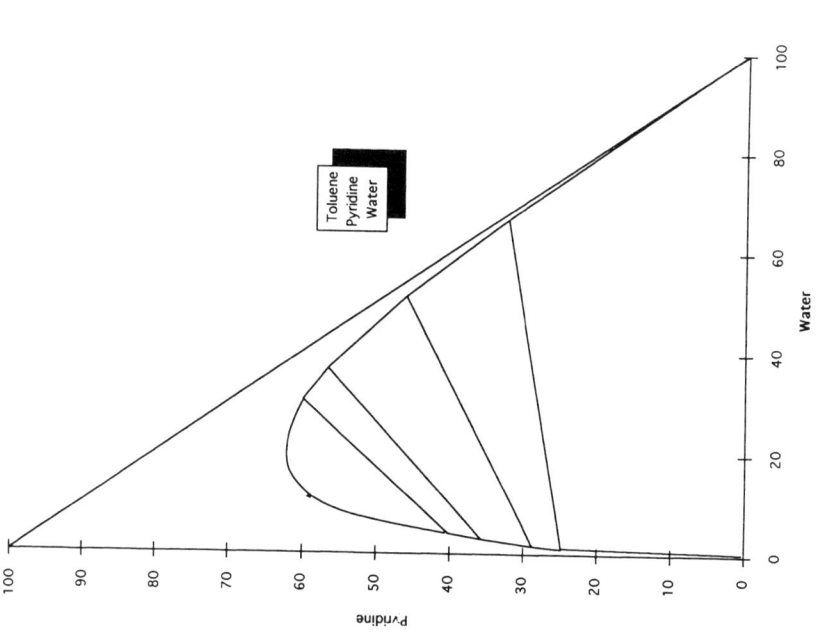

Best solvent: Pyridine. **Comments:** High content of pyridine can be reached with this system; isocratic mode.

TERNARY DIAGRAMS

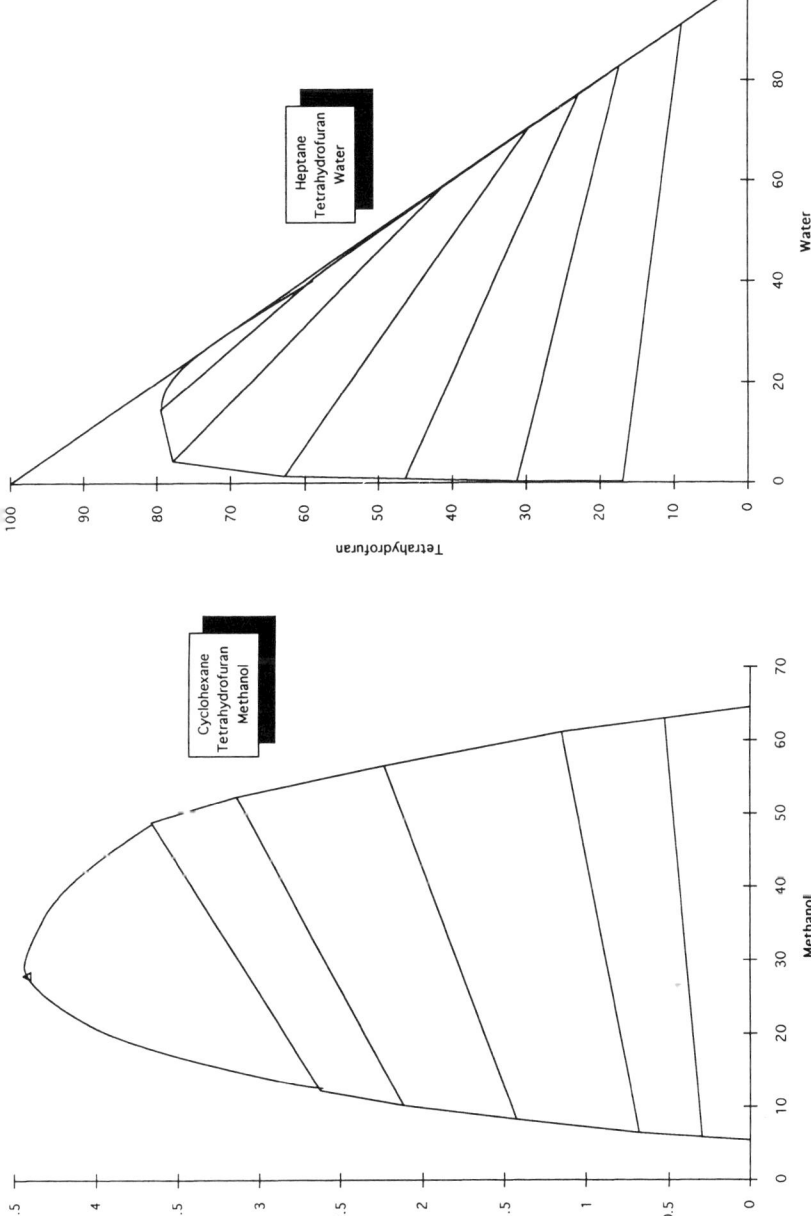

Best solvent: Tetrahydrofuran. **Comments:** Allows high content of THF; isocratic mode.

Best solvent: As only a low content of THF can be reached, the sample must be freely soluble in the binary system cyclohexane/methanol; the THF is added to modify the partition coefficients. **Comments:** Nonaqueous system, isocratic mode.

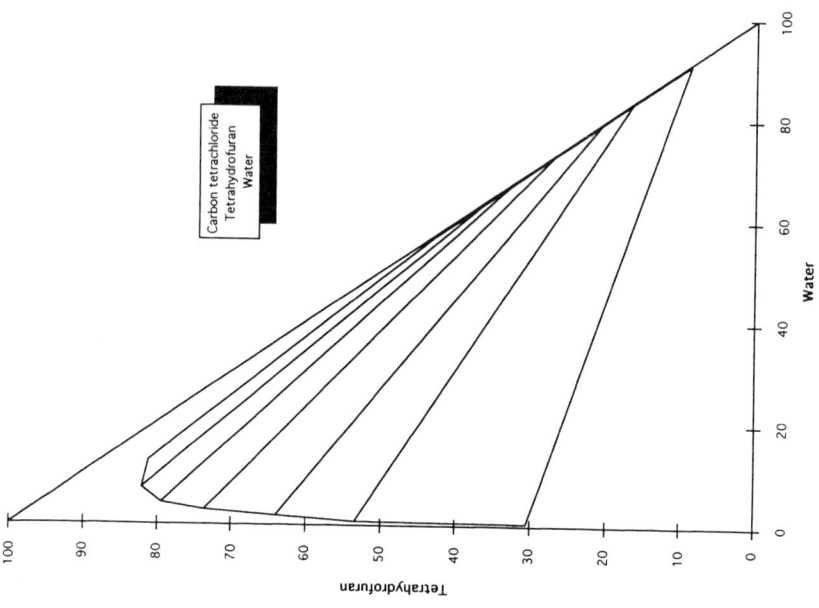

Best solvent: Tetrahydrofuran. **Comments:** Isocratic mode.

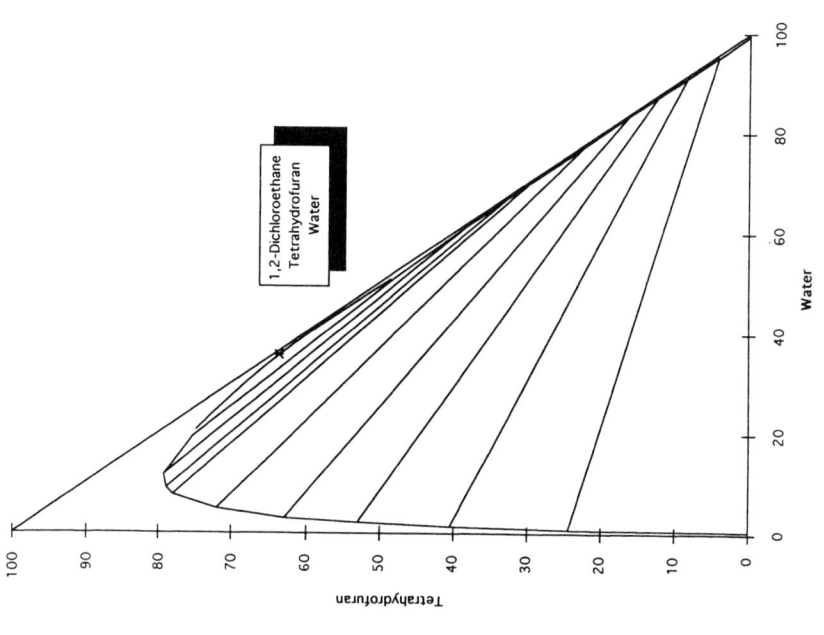

Best solvent: Tetrahydrofuran.

TERNARY DIAGRAMS

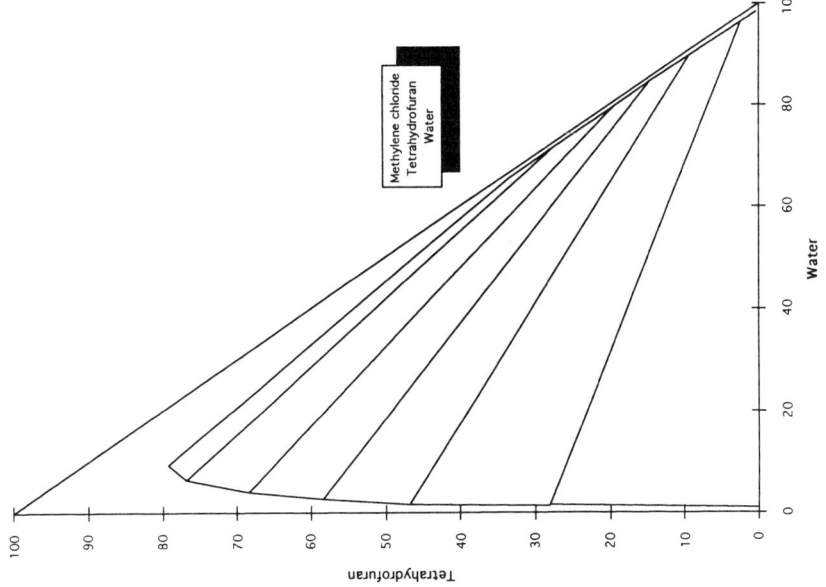

Best solvent: Tetrahydrofuran. **Comments:** Isocratic mode.

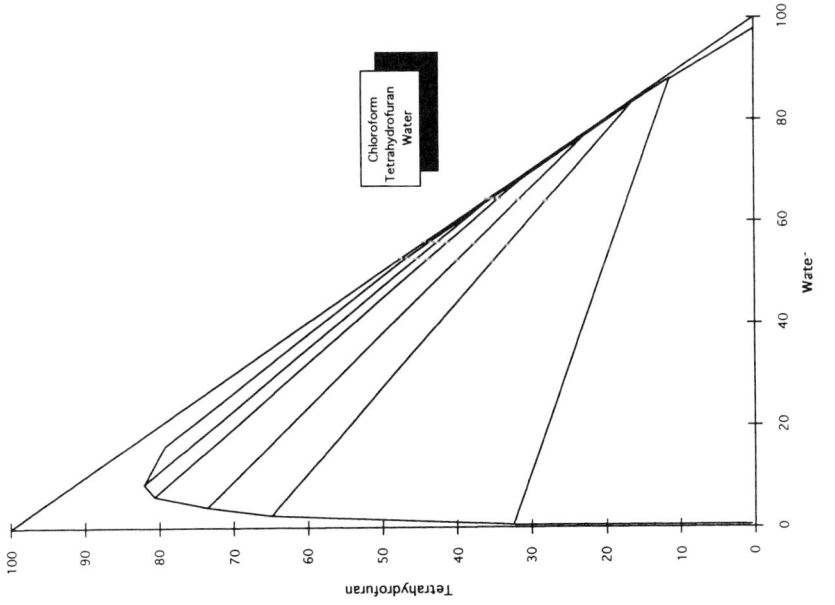

Best solvent: Tetrahydrofuran. **Comments:** Isocratic mode.

Index

Acetanilide, octanol/water partition coefficient, 206–208
Acetone, reactivity, 143
  octanol/water partition coefficient, 206–208
Acetophenone, octanol/water partition coefficient, 183, 206–208
2-Acetoxy benzoic acid, octanol/water partition coefficient, 183
Alcohol ethoxylates, 18–22
Alkylphenol ethoxylates, 18–22
Anisole, octanol/water partition coefficient, 213–215
Arsenazo III, for lanthanide detection, 225, 319

Barringtin, 123–128
Barringtonia asiatica (Lecythidaceae), 123–128
Benzamide, octanol/water partition coefficient, 183, 202–206
Benzene, octanol/water partition coefficient, 186, 191, 213–215
Benzene/1-butanol/water, 389
Benzene/2-butanol/water, 390
Benzene/formic acid/water, 367
Benzene/1-propanol/water, 392
Benzene/2-propanol/water, 393

Benzoic acid, octanol/water partition coefficient, 183, 186, 191
Benzyl alcohol, octanol/water partition coefficient, 202–206
Binodal, 71
Biphenyl, octanol/water partition coefficient, 186, 191, 213–215
Bleeding, 7
Bond number, 37
Bromobenzene, octanol/water partition coefficient, 213–215
1-Butanol/acetic acid/water, 11, 29–40, 87, 88, 94, 372
1-Butanol/methanol/water, 401
1-Butanol/1-propanol/water, 102, 392
1-Butanol/water, 29–40
sec-Butanol/water, 29–40

Capacity factor, 3
Carbon tetrachloride/1-butanol/water, 388
Carbon tetrachloride/methylene chloride/methanol/water, 149–151, 159
Carbon tetrachloride/tetrahydrofuran/water, 406
Castalagin, 112–113
Casuarictin, 112–119
Casuarinin, 112–119

409

Catalytic cracking (see Cat-cracking)
Cat-cracking, 14–18
  gradient elution for fractionation of, 15
Cerium (see Chapter 9, pp. 219–239)
  separation with DEHPA, 251–299
Cesium, separation using a crown ether, 242–251
α-Chloroacetamide herbicides (see Herbicides)
Chlorobenzene, octanol/water partition coefficient, 213–215
Chlorobenzene/acetonitrile/water, 377
2-Chlorobenzoic acid, octanol/water partition coefficient, 183
Chloroform, reactivity, 144
Chloroform/acetic acid/water, 370
Chloroform/acetone/water, 381
Chloroform/aq. HCl, separation of lanthanides, 263, 266–275, 289–299
Chloroform/1-butanol/water, 75, 388
Chloroform/dimethyl formamide/water, 403
Chloroform/ethanol/water, 396
Chloroform/formic acid/water, 368
Chloroform/isopropanol/water, 75, 394
Chloroform/methanol/water, 75, 87, 102, 400
Chloroform/methanol/1-propanol/water, 29–40, 102
Chloroform/tetrahydrofuran/water, 407
Chloroform/water, 29–40
2-Chloronitrobenzene, octanol/water partition coefficient, 183
2-Chlorophenol, octanol/water partition coefficient, 183, 186, 191, 206–208
Cocurrent chromatography, for octanol/water partition coefficient determination, 188–194, 215
Coriaria japonica (Coriariaceae), 123–128
Coriariin, 123–128
Corilagin, 112–117
Countercurrent distribution, comparison with HPLC, 332

CPC
  comparison with DCCC, 110–113
  comparison with HPLC, 4, 333, 348
  comparison with liquid membrane transport systems, 299–312
  comparison with preparative HPLC, 107–109
Crown ether, separation of alkali metal picrates with, 242–251
Cyanex 272™ (see Chapter 11, pp. 317–330)
Cyclohexane/acetic acid/water, 368
Cyclohexane/acetone/ethylene glycol, 382
Cyclohexane/chloroform/acetonitrile, 376
Cyclohexane/diethyl ether/methanol, 376
Cyclohexane/methyl ethyl ketone/water, 284
Cyclohexane/tetrahydrofuran/methanol, 405
Cyclohexane/toluene/acetonitrile, 374
Cyclohexane/water, 185

D2EHPA (see DEHPA)
10-Deacetyl-baccatin III, 335
DEHPA (see Chapter 9)
  for separation of lanthanides, 251–299
Density difference
  and pressure drop, 6 (see also Chapter 3, pp. 51–69)
  and stability (see Chapter 2)
Dibenzo-18-crown-6 (see Crown ether)
1,2-Dichloroethane/acetic acid/water, 371
1,2-Dichloroethane/aq. buffer, for separation of platinum group metals, 323
1,2-Dichloroethane/formic acid/water, 367
1,2-Dichloroethane/methanol/water, 400
1,2-Dichloroethane/tetrahydrofuran/water, 406
Diethyl ether/acetic acid/water, 370
Diethyl ether/acetone/water, 380
Diethyl ether/ethanol/water, 397
Di(2-ethylhexyl) phosphoric acid (see Bis(2-ethylhexyl)phosphoric acid)

INDEX 411

Diethyl phthalate, 41, 43
7,4' Dihydroxyflavone, 107
Diisopropyl ether/2-propanol/water, 394
Dimethyl formamide, octanol/water
  partition coefficient, 142, 202–206
Dimethylsulfoxide, 142
Droplets (see Chapter 2, pp. 25–49)
  from kinetic measurement, 328
DTMPPA (see Chapter 9, pp. 219–239)
Dual-mode CPC, 87–90
  for octanol/water partition coefficient
  determination, 185–188, 209–215
  for purification of 10-deacetyl-
  baccatin III, 345
Dysprosium (see Chapter 9, pp. 219–239)
  separation with Cyanex 272™, 319
  separation with DEHPA, 267–299

Efficiency, 3
  and chemical kinetics, 324–329
  and mobile phase volume, 42–45
  and rotational speed, 41
EHPA (see Chapter 9, pp. 219–239)
Elaeocarpusin, 112–117
(—)-Epicatechin, 103, 109–111
(—)-Epicatechin gallate, 103, 109–111
(—)-Epigallocatechin, 109–111
(—)-Epigallocatechin gallate, 100, 103, 109–111
Erbium (see Chapter 9, pp. 219–239)
  separation with DEHPA, 259–299
Ethers, reactivity, 144
2-Ethoxyphenol, octanol/water partition
  coefficient, 213–215
Ethyl acetate, reactivity, 143
Ethyl acetate/acetic acid/water, 373
Ethyl acetate/acetone/ethylene glycol, 384
Ethyl acetate/acetone/water, 383
Ethyl acetate/acetonitrile/water, 379
Ethyl acetate/1-butanol/water, 87, 389
  gradient, 92
Ethyl acetate/2-butanol/water, 390
Ethyl acetate/N,N-dimethyl acetamide/water, 402

Ethyl acetate/ethanol/water, 87, 397
Ethyl acetate/methanol/water, 401
Ethyl acetate/water, 29–40
Ethylbenzene, octanol/water partition
  coefficient, 213–215
Ethyl benzoate, octanol/water partition
  coefficient, 186, 213–215
2-Ethylhexyl 2-ethylhexylphosphonic
  acid (see Chapter 9, pp. 219–239)
bis(2-Ethylhexyl)phosphoric acid (see
  Chapter 9, pp. 219–239)
  for separation of lanthanides, 251–299
Europium (see Chapter 9)
  separation with Cyanex 272™, 319
  separation with DEHPA, 251–299
Extraction by CPC, 18–22
Extraction constants, of lanthanides, 225

Filling the CPC, 8
  control of the stationary phase
  volume, 15, 351
Flavonoids (see Chapter 5, pp. 99–131)
Flow rate and pressure drop, 59
  effect on separation of lanthanides, 231
Formamide, 142

Gadolinium (see Chapter 9)
  separation with Cyanex 272™, 319
  separation with DEHPA, 267–299
3-O-Galloylgranatin, 112–113
Geraniin, 112–117
Geranium thumbergii, 112–117
Glycycoumarin, 103, 106
Clycyrrhisoflavone, 103, 106
Gradient elution, 90–96
Granatin A, 112–113
Guavin, 112–116

Heptane/acetone/water, 383
Heptane/aq. buffer, for separation of
  metal ions (see Chapter 11, pp. 317–330)
Heptane/aq. HCl, separation of
  lanthanides, 251–299
Heptane/benzene/dimethyl formamide, 373
Heptane/1-butanol/acetonitrile, 386

Heptane/1-butanol/acetonitrile/water, 84
Heptane/1-butanol/ethylene glycol, 386
Heptane/1-butanol/water, 387
Heptane/ethanol/acetonitrile, 396
Heptane/ethyl acetate/methanol/water, 82
  for purification of 10-deacetyl-
    baccatin III, 337
Heptane/methanol, 29–41
Heptane/methanol/acetonitrile, 398
Heptane/methanol/water, 399
Heptane/methyl ethyl ketone/water, 385
Heptane/2-propanol/water, 393
Heptane/tetrahydrofuran/water, 405
Heptane/toluene/dimethyl sulfoxide, 375
Heptane/toluene/sulfolane, 375
Herbicides, C14-labeled and metabolites, 9–14
Heterocentron roseum
  (Melastomataceae), 120–125
Hexane/acetic acid/water, 369
Hexane/acetone/water, 382
Hexane/acetonitrile, 87, 146, 185
Hexane/acetonitrile/chloroform, 87, 88
Hexane/benzene/ethylene glycol, 374
Hexane/1-butanol/water, 387
  gradient, 94–96
Hexane/ethanol/acetonitrile, 395
Hexane/ethanol/water, 87, 90, 395
Hexane/ethyl acetate/acetonitrile/
  methanol, 87
Hexane/ethyl acetate/1-butanol/methanol/
  water, 82
Hexane/ethyl acetate/isopropanol/
  methanol/water, 151, 157
Hexane/ethyl acetate/methanol/water, 87
Hexane/methanol/acetonitrile, 92, 398
Hexane/methanol (4% water), 42
Hexane/methanol (*n*% water), 146
Hexane/methylene chloride/ethyl acetate/
  acetonitrile/methanol/water, 153–157, 161
Hexane/methyl ethyl ketone/water, 385
Hexane/water, 185
  and pressure drop, 61–63
Holmium (*see* Chapter 9, pp. 219–239)
  separation with DEHPA, 267–299

p-Hydroxybenzoic acid, 41
  octanol/water partition coefficient, 191
19-Hydroxy-10-deacetyl baccatin III (*see*
  Chapter 12, pp. 331–350)

Immunosuppressant, 147–162
Interfacial tension, and stability (*see*
  Chapter 2, pp. 25–49)
Iridium, separation with
  trioctylphosphine oxide, 320–326
Isoamyl alcohol/acetic acid/water, 371
Isorugosin, 116–119

Kerosene/aq. buffer, for lanthanide
  fractionation (*see* Chapter 9, pp. 219–239)

Lanthanides (*see* Chapters 9–11, pp. 219–330)
Lanthanum (*see* Chapter 9, pp. 219–239)
  separation with Cyanex 272™, 319
  separation with DEHPA, 251–299
Licochalcone A, 106
Licochalcone B, 106
Licopyranocoumarin, 103, 106
Licorice, 105–109
Licorice polyphenols, partition
  coefficient of, 103
Linear velocity, of the mobile phase in
  the channels (*see* Chapter 2, pp. 25–49)
Liquidambar formosana
  (Hamamelidaceae), 116–119
Liquid membrane transport (*see* Chapter 10, pp. 241–316)
  of lanthanides, comparison with CPC, 275–312
Lithium, separation using a crown ether, 242–251
Lutetium (*see* Chapter 9, pp. 219–239)
  separation with DEHPA, 267–299

Metholachlor, 18–22
Methyl *t*-butyl ether, 144
Methyl *t*-butyl ether/acetonitrile/water, 378

INDEX

Methylene chloride/tetrahydrofuran/ water, 407
Methyl isobutyl ketone/acetic acid/water, 372
Methyl isobutyl ketone/acetone/water, 29–40, 381
  for purification of 10-deacetyl-baccatin III, 338
Methyl isobutyl ketone/acetonitrile/ water, 378
Mobile phase volume, minimum in the channels, 26
  and flow rate, 27

Naphthalene, octanol/water partition coefficient, 186, 191
1-Naphthol, octanol/water partition coefficient, 186, 191, 213–215
Neodymium (see Chapter 9, pp. 219–239)
  separation with Cyanex 272™, 319
  separation with DEHPA, 251–299
Nitrobenzene, octanol/water partition coefficient, 213–215
p-Nitrophenol, 249
2-Nitrotoluene, octanol/water partition coefficient, 213–215
Nobotanin, 120–125
Nobotanin K, 108
Nonionic surfactants (see Surfactant)

Octanol/water, 29–40
Octanol/water/hexane, 185
Octanol/water partition coefficient (see Chapters 7–8, pp. 167–218)
Oenothein B, 103, 112–113, 118–121
Oligomeric hydrolyzable tannins, 118–128
Organophosphorous extractants (see Chapter 9, pp. 219–239)
Oxygen, reactivity, 144

Palladium, separation with trioctylphosphine oxide, 320–326
Partition coefficient, 2, 140
  in octanol/water (see Chapters 7–8, pp. 167–218)

Pedunculagin, 112–125
Peptides, gradient elution, 95
Perfluorohexane/hexane/benzene, 404
Petroleum products (see Cat-cracking)
Phase stability, and injection, 143, 340
Phase volume, ratio, 4
Phenanthrene, octanol/water partition coefficient, 191
Phenol, octanol/water partition coefficient, 186, 202–206, 213–215
1-Phenyl-3-methyl-4-benzoyl-5-pyrazolone (see Chapter 11, pp. 317–330)
Phthalimide, octanol/water partition coefficient, 191
Picric acid, 249
Plait point, 73
Plant polyphenols (see Chapter 5, pp. 99–131)
Platinum, separation with trioctylphosphine oxide, 320–326
Platinum group metals (see Chapter 11, pp. 317–330)
Polyphenols (see Chapter 5, pp. 99–131)
  toxicity, 100
Potassium, separation using a crown ether, 242–251
Praecoxins, 112–116
Praseodymium (see Chapter 9, pp. 219–239)
  separation with Cyanex 272™, 319
  separation with DEHPA, 251–299
Preparative CPC (see Chapter 12, pp. 331–350)
Pressure drop, 6
  in CPC (see Chapter 3, pp. 51–69)
  and rotational speed, 7
Promethium (see Chapter 9, pp. 219–239)
4-(2-Pyradylazo) resorcinol, for detection of Ni(II), 319

Rare earth metal ions (see Lanthanides)
Rekker-Hansch fragmental constant, 177
Resolution, 4
  and mobile phase volume, 45–48

Rhodium, separation with
 trioctylphosphine oxide, 320–326
Rotational speed
 effect on separation of lanthanides, 231
 and efficiency, 41
 and pressure drop, 6
Rubidium, separation using a crown
 ether, 242–251
Rugosin, 112–116, 123–128

Samarium (*see* Chapter 9, pp. 219–239)
 separation with Cyanex 272™, 319
 separation with DEHPA, 251–299
Sanguiin H-11, 108
Scaleup, 145–147
Scandium (*see* chapter 9)
Schimawalin, 123–128
Selectivity factor, 3
Sodium, separation using a crown ether, 242–251
Solute retention, 2
Solvent, selection, 80–87, 136–147
 reactivity, 143
 polarity, 168–172
 physiochemical properties, polarity and water solubility, 170
 effect of added solvent on separation of lanthanides, 227
 selection in preparative CPC, 336–340
Solvent systems (*see* chapter 4 & appendix III)
 classification, 78, 364
Stachyurus praecox (Stachyuraceae), 112–116
Stokes law, (*see* chapter 2)
Strictinin, 112–116
Surfactant, contaminant from waste water, 18–22

Tannins (*see* Chapter 5, pp. 99–131)
Taxotere (*see* 10-deacetyl-baccatin III)
TBP (*see* tributyl phosphate)
Tea polyphenols, 103, 109–111
Tellimagrandin, 112–119
Temperature, effect on separation of lanthanides, 229, 271

Terbium (*see* Chapter 9, pp. 219–239)
 separation with DEHPA, 267–299
Ternary diagram, 363 (*see also* Chapter 4, pp. 71–97)
Tetrahydrofuran/dimethylsulfoxide/water, 29–40, 403
Thulium (*see* Chapter 9, pp. 219–239)
 separation with Cyanex 272™, 319
 separation with DEHPA, 267–299
Tie-line, 71
Toluene, octanol/water partition coefficient, 186, 191, 213–215
Toluene/acetic acid/water, 369
Toluene/acetone/ethylene glycol, 380
Toluene/acetone/water, 379
Toluene/acetonitrile/ethanol/water, 87, 88
Toluene/acetonitrile/water, 377
Toluene/aq. HCl, separation of lanthanides, 263
Toluene/2-butanol/water, 391
Toluene/dimethyl formamide/water, 402
Toluene/methanol/water, 399
Toluene/1-propanol/water, 391
Toluene/pyridine/water, 404
Transition metals (*see* Chapter 11, pp. 317–330)
Transparent CPC, 63–66
Trapanin B, 108
Tributyl phosphate, as a solvent for lanthanide fractionation, 237, 264
bis(2,4,4-Trimethylpentyl) phosphinic acid (*see* Cyanex 272™ and Chapter 9, pp. 219–239)
2,4,6-Trinitrophenol, 249
Trioctylphosphine oxide (*see* Chapter 11, pp. 317–330)
Tripdiolide, 147–162
Tripterygium wilfordii (*see* Chapter 6, pp. 133–165)
Triptolide, 147–162

Vescalagin, 112–113
Viscosity
 and pressure drop, 6 (*see also* Chapter 3, pp. 51–69)

INDEX

[Viscosity]
  and stability (*see* Chapter 2, pp. 25–44)
Visual observation, 63–66

Woodfordia fructicosa, partition
  coefficient of tannin of, 103, 118–121
Woodfordin C, 103, 118–121

o-Xylene, octanol/water partition
  coefficient, 213–215
Xylenol orange, for detection of
  lanthanides, 253, 268

Ytterbium (*see* Chapter 9, pp. 219–239)
  separation with DEHPA, 259–299
Yttrium (*see* Chapter 9, 219–239)